THE CONSTRUCTION LAW LIBRARY FROM WILEY LAW PUBLICATIONS

ALTERNATIVE CLAUSES TO STANDARD CONSTRUCTION CONTRACTS
James E. Stephenson, Editor

ALTERNATIVE DISPUTE RESOLUTION IN THE CONSTRUCTION INDUSTRY
Robert F. Cushman, G. Christian Hedemann, and Avram S. Tucker, Editors

ARBITRATION OF CONSTRUCTION DISPUTES
Michael T. Callahan, Barry B. Bramble, and Paul M. Lurie

ARCHITECT AND ENGINEER LIABILITY: CLAIMS AGAINST DESIGN PROFESSIONALS
Robert F. Cushman and Thomas G. Bottum, Editors

CALCULATING CONSTRUCTION DAMAGES
William Schwartzkopf, John J. McNamara, and Julian F. Hoffar

CALIFORNIA CONSTRUCTION DEFECT LITIGATION: RESIDENTIAL AND COMMERCIAL
Thomas E. Miller

CALIFORNIA CONSTRUCTION LAW (FIFTEENTH EDITION)
Kenneth C. Gibbs and Gordon Hunt

CONSTRUCTION AND ENVIRONMENTAL INSURANCE CASE DIGESTS
Wiley Law Publications, Editors

CONSTRUCTION BIDDING LAW
Robert F. Cushman and William J. Doyle, Editors

CONSTRUCTION CLAIMS AND LIABILITY
Michael S. Simon

CONSTRUCTION CLAIMS ORGANIZATION SYSTEM
Robert S. Brams and Christopher Lerner

CONSTRUCTION CONTRACTING ALERT NEWSLETTER
Smith, Currie & Hancock

CONSTRUCTION CONTRACTOR'S HANDBOOK OF BUSINESS AND LAW
Robert F. Cushman, G. Christian Hedemann, and Peter J. King, Editors

CONSTRUCTION DELAY CLAIMS (SECOND EDITION)
Barry B. Bramble and Michael T. Callahan

CONSTRUCTION DOCUMENTATION (SECOND EDITION)
John A. Ricchini and James J. O'Brien

CONSTRUCTION ENGINEERING EVIDENCE
Loren W. Peters

CONSTRUCTION INDUSTRY CONTRACTS: LEGAL CITATOR AND CASE DIGEST
Wiley Law Publications Editorial Staff

CONSTRUCTION INDUSTRY FORMS (TWO VOLUMES)
Robert F. Cushman and George L. Blick, Editors

CONSTRUCTION INDUSTRY INSURANCE HANDBOOK
Deutsch, Kerrigan & Stiles

CONSTRUCTION JOINT VENTURES: FORMS AND PRACTICE GUIDE
Mitchell W. Becker and Robert F. Cushman, Editors

CONSTRUCTION LITIGATION FORMBOOK
David M. Buoncristiani, John D. Carter, and Robert F. Cushman

CONSTRUCTION LITIGATION: REPRESENTING THE CONTRACTOR (SECOND EDITION)
John D. Carter, Robert F. Cushman, and William J. Palmer, Editors

CONSTRUCTION LITIGATION: REPRESENTING THE OWNER (SECOND EDITION)
Robert F. Cushman, Kenneth M. Cushman, and Stephen B. Cook, Editors

CONSTRUCTION LITIGATION: STRATEGIES AND TECHNIQUES
Barry B. Bramble and Albert E. Phillips, Editors

CONSTRUCTION RENOVATION FORMBOOK
Robert F. Cushman and John W. DiNicola, Editors

CONSTRUCTION SCHEDULING: PREPARATION, LIABILITY, AND CLAIMS
Jon M. Wickwire, Thomas J. Driscoll, and Stephen B. Hurlbut, Editors

CONSTRUCTION SUBCONTRACTING: A LEGAL GUIDE FOR INDUSTRY PROFESSIONALS
Overton A. Currie, Neal J. Sweeney, and Randall F. Hafer, Editors

CONSTRUCTION SURETY AND BONDING HANDBOOK
Gorden Hunt

DESIGN-BUILD CONTRACTING HANDBOOK
Robert F. Cushman and Kathy Sperling Taub, Editors

DESIGN PROFESSIONAL'S HANDBOOK OF BUSINESS AND LAW
Robert F. Cushman and James C. Dobbs, Editors

DIFFERING SITE CONDITION CLAIMS
Robert F. Cushman and David Tortorello, Editors

FEDERAL CONSTRUCTION CONTRACTING
James F. Nagle

FIFTY STATE CONSTRUCTION LIEN AND BOND LAW (TWO VOLUMES)
Robert F. Cushman, Editor

LEGAL GUIDE TO AIA DOCUMENTS (THIRD EDITION)
Werner Sabo

1992 WILEY CONSTRUCTION LAW UPDATE
Overton A. Currie and Neal J. Sweeney, Editors

PROVING AND PRICING CONSTRUCTION CLAIMS
Robert F. Cushman and David A. Carpenter, Editors

SWEET ON CONSTRUCTION INDUSTRY CONTRACTS (SECOND EDITION)
Justin Sweet

TROUBLED CONSTRUCTION LOANS: LAW AND PRACTICE
Stanley P. Sklar, Editor

AVOIDING OR MINIMIZING CONSTRUCTION LITIGATION

AVOIDING OR MINIMIZING CONSTRUCTION LITIGATION

JONATHAN J. SWEET, Esq.

The Law Offices of Jonathan J. Sweet
San Jose, California

Wiley Law Publications

JOHN WILEY & SONS, INC.

New York • Chichester • Brisbane • Toronto • Singapore

This text is printed on acid-free paper.

Library of Congress Cataloging in Publication Data:

Printed in the United States of America

10 9 8 7 6 5 4 3 2 1

SUBSCRIPTION NOTICE

This Wiley product is updated on a periodic basis with supplements to reflect important changes in the subject matter. If you purchased this product directly from John Wiley & Sons, Inc., we have already recorded your subscription for this update service.

If, however, you purchased this product from a bookstore and wish to receive (1) the current update at no additional charge, and (2) future updates and revised or related volumes billed separately with a 30-day examination review, please send your name, company name (if applicable), address, and the title of the product to:

Supplement Department
John Wiley & Sons, Inc.
One Wiley Drive
Somerset, NJ 08875
1-800-225-5945

For customers outside the United States, please contact the Wiley office nearest you:

Professional and Reference Division
John Wiley & Sons Canada, Ltd.
22 Worcester Road
Rexdale, Ontario M9W 1L1
Canada
(416)675-3580
1-800-263-1590 in Canada

John Wiley & Sons, Ltd.
Baffins Lane
Chichester
West Sussex PO19 1UD
United Kingdom
(44)(243)779777

Professional, Reference and Trade Division
Jacaranda Wiley Ltd.
PO Box 174
Ryde, NSW 2113
(02)805-1100

PREFACE

This book is designed primarily to help nonattorneys in the construction field resolve disputes. Many of you will never have been involved in a lawsuit, and will only have heard about the perils of litigation from your friends in the field who have been sued or had to sue someone. This book will enable you to understand how the construction process, law, and the judicial system interact.

The text will clarify the basic elements of the legal process. It will make you aware of some of the issues, alternatives, and options you will face at the outset of a project, and others that may occur when you are already engaged in a project. With this information, you may be able to prevent disputes or unnecessary expenditures that can occur as a result of a lack of knowledge or a misunderstanding of the consequences of your acts or the acts of others.

You will be provided with tactics and strategies that may help you avoid being a party to litigation. For situations where litigation is unavoidable, the book will detail methods to minimize the disastrous effect of litigation on your business.

The effect of certain contractual provisions, and the benefits and disadvantages of their use, will be discussed. The functions served by bonds and insurance in risk allocation will also be analyzed. For example, a subcontractor may lack the leverage to negotiate terms of a contract with a general contractor. This book can help the subcontractor to evaluate the risks of particular contractual provisions, and to make an informed decision on whether to accept the contract. Once aware of certain risks, the subcontractor may be able to provide that some of the risks otherwise falling on the subcontractor will be covered by insurance or will be reimbursed by the general contractor or owner.

This book will also analyze contract provisions that may not be legally enforceable—provisions that a subcontractor can assume would not be effective. As an example, indemnity clauses that seek to transfer liability of one party to another for intentional acts are not enforceable, because the courts view this transfer as a violation of public policy.

This book cannot substitute for individual legal counseling provided by an attorney who is familiar both with the facts of your dispute and with the laws that govern the dispute. However, this book can provide you with some background knowledge that will reduce the confusion you feel when you are caught up in construction disputes and litigation.

In addition to emphasizing litigation avoidance, this book will address some of the criteria to use in retaining attorneys to work on construction matters. It will help you to cooperate with your attorney and to make informed and intelligent decisions. Most of all, this text will describe the limitations inherent in the use of traditional litigation to resolve disputes, and will provide reasons why attorneys who practice construction law may prefer alternative dispute resolution methods to conventional litigation as a means to resolve construction disputes.

Format of the Book

Chapter 1 outlines the elements and patterns that develop in construction disputes, and discusses some types of disputes that arise in the construction setting and how they may be resolved.

Chapter 2 examines the structure of the public court system in the United States and explains the differences between the federal and state systems and the considerations regarding selection of the court system that will be used to resolve a dispute.

Chapter 3 explores the existing alternatives to resolution of cases by trial, and available alternatives to dispute resolution within the public system. It describes the advantages and disadvantages of mediation, arbitration, and settlement.

Chapter 4 discusses and analyzes the construction process, the parties to construction projects, and their relationships and duties to each other. Alternatives to the traditional construction hierarchy (owner, architect, prime/general contractor, and subcontractors) are examined. These alternatives include multiple prime contracting, fast-tracking, use of construction managers, and design-build.

Chapter 5 analyzes the contract as the foundation of business relations, with particular focus on construction projects. Elements required for a valid contract are described, and examples of various construction contract situations are given. Particular contract clauses that are of primary concern are highlighted.

Chapter 6 discusses special issues that arise for parties acting as subcontractors. The complexity of the obligations of subcontractors to the owners, and of subcontractors to the other subcontractors and suppliers on large projects, is examined. In addition, the question of how contract clauses between the owner and the prime/general contractor apply to the performance of the subcontractors is discussed.

Chapter 7 examines the concept of tort, the duty and breach of duty between those parties who have no contract. Because of concerns for public health and of fairness, the government has passed laws that create duties between parties where there is no contract or where there is no express statement of parties' rights in relation to each other.

Chapter 8 details the effect and uses of insurance in the construction process. Types of insurance, and types of contractual clauses relating to insurance, are discussed.

Chapter 9 examines the bonding process and the effect of construction performance and payment bonds on contract performance and completion. Licensing bonds are also discussed.

Chapter 10 explores the role of local, state, and federal governments in relation to construction contracts and disputes. Particularly significant are laws regarding bidding procedures and regulation of storage and handling of toxic or hazardous waste.

Chapter 11 focuses on specific realistic and cost-effective strategies you can adopt to avoid construction disputes and as well as litigation, and on problems related to bankruptcies, which may arise during a project. Many steps suggested here can be taken without the services of an attorney. Others involve consulting an attorney, even before executing a contract or other agreement.

Chapter 12 provides a summary of the ideas expressed in the book.

* * *

I have illustrated a number of legal concepts by creating hypothetical cases, combined with facts from numerous cases handled in construction litigation over the last 12 years. These case descriptions occur throughout the book; they are identified numerically within each chapter: e.g. Case 1.1, 1.2, and so on.

Where the male pronoun "he" is used in the text, it refers to either men or women. I have done this to avoid the awkwardness of stating "he or she" at each use of a third-person-singular pronoun. The use of the word "he" in this context does not signify or imply that women do not work in construction and contribute to each aspect of the construction process.

I am deeply indebted to Lesly Miller, Esq., for tireless and thorough editing and proofreading; to Professor Justin Sweet, of the University of California at Berkeley, for aid and advice; and to my wife, Pamella Passaro, for allowing me the time to complete this work.

San Jose, California
December 1992

JONATHAN J. SWEET

SUMMARY CONTENTS

DETAILED CONTENTS

CHAPTER 1

INTRODUCTION TO CONSTRUCTION DISPUTE RESOLUTION

§ 1.1 Parties to Construction Contracts

In the traditional construction setting, the major participants are: (1) owner, (2) architect, (3) engineer, (4) general or prime contractor, and (5) subcontractors. The owner or occupier of land usually hires the architect and/or engineer to design a plan to build or improve a structure. The owner then hires a general contractor to execute the design. The general contractor hires subcontractors who perform specialized functions. Some of the common building subcontractor trades used in construction are: carpentry, steelwork, drywall, electrical, mechanical (plumbing, HVAC—heating, ventilation, air conditioning), and cement.

The subcontractors and the prime contractor must work closely together to successfully complete a construction project without defects. Cooperation is particularly crucial for timely completion. The potential for problems on construction projects is high. Each trade must perform its work properly and in the designated sequence. Lenders, insurers, sureties, and owners all must perform their assigned roles as well. Failure to perform properly can result in defects in the project, delay completion, and cause the parties to incur unanticipated costs.

§ 1.2 Factors Affecting Contractor's Performance

Successful completion of a construction project depends on many factors: (1) adequacy of the plans prepared by the architect and/or engineer, (2) timely acquisition of the land, (3) obtaining of necessary permits and environmental approvals, (4) reasonable cost of, and availability of, funds from the owner, or a bank, (5) contractor qualifications for bidding on particular types of jobs, (6) availability of bidding, performance, and payment bonds, (7) availability of supplies required for the contractor's particular task, and (8) efficient scheduling of the work of other contractors. Larger projects involve even more factors.

The number of parties involved in construction, and the closeness of their relationships, makes some disputes inevitable among the parties. Poor handling or resolution of these disputes can be disastrous to the timely completion of a particular project, and can have a negative overall effect on an individual contractor's business.

§ 1.3 Sources of Dispute

A completed construction project represents the combined labor of every trade on the job. A defect or problem with a construction project may be caused by the failure of several parties to do their work properly. Disputes often revolve around which party caused and is responsible for the delays in completion of construction projects. It can be difficult to determine who is responsible for these losses.

§ 1.4 Areas of Dispute

Certain factors are the source of recurring disputes on construction projects. Some of the most frequent causes of disputes and litigation are described in the following sections.

§ 1.5 —Contract Clause Interpretation

Determining the meaning or intent of the parties who agreed to a contract clause may be crucial for resolution of a dispute. Under the law of most states, if the meaning of a clause is clear on its face, a court will not examine other evidence, such as oral testimony, to determine the intent of the parties to the contract with regard to that clause. If language in a contract clause is ambiguous, it will usually be strictly interpreted against the drafter of the contract.

The parties with more bargaining power, such as owners, architects, and engineers, often use contract clauses to attempt to limit their liability for losses, delays, increased costs, or injuries that may occur on a job site. Where the drafter of the contract is in a position of greater bargaining power, a court may decide that a particular clause is invalid because it was not freely and willingly agreed to by the weaker parties.

If a court is willing to look beyond the written contract, it might examine outside evidence regarding the parties' intent, including oral testimony of the parties; correspondence or other documentation that discusses the contract and the relationship among the parties; and actions taken by the parties that support a particular interpretation of the contract.

§ 1.6 —Clauses Requiring Proof of Insurance

In many construction contracts a clause requires a subcontractor to provide evidence that he is insured against losses from liability claims. This clause may also require the subcontractor to name the general contractor or the owner as an additional insured under the subcontractor's liability policy. The presence of this type of clause means that the insurance company that wrote the policy for the subcontractor also will have to defend the general contractor from liability claims. Failure by an owner or general contractor to insist on advance proof that the subcontractor has obtained liability insurance could defeat a subsequent claim for contract breach against the subcontractor because the owner or general contractor did not fulfill this requirement.

§ 1.7 —Inspection of Work Clauses

Construction contracts often provide that the architect will inspect and approve all work performed on the project. Progress payments to contractors and subcontractors are generally withheld by the owner until the architect has approved the adequacy of each phase of the work and has verified that the work conforms to the specifications and plans. An architect who fails to inspect and approve work on schedule may be found liable to contractors for damages for the delay in payment, because the architect's failure to perform in timely fashion has caused damage to the contractors. The architect's liability may be based on a negligent failure to inspect and approve, or on an intentional or fraudulent failure to fulfill the architect's duty.

§ 1.8 —Indemnity Clauses

Most contracts contain an indemnity clause, which provides that, if one party to the contract must pay money for damages, that party will be indemnified (reimbursed) by another party. Courts analyze the leverage of the different parties and each party's relative bargaining power, and they consider public policy issues, before determining whether it is appropriate to enforce an indemnity clause.

§ 1.9 Liability to Third Parties

Subcontractors are third parties to many of the contracts on construction projects to which they are not parties. Architects and engineers are usually third parties to suppliers, adjoining landowners, and members of the general public. Subcontractors do not usually have contractual responsibilities to each other on projects as well.

§ 1.10 —Responsibility for Safety Conditions

The prime contractor has overall responsibility for the project and must provide safe conditions on the job site. He will be held liable for damage due to unsafe conditions. However, if a laborer or a member of the public is injured on a job site, and if the owner has some active involvement in the job, both the owner and the general contractor may be held responsible for the compensation for the injuries.

§ 1.11 —Payment for Supplies

Normally, the contractor or subcontractor who uses supplies on a project will be responsible for payment for those supplies. If the supplier is unpaid and provides the required notice, he may later file a mechanic's lien against the property and collect payment from the owner for the supplies. This recovery is based on the fact that the owner of the property has received a benefit from the use of the supplies on the property.

If a contractor leaves a construction job because of financial problems and fails to pay one of the material suppliers, the supplier may still be able to recover from the owner for supplies used on the project, without providing any mechanic's lien notice and without bringing suit on the mechanic's lien. If the owner completes the job, using the remaining supplies, the supplier also may be able to recover payment from the owner for the supplies actually used by the owner. The rationale is: to allow the owner to use the supplies without paying the provider would result in a windfall, or in unfair enrichment of the owner.

§ 1.12 —Implied Obligation to Cooperate

Written contracts on a construction project usually exist between the owner and prime contractor, owner and architect, and prime contractor and subcontractors. In addition to these written obligations, parties to a contract have an implied obligation, by law, to cooperate with each other.

Prime contractors are not entitled to make work difficult or impossible for subcontractors, and then to insist on and expect timely and complete performance from subcontractors. If the prime contractor does not cooperate with a subcontractor, the subcontractor may be entitled to leave the job. The general contractor may then be required to pay additional sums to have another subcontractor complete the work, and may be unable to collect those excess amounts from the first subcontractor.

Despite a lack of any written contract between subcontractors, each subcontractor may have legal rights and duties to other subcontractors. In most states, a legally implied duty exists for subcontractors to cooperate with each other in the performance of their work. A subcontractor's failure to cooperate may allow another subcontractor to recover against the first contractor for damages.

Some states that are more strict require that any obligation or duty of a subcontractor must be set out in a direct contract with the party sought to be held liable, before the subcontractor will be allowed to recover damages for failure to cooperate.

A clause in the agreement between the owner and the general contractor may provide that the subcontractors on the project must cooperate with one another. In the stricter states, a subcontractor suing another for failure to cooperate is unlikely to recover. The clause in the contract regarding cooperation is interpreted as having been intended to benefit only the parties to the contract, rather than outside parties, such as subcontractors.

§ 1.13 Disputes Based on Licensing and Public Bidding Requirements

Many professionals involved in construction, such as contractors, architects, and engineers, are licensed by states. The licensees must follow certain guidelines, or their licenses can be revoked. In addition, a person who is performing work that requires a license, and who has failed to obtain one, may be penalized.

§ 1.14 —Licensing Laws

In some states, such as New York, unlicensed contractors who sue owners are not allowed to recover payment for work performed. It is irrelevant whether the work was performed adequately, or whether the failure to obtain a license was deliberate. Even if the owner was previously aware of the contractor's lack of license and planned to take advantage of the lack of license, the contractor will still not be allowed to recover.

§ 1.15 —Special Problems Pertaining to Public Bidding

Contracts that are publicly bid have special problems. These projects are usually competitively bid. Any contractor who bids on such projects must comply with elaborate rules and procedures. Often, the public agency receiving bids retains the discretion to refuse to accept any bids, to cancel the job, or to solicit new bids.

Contractors' failures to follow required bidding procedures generate many disputes with the public agency or with other, unsuccessful bidders. A bidder who fails to comply with certain rules, such as listing of subcontractors or specifying the unit price for each material, may argue that the public agency can waive such defects as minor. The public agency is likely to maintain that it has the discretion to decide which infractions of its rules are minor and can be waived, which allows the public agency to accept or reject a particular bid. For contractors who bid frequently on public contracts, failure to comply with all the necessary procedures may have drastic consequences. Not only may a contractor be prevented from receiving the award of a particular contract or recovering for work performed, but that contractor also may be barred from bidding on future contracts with the awarding agency.

§ 1.16 Selecting an Attorney

Until recently, ethical rules prohibited attorneys from advertising their services. Within the past few years, via court decisions, legal ethical rules have been modified. Some types of advertising are allowed, as long as they are not misleading. However, restrictive rules still exist, regarding the ways in which attorneys may directly solicit clients, because of the opportunity for the attorney to take unfair advantage of clients, or because of the potential for misleading clients that is inherent in direct solicitation by attorneys.

Many county bar associations maintain referral panels of local attorneys, classified by specialty. Trade associations may maintain lists of attorneys whom they recommend. Most state bar associations have certifications of specialty in certain areas, such as tax, family law, and workers' compensation; however, construction law is not usually a certified specialty. Law libraries have available a legal directory called "Martindale–Hubbell,"[1] which lists many law firms geographically, by state and city. The listings may indicate representative clients and the areas of law in which the firm practices. The local Yellow Pages usually have a section headed "Attorneys," with names arranged alphabetically by firm name or surname; boxed listings or advertisements will indicate a practice specialty. Many people ask their friends for a referral to an attorney when they have a problem.

A personal referral, followed by an initial in-person consultation, is probably the best method for selecting an attorney. If no personal referrals are available, county bar associations will give a referral from an appropriate panel. A call to an attorney, to request client references, can provide an informed impression of the quality of work performed by the attorney.

Most attorneys allow new clients a brief interview before the clients decide whether to retain the attorney to handle a particular matter. The attorney may conduct the initial interview at no charge or may charge only a small fee. The interview provides both the new client and the attorney with an opportunity to become acquainted, before deciding on whether they will work together.

---

[1] *Martindale–Hubbell, Law Directory,* Reed Properties (published annually).

§ 1.17 —Working with Attorneys

Your attorney has two primary roles: (1) to provide you with advice on legal matters, and (2) to act as your agent or spokesperson, both in court and in dealing with other parties. Once you have retained an attorney, ethical rules prohibit other attorneys from directly contacting you. The attorney for the other side, in any litigation, is required to direct all communication to your attorney.

Once you have retained an attorney, you may worry about whether the attorney is handling your matter properly. Insist on your attorney's keeping you informed of the status of your case and responding to your phone calls or letters. The dispute is yours, even though your attorney is dealing with the legal aspects of the dispute.

You and your attorney must be able to cooperate. It is crucial that you be comfortable with your attorney and confident about your attorney's skills. After you have retained an attorney, if you believe the attorney is mishandling the lawsuit, you should request an explanation from your attorney. If you are still unsatisfied, you should retain a different attorney.

Using an attorney on a regular basis can become expensive. Most attorneys charge for their services by the hour, and their fees include charges for all time spent on matters for each client. You will be charged for phone calls, letters, travel time, and any other time the attorney devotes to your work. Hourly rates for attorneys can range from $80 per hour to $400 per hour. Variables used to determine hourly rates may include the number of years the attorney has been practicing, the area of specialty of the attorney, and the geographical area where the attorney is located.

Attorneys generally have an obligation to notify clients of the amount of their hourly rates, and to inform clients of any changes in the initial rates. In *Severson, Werson, Berke & Melchior v. Bolinger,*[2] the Severson firm used a fee agreement that did not specifically state the amount of the hourly rates it intended to charge its clients. At some point in time, oral quotes of those rates were given to clients. The fee agreement contained no language stating that the hourly rate might increase during the course of legal representation. In addition, the hourly rate could not be determined from review of the bills received by the client, Bolinger. In the *Severson* case, the court held that, where a client has agreed to pay a law firm's "regular hourly rates," such rates cannot be increased without prior notification of the client.

Some attorneys charge flat fees (one fee for handling the entire case) for their services, rather than hourly rates. Flat-fee billing is usually limited to the handling of simple matters, such as an uncontested divorce or a basic will. Some law firms will agree to charge flat fees for standard services to clients who provide a large volume of work. Construction disputes will rarely be handled for a flat fee, because of the large number of variables or unknowns

---

[2] _____ Cal. App. 3d _____, 1 Cal. Rptr. 2d 531 (1991).

involved, and because of the specialized knowledge required to handle these disputes.

Personal injury attorneys commonly accept cases for a percentage of the recovery. This is called a "contingency fee." This term means that the attorney's fee is contingent on the attorney's obtaining a recovery for you. If there is no recovery, you pay nothing to the attorney except the costs of the suit (filing fees and other out-of-pocket amounts incurred by the attorney). The attorney usually receives from 25 to 45 percent of the net recovery as a contingency fee.

One reason for the high costs of legal fees is that many people do not consult an attorney until their situation has reached a crisis level—until they have been sued or wish to sue. By this time, preventable and costly mistakes may have been made which complicate the attorney's work.

Disputes may involve small dollar amounts, which cannot justify the cost of retaining an attorney. In those situations, the damaged party may file an action in small claims court or may decide to try to negotiate a settlement.

§ 1.18 —The Litigation Process

The person who files the lawsuit is called the "plaintiff"; the person who responds to the lawsuit is called the "defendant." Usually, the basis of a lawsuit is that a party owed a duty to another party and failed to live up to that duty, which failure damaged the plaintiff. The duty may be set out in an oral or written contract, or it may be created by law.

For a lawsuit based on breach of contract, the time for filing suit is measured from the date when the obligation set out in the contract was due, and performance was not rendered. In many states, a suit based on an oral contract must be filed within two years of the date of the "breach" (the failure to perform under the contract). For a written contract, a suit must be filed within four years or longer from the date of the contract, depending on the specific state laws or federal statutes that apply.

Tort claims (non-contract-based claims) involve damage to person or property. Where a tort claim is based on an allegation of negligence, some states require that suit be filed within one year from the date of the negligent act.

In some jurisdictions, although a suit may have to be filed to preserve legal rights, the parties still can pursue settlement while the suit is pending. The action of filing the lawsuit does not necessarily require that the suit be actively litigated. However, the federal courts, and some state courts, require the parties either to settle or to actively prepare for trial within a short time of filing suit. Statutory deadlines must be met for filing and pursuing lawsuits, regardless of whether settlement still may be a possibility.

Some people blame their attorney if they do not win a case. If your attorney is fulfilling his role properly, he will have advised you of the strengths and weaknesses of the case at the outset of litigation. Your attorney does not create the facts of your case, and you may lose because the other side has stronger facts. In addition, many factors outside your attorney's control can cause or contribute to an unsatisfactory conclusion to the dispute. However, if you feel your attorney has acted improperly or unethically, you may complain to the state bar association.

An attorney's failure to handle a suit properly is called professional malpractice or professional negligence. If you have evidence that your attorney has mishandled your case and that the mishandling has caused you damage, you may sue your attorney for malpractice.

§ 1.19 —Client's Role in Discovery Process

Both you and your attorney bear responsibility for ensuring that the attorney–client relationship is successful. You must cooperate with your attorney, and your attorney should be responsible to you and responsive to your needs. In addition to periodic status reports, a part of your attorney's responsibility is to give you approximate time frames in which you can expect events in the litigation to occur. If facts are discovered that dramatically shift the likelihood of winning a case, your attorney must advise you of these facts as soon as possible, to avoid later disappointment and to prevent needless expenditure for legal fees.

Lawsuits require that you provide your attorney with information and with documents. The pretrial investigation process, called "discovery," consists primarily of the exchange of information and documents between the opposing sides to a lawsuit. It is essential that you cooperate with requests by your attorney for information or documents that your attorney must provide to the opposing side. The documents must be provided by your attorney to the opposing side within the time stated by the opposing attorney, a time usually dictated by court rules. Failure of your attorney to meet the deadlines for document production may result in the court's imposing fines or penalties on both you and your attorney.

Pretrial oral testimony, called a "deposition," often is used to obtain information from parties or witnesses. Few clients or witnesses enjoy testifying at a deposition. The deposition is an opportunity for the opposing attorney to assess your believability and to test your memory about the events important to the dispute.

The testimony you give at your deposition is important to your case. However, you are not giving a poor deposition if you cannot positively answer every question asked at the deposition. You need to be aware that "I do not know" or "I cannot remember" are acceptable answers and, in fact, are often likely to be the only honest answers. Most attorneys will sit down with you and explain the deposition process in advance, so you will know what to expect.

It is important that you retain, for your attorney's review, all documents that pertain to a legal dispute. Having the documents available for review will allow you and other witnesses to testify more accurately at deposition or at trial. Although both documents and oral testimony are evidence that can be used in court, your memory of details may fade, and the documents will store information in a fixed, relatively permanent form. If you store your documents using a system that allows quick and easy retrieval, you will make your attorney's work easier—and perhaps reduce your legal fees.

§ 1.20 Methods of Dispute Resolution

Different methods may be used to resolve disputes that arise regarding construction projects. Prelitigation procedures may be set out in the contract documents, described in union agreements, or set by statute. Alternatives to conventional litigation also exist.

§ 1.21 Use of the Public Court System

Traditionally, when negotiations break down between parties to a dispute, one of the parties files a lawsuit against the other. With many construction disputes, there are valid reasons why this should not be the first step toward resolution.

§ 1.22 —Public System Management of Construction Disputes

The present public court system is ill-equipped to handle construction disputes, although some court systems are attempting to educate litigants (see **Appendix M**). Arbitration clauses are found in most construction contracts because the parties to a dispute have at least one common goal: They wish to resolve the dispute. They may not agree on the method to resolve the dispute, or on the figure at which the dispute should be resolved, but most people do not want to spend time, energy, and money in pursuing lawsuits.

Even for simple lawsuits, the court system is expensive and slow. Because construction disputes tend to be more complex, the delays before trial may be longer than for the average commercial dispute. More complex cases require longer trials. The longer the parties estimate the trial will last, the further ahead in time the courts must schedule trials.

Most states require that the plaintiff bring the matter to trial within a specified period of time. For instance, in California, all suits must be brought to trial within five years from the date of filing, unless the parties to the litigation agree to extend the five-year period. Many years may pass before a civil matter actually comes to trial. One of the reasons that civil cases are frequently delayed is that criminal matters have priority over civil cases. (Most construction disputes are civil cases.)

Many states require that the parties go through nonbinding arbitration and numerous settlement conferences before the case is finally brought to trial. These requirements make it so expensive to bring civil cases to trial that many parties are discouraged from going to trial, unless very large amounts of money are involved or any compromise of the dispute is impossible.

§ 1.23 —Relief Available from Conventional Litigation

Money is the usual award or "judgment" obtained in a lawsuit. Occasionally, parties may seek an order from a court that a party perform, or not perform, a particular act. This order is called an "injunction." For instance, where a bidder protests the award of a contract to another bidder, the protesting party might ask the court to order that the awarding agency not be allowed to award the contract to another bidder, and that the contract be awarded to the protesting bidder. The standards that the courts apply to requests for injunctions are high, and money awards are by far the more common relief requested and obtained from courts.

One of the important considerations for you, in deciding whether to file a lawsuit, is whether an award of money will compensate you for the loss you have suffered. Not all losses can be quantified into dollar amounts. In fact, the negative effects on your business, in addition to the cost of litigation, may be strong arguments against proceeding with litigation to resolve construction disputes.

Often, because of your anger with another party with whom you have a dispute, you may wish to have revenge against the other party. The civil legal system is primarily concerned with compensating those who are damaged, not with punishing the "bad" behavior of defendants. The award that a party to litigation will receive is generally calculated to be the amount of money that will place the damaged party in the position it would have been in had the defendant acted in accord with applicable standards, rather than an amount that will "punish" the defendant.

§ 1.24 —Drawbacks to Being a Litigant

Before agreeing to work on a project, many of the potential parties will investigate the reputation of the others who will be participating on the project. Discovering that one of the parties on a prospective project has previously been involved in litigation may cause concern. Potential bidders may be less willing to participate in the project, because they may believe that the person who has been in litigation in the past is more likely to use litigation to resolve disputes that may arise from the project.

Mutual trust and respect between parties to a construction contract are critical. Even if projects encounter delays or problems, if these situations can be resolved without a lawsuit, the parties are more likely to be willing to work together again on future projects. Because the community of people in the construction business is fairly small, parties are likely to encounter again those with whom they have worked before. The desired mutual trust and respect are more likely to exist where parties have not been opponents in a lawsuit.

To avoid many of the drawbacks that accompany litigation, attorneys frequently advise their clients to agree to a compromise figure, or to "write off" the entire sum involved in the dispute, rather than to file suit. After going through the entire litigation process, the amount originally disputed may seem small when it is compared to the money, time, and energy spent in pursuit of that amount. Unless legal fees are specifically provided for by contract or statute, legal fees incurred in bringing and pursuing a lawsuit usually are not recoverable.

It should also be noted that obtaining a judgment or award from a court does not guarantee a party the actual recovery of money. If the other party lacks assets, or has no insurance available to pay for the judgment, winning the lawsuit may be an empty victory.

Although there are times when litigation is necessary, it has many drawbacks. Some people who become involved in litigation are stubborn and refuse to agree to any compromise. They assume that they benefit automatically if a litigation tactic or maneuver damages the other side, or if the other side is forced to incur costs.

§ 1.25 —"Early Dispute Resolution"

In some states, the public court system has initiated early dispute resolution programs and nonbinding judicial arbitration as a part of the court procedures, to dispose of cases without trials (see **Appendix M**). Because of the high cost of using the public court system, and the delays attributable to a shortage of judges, courtrooms, and other resources, many parties have shifted to use of private organizations that act as arbitrators, mediators, or judges, in order to resolve construction disputes.

§ 1.26 Alternatives to Conventional Litigation

The current trend in dispute resolution is toward mediation and arbitration, as opposed to the more combative conventional courtroom litigation. This trend is a response to the high costs and disadvantages of conventional litigation. Most people prefer not to waste time and money on traditional litigation if the dispute can be resolved by some other method. Arbitration, mediation, and private trials are all services provided by most of the private organizations.

The American Arbitration Association is the oldest of these private organizations (see **Appendices I–L**). Many others have developed over the years, for use by those who are dissatisfied with the public court system. The primary barrier to using these private organizations for alternative dispute resolution is that all parties must agree to their use. High cost is also a disadvantage. Fees of $250 to $300 per hour for the use of these organizations are not uncommon. If the parties will agree to accept as binding a decision from one of these private organizations, the decision made by the representative of that organization is final. This allows the parties to be sure that the dispute resolution procedure they have chosen will end the dispute.

§ 1.27 —Arbitration

With arbitration, an impartial third party hears evidence in the form of testimony, reviews documents, and hears arguments by the parties or their attorneys. The arbitrator then gives the parties his or her decision. If the arbitration is binding, the parties must accept the arbitrator's decision. If it is not binding, the parties then may file suit, or, if suit has already been filed, proceed with the litigation process. Arbitration is the oldest form of alternative dispute resolution, and remains the most common, particularly with construction disputes.

An arbitration clause requiring the parties to a contract to submit the dispute to arbitration will not be triggered automatically when a dispute arises. If one party sues because of a dispute, and the contract contains an arbitration clause, the party who prefers arbitration to litigation must demand arbitration. If a lawsuit has been filed, the arbitration request will force suspension of the litigation until the arbitration hearing has occurred and a decision has been rendered. Frequently, a contract will provide that the rules and procedures of a particular organization, such as the American Arbitration Association, will regulate any arbitration held pursuant to the contract (see **Appendix I**).

In judicial arbitration, if a party is dissatisfied with the arbitration result, that party may request a trial. Judges will often try to persuade the parties to agree to make the judicial arbitration binding, with the parties waiving their right to a trial following the arbitration. If the judicial arbitration is not binding, a party who is unhappy with the arbitration result may seek a trial. Even after a trial, he may appeal.

Problems may arise where not all parties to a dispute are subject to arbitration. For example, in a construction dispute, one set of claimants may have a contractual obligation to arbitrate before filing suit, while other disputing parties are not obligated to arbitrate. In such mixed cases, a court may be faced with a decision on whether to delay the litigation, in order to allow the arbitration to proceed to conclusion before allowing the lawsuit to proceed.

The argument in favor of delaying the litigation is that the arbitration may resolve the entire dispute, making any trial of the dispute unnecessary. On the other hand, if all issues are not resolved by arbitration, using the arbitration procedure will not expedite the dispute resolution process; after the arbitration, a trial still will be necessary. Most states have a policy of encouraging arbitration and out-of-court settlement of disputes. This policy has to be balanced against the constitutional right of parties to a trial on the merits of their dispute.

§ 1.28 —Mediation

Mediation is an informal meeting between the parties to a dispute and an impartial third party, with the goal of resolving the dispute. The process is fluid and usually examines many practical factors, such as the availability of insurance or of assets for payment, which are not relevant to an arbitration or trial. Mediation is most effective where the parties are able to cooperate sufficiently to negotiate a settlement, but need some direction.

Mediation is usually privately arranged by the parties or through private organizations. Some states, such as Michigan, have a mediation process incorporated into the traditional court system. Penalties are imposed on parties who refuse to accept the mediator's evaluation of the case.

§ 1.29 —Dispute Resolution Through Private Organizations/Alternative Dispute Resolution

The alternative dispute resolution (ADR) services offered through private organizations use retired judges, or other persons who have special expertise in given areas of the law, to help parties resolve the disputes. These organizations allow parties and their attorneys to schedule arbitrations, mediations, or private trials at their own convenience. The public court system lacks this flexibility and responsiveness.

Some people are concerned, however, that widespread use of these organizations results in the application of a different set of standards for those with the money available to hire these organizations. Many judges are retiring early to work for the private organizations. They can work fewer hours, and make more money, by leaving the public system. Some people fear that the quality of judges in the public system will suffer, and that fewer people will wish to remain as judges in the public system.

An opposing argument holds that taking complex and time-consuming construction cases out of the public court system will allow the public system to deal more expeditiously with simpler cases. Removal of construction cases makes more courtrooms and facilities available for those without the money to use the private systems.

§ 1.30 Conclusion

Filing lawsuits and pursuing conventional litigation is often an expensive and inefficient method of solving construction disputes. Disputes can be avoided by carefully reviewing and understanding the legal effect of the documents pertaining to each project.

Unfortunately, many contractors hesitate to hire attorneys to review the documents before agreeing to work on a project, because of the cost of the attorney's time. Attorneys say that many of the lawsuits they handle could have been avoided if the attorneys had been consulted in advance. Awareness of the significance of particular contractual clauses, of the importance of securing insurance, and of securing performance or payment bonds, may save a contractor from personal liability for damages arising from a dispute.

Once a dispute arises, an open attitude regarding compromise and a willingness to be flexible may prevent many lawsuits.

Resolution of construction disputes is handled most efficiently between the parties themselves, or by their attorneys. If informal resolution is not possible, the next best method may be private arbitration or mediation. Mediation or arbitration may be unavailable unless all parties will agree to use of these means to handle the problem, or unless the contract provides for one or more of these alternative dispute resolution procedures. The cost of using these organizations may also be a barrier.

Your selection of an attorney who understands the construction business, or is willing to learn about the construction process, may be one of the most critical steps you can take to avoid or minimize litigation. You must be sure that your attorney is aware of your objectives and cost constraints. It is also important that you listen to your attorney's recommendations, and that you avoid insisting on revenge against those whom you blame for the problems that have arisen.

CHAPTER 2

OUTLINE OF THE AMERICAN LEGAL SYSTEM

§ 2.1 Rules for Regulation of Behavior

It is generally accepted that society needs laws to regulate the behavior of its members. For the smooth operation of the construction process, parties need to understand the nature and extent of their obligations to each other. It is equally important that parties be aware of the probable consequences of failing to meet their obligations.

§ 2.2 Structure of American Government

There are three branches of the American democratic system of government: (a) executive, (b) legislative, and (c) judicial. Each of the branches exercises power over certain areas of government and acts as a check on the power of the other branches.

§ 2.3 —The Executive Branch

On the national level, the president and his cabinet members comprise the executive branch. The governor and his cabinet perform the same function on a state level. On a local level, the executive branch may include a mayor, city council, or board of supervisors.

§ 2.4 —The Legislative Branch

Elected representatives, such as senators, representatives, and assembly members, enact laws called statutes, which directly regulate construction. Once a legislative body passes a statute, administrative agencies interpret and enact implementing rules and regulations.

§ 2.5 —The Judicial Branch

The American legal system has two primary court systems, the federal system and the state system. Each has jurisdiction (control or power) over different types of disputes.

§ 2.6 The Common-Law System

The courts and judges make findings regarding the facts of cases placed before them, and interpret the applicability of statutes passed by legislative bodies to the facts before them. If no statute applies, the rule developed in previous cases with similar facts is applied. Using the rule of earlier cases to decide later cases is called **applying precedent**. This legal system, derived from the customs of the English **common-law system**, is used in every state except Louisiana. Louisiana uses the "civil law" system, derived from France's legal system.

In the common-law adversary system, each party has his own independent attorney. This concept of the independence of the attorneys is critical to the common-law system. There is an impartial judge who hears the evidence and decides the matter. In other legal systems, the judge may be a civil servant who is a representative of the state, and who has no independence from the government. Both attorneys or legal representatives may also be employees of the state, having far greater loyalty and commonality of interest with the government than with their clients. The adversary system used in the United States strongly relies on the fact that American attorneys usually have loyalty only to their clients.

U.S. trial courts have the responsibility for administering remedies for violations of statutes or of administrative regulations. Appellate courts review the decisions of the trial courts to ensure that no errors of law, or abuses of discretion, have been committed by trial judges. Appellate court decisions are published in books, and those books are available to judges and attorneys for guidance.

Generally, judges are required to follow the decisions of judges who serve at a higher level within the same court system. A municipal or superior court judge is required to follow the decisions of the state appellate court, its state supreme court, and the decisions of the U.S. Supreme Court. The state judge is not required to follow the decisions made at the trial level by other judges within his or her system, or by judges of other states, or by federal judges. The one exception to this rule is decisions of the U.S. Supreme Court, which must be followed by all courts within the United States.

Appellate judges in different districts within the same state court system may make inconsistent rulings. When this conflict arises, the state supreme court is likely to accept a case for decision in that area, to settle the inconsistency between its appellate courts. Federal trial and appellate level decisions are not binding on judges in the state court systems. Although decisions from other jurisdictions may not be binding, they may provide guidance for a state that has not yet faced a particular issue, and where no prior case or statutory authority exists. If a judge, or an attorney for a party, does not want a particular precedent to be followed, he or she will argue that the facts of the case before the court are different from those of the case that states the rule.

For example, an appellate decision articulates a rule that a contract will be interpreted against the person who drafted the contract, and in favor of the other party.[1] The prior case being cited as precedent offers clear facts that the drafting party had his attorneys draw up the contract, and the other party did nothing but sign the contract. However, assume that, in the case being argued, Brown typed up the contract, but Green wrote paragraphs that were inserted and altered the language of the paragraph in dispute. In this situation, it is most likely that the case rule (the contract will be interpreted against the drafter) will not be followed. The facts are distinguishable because Brown and Green prepared the contract together, as opposed to one party preparing the contract and the other party blindly signing it. In this case, there is no reason to place the full burden for vagueness in the contract on one party.

In some situations, the facts of a case cited may be different, but the nature of the difference will not prevent application of the rule. A case might cite a rule that an owner of an automobile has a duty to maintain his vehicle properly. If the poor state of maintenance of the vehicle causes an accident and injury, the owner of the vehicle will be legally responsible to the victim. The owner cannot legally pass that responsibility to anyone else, such as a mechanic who services the vehicle. The owner may have a right of indemnity from the mechanic because the mechanic's act, or failure to act, caused the mechanical defect, but the owner remains primarily responsible to the injured person.

In the case that established the precedent, the vehicle was ten years old. The vehicle in the instant case is only six months old. This factual variation in the age of the vehicle does not prevent application of the rule. The owner of the vehicle remains responsible for the lack of maintenance of the vehicle, regardless of its age, when that lack of maintenance has caused an injury.

It is common for judges to refuse to follow the rule of a case where a relevant fact makes it inappropriate to apply the earlier rule. In addition, changes in society may demand that the rule of an earlier case not be followed.

Years ago, a person could bring a suit for a defective automobile only against the party from whom the injured person purchased the vehicle. In most cases, that would have been the automobile dealer. At a certain point, the U.S. courts decided that the injured person should be able to sue the manufacturer even though there was no direct contractual relationship, called **privity of contract**.

The courts eventually concluded that it was more important to compensate the injured victim than to shelter the manufacturer, by relying on the lack of contractual relationship. The rationales used by the courts for this change in law were that manufacturers could better spread the cost of making automobiles safer and could better shoulder the financial responsibility of paying those injured by the defective cars. This change in law provided the manufacturers with a strong incentive to make products as safe as possible, to avoid being held liable for such injuries.

Today, insurers maintain that they cannot accurately price the cost of the risk for defective products and that consumers are not willing to pay corresponding

---

[1] *See* Williams Engineering Inc. v. Goodyear, 496 So.2d 1012 (La. 1986).

higher insurance rates. Insurance companies and manufacturers have sought to change the laws to limit the amounts recoverable by victims injured by products. Manufacturers and insurance companies claim that juries, swayed by sympathy for injured victims, are handing down excessive verdicts and causing financial crises for business. Manufacturers and insurers claim that these excessive judgments force them to raise prices for their products or services. They claim that, if they are not able to alter the law, or to charge higher prices, they cannot make an acceptable profit. In response to industry pressure, some courts have limited the rights of the persons injured by products, and have attempted to limit the coverage of insurance for those risks the courts feel are unreasonable, or for those risks the insurance was not clearly designed to cover.

§ 2.7 —Entitlement to a Jury Trial

In criminal matters, a defendant is constitutionally entitled to a trial before a jury. The fact that the liberty of the criminal defendant is at stake is one of the reasons that the courts provide what is considered to be this higher level of protection. For this same reason, a criminal defendant is constitutionally guaranteed the right to be aided by an attorney. If the defendant cannot afford a defense attorney, the court will appoint and pay for a defense attorney.

In civil (noncriminal) matters, a party is entitled to a jury trial only if the relief sought is **legal** relief, the award of money. Where a party asks the court for something other than the award of money, the relief sought is considered **equitable**. This type of court order, an injunction, calls for a party to do or to abstain from doing a particular act. No entitlement exists to a jury where the only relief sought in a lawsuit is equitable. Cases seeking equitable relief only will be heard by a judge.

Originally, in the United States, there were courts of equity, which handled injunctions, and courts of law, which handled legal, or money, awards. Over time, the two systems have essentially merged. One court system administers both types of remedies.

For example, if a rejected low bidder filed suit against an awarding agency for money damages, the low bidder would be entitled, at trial, to a jury. The low bidder might seek to recover the costs of preparing and submitting the bid, as well as lost profits. If, however, the low bidder requests an order from the court to force the awarding agency to grant the low bidder the contract, this requested injunction would constitute equitable relief, to be heard and decided by a sole judge. In a case involving a request for injunctive relief, the first step would be to seek a **preliminary injunction**, an injunctive order granted prior to trial. To obtain a preliminary injunction, the disappointed low bidder must prove, to the satisfaction of the court, the need for the court to grant the injunction immediately, without a trial, in order to preserve things as they stand until the trial. The low bidder must prove to the court that the low bidder would suffer irreparable loss if the injunction is not granted, that an award of money is not a sufficient remedy, and that there is a strong likelihood that the low bidder will win at the trial of the matter.

Although parties may be entitled under the above rules to a jury trial, if both parties waive the right to a jury trial, the case will be heard only by a judge. Among the reasons for waiver are the probabilities that the trial of the matter will occur more quickly, and that the trial may take less time when it does occur.

§ 2.8 Function of Attorneys

An attorney performs two primary functions for clients: (1) the attorney acts as the client's representative in dealing with other attorneys, the court system, arbitrations, administrative hearings, or other proceedings, and (2) the attorney advises the client on legal matters.

§ 2.9 —Courtroom Advocates

Most court systems allow parties to choose to represent themselves in court. There are two reasons why this self-representation is not wise: (1) the party to the dispute is probably unfamiliar with the courtroom procedures required to put his or her case on at trial; and (2) the party usually lacks the emotional or psychological distance to present the case objectively and effectively.

Judges and juries are aware that attorneys are hired by clients to represent the clients and to present the clients' point of view in court. For the client, it is valuable to have, as the presenter of the case, a person who is distinguishable from the one asking for relief, or from the one claiming that no relief should be granted. If the attorney performs properly, a jury should be impressed by the attorney's sincerity and by the apparent fact that the client has convinced the attorney of the validity of the client's position.

Another reason to use an attorney is that, because of an emotional distance from the facts of the case, the attorney will be able to assess the strengths and weaknesses of the case more objectively. This does not mean that persons who represent themselves will never win or cannot do a good job; it is an explanation of why persons who represent themselves will have an uphill battle and will be laboring under a disadvantage.

§ 2.10 —Legal Advisers

Ethical rules for attorneys state that your attorney should be concerned about your moral responsibilities, as well as with the legal implications of your actions. As a practical matter, often it is difficult for an attorney to advise you regarding nonlegal concerns. If you describe a proposed course of action to your attorney, you are probably only asking for the attorney's knowledge regarding the legal ramifications of following that course of action. You are not usually interested in whether the attorney approves of your actions, or whether the attorney believes the actions are admirable or morally correct.

For example, you inform your attorney that you have recently hired a former employee of a competitor. The former employee of the competitor is willing to divulge confidential information to you about the clients, techniques, and marketing approach of the ex-employer. You can use this information to your advantage and to the detriment of your competitor. Even though taking these steps may not be illegal, it may be morally wrong.

In other situations, you may feel that your financial future rests entirely on taking a step that is illegal or could subject you to criminal or civil liability if you are caught. However, the potential profit in cutting that corner is so great that it appears to you to outweigh the risk.

Although legal rules are based on right and wrong, many actions that are morally questionable are not illegal. Still other acts that are both wrong and illegal carry only a small risk of being caught. Most states have rules of professional responsibility for attorneys that require that your attorney attempt (1) to explain the legal consequences of your actions to you and (2) to convince you not to do things that are wrong.

§ 2.11 —Selection of an Attorney

If possible, you should select an attorney who has some experience in handling construction cases. The interrelationships of the parties, and the continuity of business relationships in construction, are far more important than they may be in other business areas. It will be critical for you to select an attorney with whom you are comfortable and whom you respect.

Your initial perception of the attorney is likely to be a fairly accurate reflection of how he is perceived by other attorneys, or by a court. The focus of the attorney should be on your problems and their possible solutions. If you do use an attorney who lacks prior experience with construction matters, the attorney's willingness to be educated by you about the construction field is a good indication that he possesses the necessary flexibility to deal with your case effectively. A large part of being a good attorney is being able to quickly grasp the essentials of other fields. In order to handle a trial, or other legal proceedings, the attorney must have or develop a fundamental grasp of the essentials of the issues. In construction, this means the attorney must learn about engineering, architecture, physics, banking, and other areas, so that he can understand the problems that arise.

When you are considering hiring an attorney, it is essential that you select a person (1) with whom you feel comfortable, (2) who is knowledgeable about construction, or willing to learn, and (3) who is responsive to your needs as a client.

§ 2.12 —Attorney Fee Agreements

Some states require attorneys to have written fee agreements with clients. Even where these are not required, it is a wise practice to have such agreements. Both you and your attorney should have a clear understanding of your respective rights and obligations. A written agreement that addresses these issues is the most effective way to ensure this mutual understanding.

Usually, a fee agreement specifies the basis for the attorney's fee or billing and sets the amount of the retainer (advance payment). **Figure 2–1** is a sample fee agreement.

After reviewing your written fee agreement, if you are still unsure about your responsibilities and those of your attorney, you should ask your attorney for clarification. From the outset, it is important that you feel comfortable about asking questions about any of the issues that arise when you are dealing with legal matters. Your attorney should be patient and thorough in explaining both the agreement and the issues that may arise in the course of your legal representation.

§ 2.13 Jurisdiction of the Courts

Different court systems have jurisdiction (authority or power) over different types of disputes. Jurisdiction may depend on the citizenship of parties, the amounts in dispute, or the specific laws that are claimed to have been violated by the parties.

§ 2.14 —The Federal Courts

The federal court system has three levels: (a) the federal district courts, (b) the circuit courts of appeal, and (c) the United States Supreme Court. The district courts are the trial level courts, the circuit courts of appeal are the first level of appeal, and the U.S. Supreme Court is the last level of appeal. Essentially, the standards followed in the state court system apply to the power of the federal appellate courts to review the decisions of the district courts: the trial judge must have made an error of law or **abused judicial discretion**. An abuse of discretion might be found if the judge made a finding that was totally unsupported by a reasonable interpretation of the evidence. For example, assume no evidence exists to support a finding that a supplier was paid for goods. In spite

We have agreed to handle the following matter on your behalf:

This written agreement specifies our responsibilities to each other, including our fee arrangement.

1. *Attorney's Responsibilities:*

It is our duty to evaluate the facts in your matter and to handle your matter as those facts and your goals suggest it should best be handled. It is our duty to explain the law(s) which governs your matter to the extent you request.

It is our further duty to advise you if, at any time, it appears to us that the law will not allow us to achieve your goals and to explain why.

2. *Client's Responsibilities:*

As our client, your responsibilities include:

A. Complying with the fee arrangements set out in this Agreement;

B. Advising us of the complete facts which you believe to be harmful to your case;

C. Promptly furnishing all papers and documents we request;

D. Keeping all appointments or rescheduling them before the scheduled time;

E. Keeping us informed of your current address and telephone number;

F. Cooperating with us to develop and prove your case;

G. Advising us what you know or believe others involved in the case will claim the facts to be, whether or not these claims are true.

3. *Payment:*

Our fees in this matter will be paid on an hourly basis in the amount of $_________ per hour. At the commencement of work, you will provide us with a retainer in the amount of $_________. The retainer will be placed in our Client Trust Account, and as work is performed, payments for that work will be deducted from the Client Trust Account. When the retainer balance drops below $_________, we will request a supplemental retainer. We will provide you with monthly statements regarding the legal services provided, even where the funds used for payment originate from the Client Trust Account. If the amounts incurred in any given month exceed the funds held in the Client Trust Account, we will expect payment of those amounts within thirty (30) days of the amounts billed.

4. *Costs:*

You understand and agree that there may be costs incurred in this matter, including but not limited to fees for filings with the Court or the State, process servers, court reporters, subpoenas, postage, long-distance telephone calls, investigators, and expert witnesses. We will deduct these costs, as they are incurred, from the funds deposited on your behalf in the Client Trust Account. In the event that the funds held

Figure 2–1. Attorney and client fee agreement of representation.

in the Client Trust Account are insufficient to cover the costs, we will expect reimbursement of those costs within thirty (30) days. We may submit invoices for costs in excess of $ __________ directly to you for payment.

5. *Client's Right to Terminate Attorney:*

You understand that you have the right at any time to request that we discontinue representation of you in this matter, and that upon notification we will withdraw from the case and provide you with copies or originals of your documents and files pertaining to this matter.

6. *Withdrawal of Representation by Attorney:*

You understand and agree that the occurrence of any of the following acts or omissions entitles us to withdraw from representing you in this matter:

A. Failure to comply with our requests for proceeding in this matter, including, but not limited to failure to supply prompt responses for Interrogatories or to supply any information or documents requested by us;

B. Failure to notify us in writing of any new address and/or telephone number;

C. Failure to disclose material facts to us, or;

D. Any misrepresentation of material facts regarding your case.

Upon occurrence of any of the above, you agree, upon our request, to execute a substitution of attorney, or substitution of client in propria persona (client acting as own attorney).

Should you fail or refuse to sign such a substitution, you agree that we shall be compensated by you at the hourly rate specified in this agreement for the effort and time required by us to prepare and attend a motion to withdraw as your attorney.

7. *Right to Independent Legal Advice Regarding this Agreement:*

This document creates legal obligations on both of our parts. You have the right to be advised by an independent attorney with respect to this agreement if you desire.

BY SIGNING THIS AGREEMENT, YOU UNDERSTAND AND AGREE TO ACCEPT ITS TERMS AND YOU ACKNOWLEDGE RECEIPT OF A COPY OF THIS AGREEMENT.

Dated: ______________________,

By: ______________________

____

CLIENT

Figure 2–1. *(Continued)*

of this, the judge's ruling incorporates the "fact" that the supplier was paid, as though this finding is based on the evidence.

Questions sometimes arise regarding the choice of court in which a lawsuit should be filed. Usually, each U.S. citizen is also a citizen of a state or dependency. The most common reasons for filing a lawsuit in federal court are: (1) the other party to the dispute is a citizen of a different state or country, (2) the lawsuit is brought for violation of a federal statute or constitutional provision, and (3) the United States is a party.

If Frank, a citizen of New York, wishes to sue Bill, a citizen of Montana, he probably would sue in federal court because they are citizens of different states.

If Frank sues Jane, believing that Jane fired him because Frank is Black, the suit would be likely to be filed in federal court because discrimination in employment is governed largely by federal statutes. The fact that both parties are citizens of the state of New York does not determine which court has jurisdiction if the basis for the suit is violation of a federal statute. The suit would therefore be filed in federal court.

If Frank sues Jane because Jane would not allow him to pass out leaflets stating that Jane is a racist, in front of Jane's business, Frank's suit would be likely to be filed in federal court because the right to free speech is protected by the First Amendment to the U.S. Constitution. However, some federal statutes provide for concurrent jurisdiction, which means either federal or state court may be used.

§ 2.15 —The State Courts

Within a state, the trial level courts are municipal, superior, and, in some states, justice and small claims courts. Municipal, justice, and small claims courts generally are limited to handling disputes involving lesser dollar amounts. These amounts vary from state to state. A typical limit for municipal courts is $25,000. The small claims division of municipal courts, also called small claims court, is typically limited to an even smaller amount, approximately $5,000 or less.

Parties are required to represent themselves in small claims court; attorneys are not allowed to represent parties. Procedures in small claims court are simple, and the process is a cost-effective method for ordinary citizens to dispose of disputes. Even if a dispute involves a sum of money in excess of the small claims jurisdictional limit, a party can limit his or her recovery to the small claims limit and proceed in small claims court.

Disputes between citizens of the same state, involving amounts in excess of the municipal court limits, are heard by the superior courts, unless a suit involves a federal statute (legislatively enacted law) or a federal constitutional issue. Most lawsuits are filed in superior court. The first state appellate level is the court of appeal. The state supreme court is the highest court of appeal in most states. (New York has essentially the same system, but the trial level court is called the Supreme Court, there is an intermediate level of appeal called the Superior Court, Appellate Division, and the final level of appeal is called the New York Court of Appeals.)

If a case is not settled after being filed, it is brought to trial. Following trial, if any party is dissatisfied with the result, that party may file an appeal if the necessary steps are taken within the time set by statute. As a rule, the appellate court will not retry the facts of the case. The appellate court usually is restricted to reviewing the trial court's decision on issues of law and procedure, and to establishing that the trial judge did not abuse judicial discretion by making a decision totally unsupported by the evidence in the case.

For example, interpretation of whether a particular clause in a contract is an issue of law can be examined by an appellate court. On the other hand, the finding of a trial court that wood of a particular quality was used for a construction job would be a finding of fact, which the appellate court could not reverse or overrule unless the trial court committed gross error or abused its discretion.

If a party is dissatisfied with the result in the appellate court, it can seek review by the state supreme court. Generally, a state supreme court will only review a matter if it is necessary (1) to secure uniformity of decisions or to settle important questions of law, or (2) because the court of appeal lacked the jurisdiction (power) to render its decision.

Of the lawsuits that are filed, few actually come to trial. Of those brought to trial, fewer are appealed. Only a small number of cases that are appealed are reversed, and an even smaller number are given review by the state supreme court. When a state supreme court accepts a case for review, it may affirm (agree with) the appellate court's decision, or it may reverse and remand (send back for retrial). On occasion, portions of the appellate decision may be upheld, and portions may be reversed and remanded.

In the construction business, contractors usually carry workers' compensation insurance to cover the employer's liability for injuries to workers. The workers' compensation system was developed to provide expeditious help for workers who were injured in the course of their work. Workers' compensation is subject to a legal system that is completely different from the state or federal civil court system. It has its own judges, rules, and formulas for compensation. The system was intended to pay for the cost of medical bills and disability for workers injured on the job, regardless of whether the workers were negligent in causing the injury. The exclusive remedy, against the employer, of a worker who is injured on the job is usually through the workers' compensation system. There are few exceptions to this rule.

§ 2.16 Pleadings

Parties set out their contentions in pleadings, which are official documents filed with the court. A complaint, which sets out the claims of the plaintiff against the defendant, is the first document filed. In the complaint, the plaintiff sets out the facts and legal theories that form the basis for his request to the court for relief.

An introductory portion of the complaint usually states the identity of the parties to the suit, and their relationship to each other. The first cause of action (legal theory) stated in the complaint describes the facts that give rise to the lawsuit, and the reason the plaintiff believes he or she is entitled to relief.

Alternative legal theories can be presented in other causes of action listed in the complaint. For example, if a suit is filed by a general contractor to enforce a mechanic's lien (legal right to bring an action against property) against an owner's property, the complaint will allege (1) that the work was performed by the contractor, according to a contract, (2) that the contractor has complied with all the provisions of the contract, and (3) that the owner has not paid all the money the contractor claims is due. An alternative legal theory would be that the owner must pay the contractor for the reasonable value of the work, whether or not a contract existed, because the contractor did work that benefited the owner.

The plaintiff files the complaint with the court, and then has the defendant personally served with the complaint, together with an order from the court, called a **summons**. The summons orders the defendant to answer the complaint within a specified period of time, usually 30 days, or the plaintiff may request a judgment from the court by default. In the answer to the complaint, the defendant will usually deny the claims of the complaint and will set out legal defenses to the claims of the plaintiff. These defenses might include allegations that the plaintiff waited too long before filing the suit, that the plaintiff has no valid claim even if all the facts described in the complaint are true, that the defendant is not responsible for the plaintiff's damages because the damages are the responsibility of some third party or because the plaintiff caused the claimed damage. In some states, the plaintiff then must reply to the answer of the defendant.

At the time of filing the answer, the defendant may also file a pleading called a **counterclaim** or **cross-complaint** (a pleading that sets out claims against some third party, which are connected to the plaintiff's complaint and claimed damages). If the defendant has affirmative claims against the plaintiff, as opposed to defenses, they are set out in the answer to the complaint. The cross-complaint claims that the cross-defendants named in the cross-complaint are responsible to the defendant (cross-complainant). In most states, the cross-defendant must answer the cross-complaint.

The most commonly filed cross-complaint asks for indemnity, stating that the cross-complainant (and defendant) is entitled to reimbursement from the cross-defendant if the cross-complainant has to pay any money to the plaintiff in the original suit. For example, if a construction worker is injured on a project because the scaffolding on which he was standing collapsed, he may sue the party that erected the scaffolding. That party may cross-complain against the manufacturer of the scaffolding, claiming that the scaffolding was defective and that the defective condition was the sole or partial cause of the worker's injuries.

The purpose of the complaint and answer (and cross-complaints and their answers) is to frame the issues of the lawsuit. The pleadings also attempt to eliminate issues and facts about which there is no disagreement. As an example, the parties may agree, as set out in the answer and complaint, that the contract was executed on a certain date. Although the parties may disagree about the terms of the contract, the date of execution, or the fact that a contract was executed, are less likely to be disputed issues, and litigation can be avoided regarding those issues.

If a defendant is served with a summons and complaint and fails to file an answer, the plaintiff may request that the court enter the defendant's default. The plaintiff wins, without proving liability, because the defendant did not appear before the court to contest the plaintiff's claims. If the default is entered, the plaintiff usually still must prove the amount of damages, at least to the judge. Judgment will then be entered for the plaintiff, and the plaintiff may attempt to collect that judgment by foreclosing on real property or by obtaining the legal right to secure other assets of the defendant.

Unfortunately, the judgment is only a statement from the court that the plaintiff has a right to collect money *if* the plaintiff can find the money. The court will not put the defendant in jail to force the defendant to pay. If the defendant has no visible assets, the plaintiff has won only a piece of paper.

§ 2.17 Means Available to Secure a Judgment before Trial

If a lawsuit is filed and the defendant appears, it will be some time before the matter is tried and judgment entered, unless the case is settled before trial. During the pretrial period, the plaintiff may fear that the defendant will hide or attempt to hide assets, or that the defendant will transfer the assets to others, in order to conceal the existence of assets that could be used to satisfy a judgment. Many states have procedures that a plaintiff, or other party, may use to protect against this possibility.

The plaintiff may apply to the court for a **writ of attachment** (a court order that seizes funds or assets). This method may be used if the plaintiff can prove to the court that the plaintiff is likely to win at trial, is in danger of suffering irreparable harm if the writ is not granted, and can demonstrate a likelihood that the defendant will transfer or hide the assets. The plaintiff seeking the writ of attachment may be required to post a bond, to cover the risk that the plaintiff will lose at trial. The bond will then be used to compensate the defendant for any damage he can prove he suffered because of the attachment if the plaintiff loses. To avoid the attachment of assets before trial, the defendant may instead post a bond to provide a source for satisfaction of a future judgment.

§ 2.18 Pretrial Litigation Procedures

In the federal court system, and in most state court systems, pretrial procedures exist to enable your attorney to develop your case and prepare it for trial. These procedures are called **discovery**.

§ 2.19 Discovery

Discovery procedures are intended to allow an orderly exchange of information and documents. The general rule in discovery is that, on demand, each party must produce relevant evidence, that is, evidence that applies to issues involved in the case or is likely to lead to the discovery of admissible evidence. Relevant evidence has a tendency to prove or disprove a fact that is at issue in the lawsuit. This is a basic concept, but it can become difficult when applied to the facts of a particular case.

For example, the fact that there was a full moon the night a construction worker shot his coworker may not be relevant to prove the worker was crazy, although some studies draw a correlation between insane acts and phases of the moon. The existence of the full moon *is* relevant, however, to prove that a witness was credible when he said he could identify the worker because of the moon's bright light.

Because the standard for admissibility of evidence is subjective, and no judge is present at depositions, production of documents, or site inspections, the cooperation of attorneys is important. At a deposition, if the attorneys disagree about what questions should be asked and answered, a judge or magistrate may be called to resolve the issue. Usually, however, attorneys try to solve the problem themselves, or they defer questioning in the disputed areas until a later time.

§ 2.20 —Privilege

Evidence that is relevant may still not be required to be produced if it is **privileged**. Privileged material is confidential material to which only the client and his attorney have access. The most common privilege applies to confidential conversations or other communications between a client and his attorney, for the purpose of seeking or offering legal advice. No part of these communications needs to be revealed in discovery. Exceptions to this rule are where the privilege is waived because third parties are present or have access to the information, or where no attorney–client relationship existed between the attorney and the individual at the time of the communication.

Generally, for an attorney–client communication to be privileged, third parties who are not essential to the communication must be excluded. If a translator is necessary at a client interview because the client does not speak English and the attorney does not speak the client's language, the translator's presence does not destroy the privilege. However, if a friend of the client is present during the attorney–client conference, that person's presence can destroy the privilege because the attorney and the client are not treating the communications as confidential. If a person inadvertently overhears the attorney–client conversation, for example, over an office partition, the privilege is not destroyed. The critical issue in determining whether the privilege applies is: Did the attorney and the client intend that the communication be confidential?

Attorney–client or client–attorney letters may lose their privileged status if third parties are sent copies of the correspondence. To protect the privilege, it is important that a client not send copies of attorney–client letters to outsiders. If an individual communicates with an attorney he has not retained or is not consulting for legal advice or representation, no privilege applies to this communication. If an attorney communicates with an individual with whom the attorney has no attorney–client relationship, their communications are not protected by attorney–client privilege.

Attorney–client privilege is absolute; it cannot be invaded by an opposing side. The rationale for protection of attorney–client communications is to encourage clients to provide all relevant information to their attorneys, without fear that the information will be disclosed to others. This purpose would be destroyed if, under certain circumstances, the privilege could be invaded.

Another important privilege attaches to the **work product** of an attorney, or to the work product of the attorney's employees or agents obtained in the course of representation of the client. Examples of work product include memoranda that set out an attorney's evaluation of evidence, or his or her analysis of law pertaining to a case. A party to litigation is not entitled to use the discovery process to request information that discloses the opposing side's attorney's work, when that information is equally accessible to each attorney. If the work product were not protected, no attorney would need to perform his or her own investigation. The attorney could wait for the other side to do the work, and then request the fruits of that labor.

The work product privilege is not absolute, unlike the attorney–client privilege. If an attorney can demonstrate that the only way he can obtain certain information is by requesting documents or materials from the opposing attorney, the opposing attorney may be required to provide the information even though the request seeks attorney work product material. For example, assume one party's attorney interviewed a witness to an accident, and tape-recorded the interview. The witness died before the other party's attorney had a chance to interview the witness. Because there is no way for the second attorney to obtain the facts from that witness, except from the tape recording, a court would probably order that a copy of the tape be produced to the second attorney.

Attorneys are very cautious about inadvertently producing attorney–client or work-product privileged information. Accidentally producing privileged material may result in the waiver of attorney–client privilege regarding *all* documents in the case.

§ 2.21 Methods of Discovery

The most commonly used methods of discovery in construction litigation are (1) depositions, (2) interrogatories, (3) requests for documents, and (4) site inspections or testing.

§ 2.22 —Depositions

A **deposition** is a procedure in which pretrial testimony is given under oath and transcribed by a court reporter. It is an opportunity for an attorney to ask questions of an opposing party or of an independent witness, prior to the trial, so that the attorney will know what information that person has and how that person will testify at trial. The deposition usually takes place in the office of the attorney requesting the deposition, although it may be taken at other locations for the convenience of the attorneys or the witness. Usually, the deposition is taken in a conference room, with the attorney who requested the deposition seated on one side of the table, and, if the witness has an attorney, the witness and his attorney seated on the other side. If there are other parties to the lawsuit, they may also be present and may have their attorneys present as well.

If the attorney for the witness believes a question is improper or unclear, that attorney may state objections. If attorneys representing other parties or witnesses are present and wish to state objections, they may also do so. After the attorney who requested the deposition has finished asking his or her questions, the other attorneys may ask relevant questions. As a practical matter, the questions of the other attorneys usually cover only the areas the attorneys felt were not explored by questions already asked of the witness. The attorney for the witness may also ask the witness some questions. This is rare, however, and usually occurs only if the attorney felt that the answers of the client were incorrect or incomplete.

The court reporter at the deposition will be a person who has undergone special training and is licensed by the state to administer the oath to the witness to tell the truth, to take down testimony, and to transcribe that testimony. The transcript includes the questions of the attorney who requested the deposition, the objections and questions of the attorney for the witness, and the objections and questions of any other attorneys present.

A deposition that involves many objections and discussions between counsel will proceed more slowly and will be more costly for the client. For most intents and purposes (except, as discussed below, at video depositions), objections to questions asked at a deposition should be kept to a minimum. Under certain circumstances, attorneys use objections or comments to questions to signal the client that a question is dangerous or harmful to the client's case, and that the client should not answer a question when it is phrased in a particular way.

Some attorneys will spend a number of hours with you, if you are going to give a deposition. They will explain the deposition process to you, and may rehearse you by asking the questions you are likely to be asked at the deposition and coaching you on how to answer. Other attorneys will spend little time with you, advising only that you tell the truth and not volunteer any information. These attorney–client meetings are protected by the attorney–client privilege and the substance of the meetings will not be disclosed to the opposing side. Extensive predeposition discussions usually do not take place with independent fact witnesses, because any statement made to the witness by an attorney prior to the deposition may be explored at the deposition.

Meetings with expert witnesses are usually more extensive, with the understanding that the conversations will be explored in depth by the opposing attorney at the deposition.

At the outset of the deposition, the attorney who requested the deposition usually cautions the witness that (1) the transcript will be available for demonstrating inconsistent testimony at the time of trial, and (2) testimony at the deposition is given under penalty of perjury. It is important for the witness to exercise care in answering questions at the deposition and to review the deposition before the time of trial.

At trial, the attorney may ask a question that was asked at the deposition. If the response of the witness is substantially different from that given at the deposition, the attorney is likely to read the question and answer from the transcript of the deposition, to try to demonstrate to a judge or jury that the witness was either lying at the deposition, or is lying at the trial. In either case, if this argument is successful, the believability of the witness is impaired. If the attorney attempts to use this tactic and the inconsistency can be explained by the witness, or the inconsistency is minor, this tactic may backfire against the attorney. The jury is likely to resent the attorney's attacking the witness on a trivial matter.

The most significant advantage to taking a deposition from a party or witness is that it provides the attorney with the opportunity to observe the witness in a setting similar to that of a courtroom, even though there is no judge or jury present, and the witness must testify with little or no help from his attorney. The deposition provides the attorney with factual information, but it also gives the attorney a preview of how convincing the witness will be at a trial. Parties often attend depositions, to help their attorneys formulate questions to be asked of the witnesses giving depositions.

Disadvantages of taking depositions include the fact that the procedure is expensive. The attorney charges for all time spent on preparation for, and attendance at, the deposition. The attorney, or his employee, usually prepares a summary of the deposition. The cost of preparing the transcript is also expensive. For a three-hour deposition, several hundred dollars is a common charge. Another disadvantage is that the deposition provides an adverse witness with an opportunity to practice his testimony. By the time of trial, the presentation of that witness may be more polished and effective.

Video depositions are also allowed in some states, such as California. Video depositions are usually taken when there is a possibility that the witness may not be available for trial. Most states require that a live witness be produced at trial, if at all possible. However, under certain circumstances, the court may allow use of a videotaped deposition, even where the witness is technically available to testify at the time of trial. The most common situation would be taking the testimony of an expert witness who lives in another part of the country, when it would be inconvenient or very costly to have the expert on call during a trial. In addition, if the opposing attorneys agree to use the videotape of a deposition in place of live testimony, a court may permit the tape to be used at trial.

One problem in taking a video deposition is that the standard for admissibility of evidence at a deposition is more relaxed than the standard that applies at trial. For this reason, all objections to the evidence at a video deposition must be made at the time of the deposition, rather than reserving those objections until the time of trial. The video deposition usually takes longer and is therefore more costly, because the basis for payment for deposition transcripts is their length. A video deposition also is more costly because of the need to hire a special video operator, who must be specially trained and licensed by the state.

§ 2.23 —Interrogatories

Interrogatories are written questions prepared by one attorney and sent to the opposing attorney. The responding party must provide written answers under oath, from memory or based on a review of the documents available to that party. The party's attorney is required to help in providing information. Interrogatories are an economical and effective method to obtain discovery from parties. Usually, interrogatories cannot be used to obtain information from outside witnesses, although outside witnesses can be asked to give depositions or to provide documents.

With interrogatories, an attorney can ask the opposing party detailed questions. The questions usually are intended to pin down the opposing side to a particular position, or to force that party to provide factual information that will help the attorney asking the questions to prepare his case. An attorney will often object to an interrogatory on the basis that the question calls for work-product or attorney–client privileged information. In responding to interrogatories, the attorney must reveal those facts of which he is aware, but work-product privilege applies to prevent a forced disclosure of his legal theories regarding the case. The standard regarding what information must be disclosed, and what information is privileged, is subjective. It varies greatly, depending on the judge who is determining which questions must be answered.

Some jurisdictions have controls to prevent an attorney from using the interrogatory process to cause needless effort and expense to an opposing party. Some states, such as California, have instituted numerical limits on the number of interrogatory questions that may be asked. A law passed in 1986 in California limited the number of interrogatory questions that could be sent in a superior court action to the opposing party to 35, unless the attorney for the questioning party has represented to the court that more interrogatories are necessary because of the complexity of the case or for another valid reason. If an attorney representing a responding party feels that the interrogatories demand an excessive amount of material or that certain questions seek irrelevant or prejudicial information, that attorney can seek relief from the court.

When used properly, interrogatories can help parties to locate important witnesses and documents. If interrogatory responses are received prior to the deposition of a responding party, questioning at the deposition can be focused directly on the critical facts and issues.

You must cooperate fully with your attorneys in answering interrogatories. Incomplete responses to interrogatories can cause substantial problems later on in the lawsuit. Important trial testimony of yours can be destroyed if interrogatories were answered incorrectly or incompletely.

§ 2.24 —Document Production

The third primary pretrial discovery method is a party's request for documents from the opposing parties or from third parties. Construction projects, in particular, involve substantial numbers of documents. Important documents, such as contracts between the parties, are usually in the possession of both sides. However, many other documents, such as work diaries, change orders, insurance policies, and plans and specifications may not be in a party's possession. These documents may be requested and used to analyze the strength or weakness of a party's case.

§ 2.25 —Site Inspection/Testing

Preparation of construction cases often requires that parties inspect the site of the construction. Where claims are made for defective construction, defendants will want to view the structures and look for evidence of the defects. Testing may be performed using various instruments. For example, if a part of the claim is that water intrusion is occurring at the structure, the contractors who are being sued may wish to check the ceilings, walls, or floors for signs of dry rot or wet rot, and to look for water stains or other evidence of leaks.

Some testing can be done without removing or destroying any part of a structure. In other circumstances, professionals attempting to assess the problems with defective construction may wish to perform what is called **destructive** testing. This testing requires removing parts of roofs, cutting into walls, or doing other investigative work that causes damage to part of the structure. If destructive testing is performed, the structure must then be restored to the condition it was in prior to the testing.

§ 2.26 The Jury System

The United States inherited the concept of the jury system from England. In the original jury system, a group of individuals who actually were acquainted with the parties, such as neighbors of the parties or members of the same community, would determine the outcome of the dispute. This concept of the jury has been altered. The present American system demands that none of the jury members be personally acquainted with the parties, the attorneys, or the judge. Under the current approach, the jury is made up of ordinary members of society, who have no legal expertise or special training. Although citizens have the duty of serving on juries, many people view it as a poorly compensated nuisance.

Criticism of the jury system often centers around the assertion that members of a jury are unqualified because they lack special training or expertise. Opponents of the jury system claim that what they view as excessive or erroneous jury verdicts for plaintiffs are indications that the jury failed to understand the law or the facts. Those who support the jury system feel that juries have the ability to penetrate the arguments and tactics of counsel and to make fair and reasonable decisions.

§ 2.27 Trial Procedures

A plaintiff in a civil (noncriminal) case usually bears the burden of proving his case by a **preponderance of the evidence**. This means that, to prove his case, the plaintiff must provide evidence that is more convincing than the totality of the evidence opposing his case.

In a criminal case, the burden of proof is greater. The prosecution must prove its case against the defendant **beyond a reasonable doubt**. This is a far more difficult burden of proof. If the judge or jury has *any* reasonable doubt about the guilt of the defendant, they must acquit. The defendant is not required to produce evidence in either civil or criminal cases.

Most lawsuits are composed of a series of issues or facts to be proven at trial. For example, in a suit by a supplier asking that he be paid for goods he has delivered, the supplier will need to prove the following issues at trial: (1) an agreement or contract existed to supply certain goods; (2) the proper goods were delivered; (3) all other terms of the contract were complied with by the supplier; and (4) the buyer failed to pay for the goods. To prove these issues, the supplier may use a copy of the contract, an admission under oath that the buyer executed and agreed to the terms of the contract, statements or documents that tend to prove the goods were delivered, and evidence that the buyer failed to pay for the goods.

When a case is tried without a jury, the judge decides all issues. When a matter is tried before a jury, there are two decision-making entities: the judge, who decides what law applies, and the jury, which makes a finding regarding the facts, based on the evidence presented. At the beginning of the trial, the court receives a group of potential jurors. From this group, the typical jury is chosen: twelve regular jurors and two alternates. In some systems, fewer jurors may be chosen.

Before the jury is finally selected, questions are asked of the potential jurors, to attempt to avoid the possibility of bias or of prejudgment of the case by jurors. In the federal court system, the judge conducts this process, which is called **voir dire**. In the state systems, some of the questioning is often conducted by the attorneys representing the parties to the dispute.

Each attorney attempts to determine, by the responses of the jurors to questions, whether the juror will favor or be hostile to the attorney's client. Psychologists are often consulted by attorneys to aid in jury selection.

The rules of legal procedure require that the **trier of fact** (the jury in a jury trial, the judge where there is no jury) must wait until all the evidence is presented before reaching a verdict. However, some studies suggest that members of juries form their opinion about which way they will decide a case within the first three minutes of a trial. It may take very strong evidence to alter the jury's first impression.

In a civil jury trial, after voir dire is conducted, the jury is sworn in by the judge, and the jurors promise to follow the judge's instructions. At that time, the judge will give the jury certain preliminary instructions about their conduct as jurors during the trial, together with general rules about how evidence will be taken during the trial. The jurors will be cautioned not to speak with anyone about the facts of the case until after they have reached and reported their verdict to the court. The jurors are also cautioned not to speak with the attorneys representing the parties during the trial.

Each party's attorney makes an opening statement. Most trial attorneys are convinced of the critical function served by the opening statement. This is the first opportunity for the attorney to present the jury with the attorney's version of the facts, and their significance. The attorney will outline all the evidence that will be offered at the trial and will advise the members of the jury what weight and effect they should give to the testimony of different witnesses and to different documents. Each party's attorney will seek to demonstrate to the jury that his or her version of the facts is correct, and that his or her client deserves to win.

The evidence offered at trial will generally consist of testimony (oral evidence) and exhibits (documents or objects such as photographs, diagrams, movies or videotapes, and scale models). All of these items are components of the evidence a judge or jury may use to decide a case. Arguments and statements of attorneys are not evidence; they are not to be taken into consideration except to the extent that they refer to testimony or documents or objects that are evidence.

If one of the issues to be proven at trial is that a contractor built townhouses that leak, the following items might be offered as evidence to prove that defect:

(1) photographs of water damage to the interiors of the townhouses, (2) testimony by owners or residents of the townhouses regarding leaks and the damage caused by the leaks, (3) testimony by experts that the method of construction, or its execution, was not up to the applicable standard of care for this type of construction, and (4) diagrams showing the appropriate method of construction. All of these are different types of evidence that the attorney for the homeowners might use to prove a contention that the townhouses were built defectively.

As a general rule, the attorney for the plaintiff puts on his or her opening statement first. Next, the attorney for each of the defendants makes his or her opening statement. The plaintiff's attorney then puts on his witnesses and offers into evidence documents or other exhibits to prove his case. The attorneys for the defendants are each entitled to ask questions designed to attack the testimony and credibility of the plaintiffs' witnesses, in **cross-examination**. After cross-examination, the plaintiff's attorney is entitled to ask questions of the witness in the areas covered by the cross-examination. Often, the purpose of these new questions, called **redirect examination**, is to clarify the testimony of the witness or to rehabilitate the damage done to a witness's credibility by cross-examination.

After the plaintiff's attorney finishes putting on his evidence, each defendant will present his witnesses and relevant documents or objects, as evidence to defend his position. In some circumstances, defendants may have no witnesses, and their entire case may consist of cross-examination.

Expert witnesses often are used in trials. They are persons with special expertise, such as doctors, engineers, or others, who offer opinions about matters that are relevant in proving or defending a case. They are the only witnesses allowed to testify who are not required to have firsthand knowledge regarding the facts of the case, and these expert witnesses usually are hired specifically to testify in court. A judge has the discretion as to whether to allow testimony of experts regarding particular issues. The basic standard applied is whether the expert is going to testify about a factual matter that is too complex for the average person to understand, and whether the expert's testimony will help the judge's or jury's comprehension regarding that matter.

In certain situations, a group of plaintiffs or defendants may be represented by a single attorney. This is permissible if the parties have no **conflict of interest** (a difference of needs and interests of each client, including the possibility of claims by plaintiffs or defendants against each other). In theory, it would be impossible to represent fully the interests of each client where their interests may be opposed. In reality, if the claims against each other or the differences in their interests are insignificant, the plaintiffs or defendants may agree to forgo those claims and use a single attorney to reduce the cost of the litigation.

During the entire period of the trial before jurors deliberate, they are ordered by the judge not to discuss the case with anyone, not even with the other jurors. The jurors are also cautioned to postpone any conclusions about the evidence, and not to decide how they are going to vote until they are actually deliberating over the verdict. If there is substantial publicity, or if the jurors are likely to be influenced by the press, or other contacts concerning the case, a judge may order that the jury be **sequestered** (isolated from outside contact by such means as staying in a hotel, under supervision by a bailiff) until a verdict is reached.

After all admissible evidence has been presented to the judge, or to the judge and jury, each party makes a closing statement. The closing statement by the attorney summarizes the evidence put before the jury or judge at the trial, and attempts to convince the judge or jury of the significance of certain pieces of evidence. The attorney's closing statement is the only time during the trial when the attorney is entitled to argue his client's case, drawing conclusions from the evidence presented during the trial.

After the counsel for the defendants offer their closing statements, the plaintiff is entitled to a brief rebuttal of the closing statements made by the defendants. The plaintiff is given this apparent advantage because the plaintiff bears the burden of proving his case. Counsel for the defendants does not need to disprove the plaintiff's claims. If the plaintiff's attorney fails to convince the jury of his client's position by a preponderance of the evidence, the plaintiff loses, even if the defendant has put on no evidence.

A common analogy used by trial attorneys to explain the significance of the preponderance-of-the-evidence burden of proof is: if the two sides of the scales of justice represent the two sides in the trial, the plaintiff need place only one piece of evidence, which is more convincing, on his side of the scale, in order for the plaintiff to win.

After the closing statements are completed, the judge instructs the jury. Instructions to the jury consist of the judge's reading a series of rules of law. Jurors are not allowed to take notes during the instructions, but they may be given copies of the jury instructions when they are deliberating. The jury instructions include instructions covering rules of law that apply to trial evidence generally, as well as those rules that cover the specific issue brought up in the particular trial.

During jury deliberations, the jurors are put in a room by themselves to discuss the case, and they are instructed to try to agree on a verdict. In a civil case, the agreement of a majority of the jurors is required for a decision. If there are twelve jurors, the required majority is nine. (This differs from a criminal case, in which a unanimous decision is required of the jury.) If the majority of a civil jury cannot agree on a verdict, the judge may ask the jurors to return and deliberate further. If, after further deliberation, a majority still cannot agree on a verdict, the judge will have to dismiss the jurors and declare a **mistrial**. No set time limit exists for delivery of a verdict. The time allowed varies depending on the judge's patience and the jury's perception that a verdict may be reached with further deliberation.

If a mistrial is declared, it is as if no trial had ever taken place. If the plaintiff wishes, he or she can request another trial. The plaintiff may decide against seeking a new trial, concluding that the fact that the plaintiff could not receive the vote of a majority of the jurors may signify he or she will have no better luck with a new jury.

§ 2.28 The Appellate Process

If an attorney wishes to appeal a trial court's verdict, he may file a notice of appeal and submit a written brief to the appellate court. To succeed, the person appealing, whether plaintiff or defendant, must convince the appellate court that the trial court made an error of law, which demands reversal by the appellate court. The party responding to the appeal, who is satisfied with the trial result, tries to convince the appellate court that there were no errors, or that, if any errors were made, they were minor errors and do not demand reversal of the trial court's decision.

The appellate court may refuse to review the case. If the appellate court accepts the matter for review, no new evidence is submitted. The appellate court relies solely on the record from the trial court and on the briefs written by the parties to the appeal. The trial record consists of (1) the trial transcript, a word-for-word record of the testimony and argument offered at trial, and (2) the exhibits put into evidence at trial. The court may request oral argument from the parties. At oral argument, the attorneys for the parties usually summarize their written arguments and answer questions put to them by the appellate judges.

The appellate court may summarily deny the appeal without a full written opinion, or it may issue a decision with a summary of the facts of the case and a full statement of the reasons for its decision. If the ruling of the appellate court is based on a statute, that fact is stated in the opinion. If the court is relying on a rule set out in a prior case, that is also stated, with the reasons given for the court's reliance on prior authority.

§ 2.29 The Bankruptcy Court System

Bankruptcy is the present inability of a person to meet all of that person's obligations as they come due. A **voluntary bankruptcy** proceeding may be initiated by the debtor to obtain protection from creditors. **Involuntary bankruptcy** may be brought by creditors of the bankrupt who fear losses due to the debtor's apparent inability to satisfy all creditors.

The bankruptcy court system has priority over state and federal civil court systems. It is a part of the federal system; there is no state bankruptcy court. The bankruptcy court system provides a debtor with immediate suspension of all claims against him. It issues a federal injunction that stops any debt collection efforts, litigation, or other means of acting against the debtor.

Once the bankruptcy process is concluded, the debtor is free of all claims that could previously have been brought against him, with few exceptions.

During the process, the bankruptcy court, acting through a trustee who administers the bankrupt's **estate** (assets and liabilities), may set aside prebankruptcy payments, reject or assume executory (not yet fully performed) contracts, or assign such contracts to third parties.

There are different types of bankruptcies. In a Chapter 7 liquidation, the bankruptcy trustee liquidates the assets of the debtor. Secured creditors—those who have a trust deed against property or another form of security against specific personal or real property of the debtor—are satisfied first. Those with administrative priority, such as the trustee and others who are owed fees, also have priority over unsecured claims. After these claims are satisfied, there is usually little or nothing left to pay unsecured creditors.

In a Chapter 11 reorganization, the debtor continues to operate. The debtor is required to prepare a plan of reorganization, which is presented to the creditors and must be approved by the court. Chapter 11 is intended to allow the debtor to pay its present debts out of future income, to salvage something from its debts, and perhaps to continue operations during the bankruptcy. Typically, however, Chapter 11 bankruptcies are converted to Chapter 7 liquidations, or the debtor defaults under the reorganization plan, and the court converts the bankruptcy into a Chapter 7 liquidation. A simplified reorganization for individuals is available under Chapter 13.

Once a bankruptcy is filed, nearly any other legal action against the debtor to obtain funds or assets is foreclosed, unless the bankruptcy court grants explicit permission.[2] In the event of a subcontractor bankruptcy, retainages—the unpaid contract balances still held by the prime contractor—may be claimed as the property of the subcontractor's estate in bankruptcy. In the event of bankruptcy by a prime contractor, amounts held by the owner and owed to the prime contractor may be sought by the bankruptcy trustee.

In a reorganization under Chapter 11, a subcontractor-debtor may even insist that the prime contractor pay monies owed directly to the subcontractor.[3] Failure to do so may result in liability of the prime contractor for willful violation of the federal court injunction, **the automatic stay**.[4]

Construction contracts are usually considered **executory contracts** under the bankruptcy laws. This means that all parties have not fully performed their obligations under the contract. Construction contracts may be assigned to a third party by the trustee in bankruptcy, even if the subcontract provides that it may not be assigned.[5] This variance from enforcement of the agreement between the parties regarding assignment is allowed because the bankruptcy court is concerned about protecting the rights of all creditors, not just the prime contractor, and such protection may require assignment of certain contracts.

---

[2] 11 U.S.C.A. § 362(a) (West 1991).

[3] 11 U.S.C.A. § 542 (West 1991).

[4] *See In re* Abrams, 127 Bankr. 239 (Bankr. 9th Cir. 1991).

[5] 11 U.S.C.A. § 365(f)(1) (West 1991).

If a party to a construction contract files bankruptcy after litigation has begun in the civil court system, the dispute may end up being heard entirely in the bankruptcy courts. The bankruptcy courts have jurisdiction to hear all disputes "related to" the bankruptcy proceeding if they involve the debtor. Bankruptcy courts broadly interpret this power as including the power to resolve disputes over whether the debtor owes money to, or can recover money from, other parties.[6]

Corporations and partnerships can discharge nearly any type of debt as part of a confirmed Chapter 11 plan that is not simply a liquidation.[7]

Certain types of debts cannot be discharged by individual debtors under either Chapter 7 or Chapter 11.[8] Creditors can prevent the discharge in bankruptcy of debts in the construction context if the debt was procured by false pretenses, by a false representation, by actual fraud, or by the tendering of a false financial statement.[9] Also nondischargeable are debts due to willful or malicious injury to the property of another.[10]

For an individual or a company with substantial debts, the bankruptcy process gives the debtor a chance to wipe out its debts by placing all of the debtor's assets in the bankruptcy court. If all of the claims of the creditors have been discharged in the bankruptcy process, the debtor can emerge from bankruptcy with a fresh start.

§ 2.30 Pros and Cons of Pursuing Traditional Litigation

Litigation in the American legal system is a slow and expensive process. You are looking for ways to operate your business more profitably, and to reduce the cost of litigation. If your relationship with other parties to a dispute remains fairly cooperative, alternatives to traditional litigation are available to you. Unfortunately, once a lawsuit has been filed, it is hard to persuade the parties to cooperate and to agree to alternative procedures, which require the consent of all parties. Usually, suit is filed because all voluntary attempts at resolution have failed.

Even after litigation is filed, the courts have attempted to implement new court procedures to streamline the litigation process and reduce the cost and the delays inherent in the system (see **Appendix M**).

---

[6] 28 U.S.C.A. §§ 157(a), 1334(b) (West 1991).

[7] 11 U.S.C.A. § 1141(d) (West 1991).

[8] 11 U.S.C.A. §§ 1141(d)(2), 523(a) (West 1991).

[9] 11 U.S.C.A. § 523(a)(2) (West 1991).

[10] 11 U.S.C.A. § 523(a)(6) (West 1991).

§ 2.31 Federal Court Dispute Resolution without Trial

Failure of all other methods to resolve construction disputes will normally result in filing a lawsuit. The public court system is short of resources and not particularly well-adapted to resolving construction disputes, which are technical and cumbersome, and whose resolution may require specialized knowledge. Complex, multiparty cases such as the claims of workers who were exposed to asbestos and are suffering from health problems, as well as ever increasing criminal matters, crowd the dockets of many court systems. Parties to construction disputes may be required to wait lengthy periods of time before their cases come to trial.

Traditional litigation entails substantial expense for parties, because the discovery procedures and other pretrial litigation procedures outlined in this chapter require extensive attorney time. Traditional litigation also requires that parties spend a large amount of their time attending or giving depositions, preparing or answering requests for documents, or preparing or answering interrogatories. After the passage of several years, and the expenditure of many thousands of dollars in legal fees, the typical construction case may be ready for trial.

The concept of case management by the court system was recently initiated with the federal system (see **Appendix M**). Under the traditional court system, the attorney controlled the pace of litigation. Even if a matter was not yet ready for litigation, the attorney could file an action. As backlogs increased, the attorney could pick and choose which matters he or she would actively pursue, and which matters would remain in limbo. The complaint might not be served on the opposing parties; or, the complaint would be served but active discovery would be postponed.

Under the new federal system, a single judge is assigned to handle every aspect of the case except settlement conferences. The judge handles motions and discovery disputes, and will be the trial judge if the case does not settle. Shortly after the case is filed, typically within 90 to 120 days, there is a conference between the judge and the parties. The parties are expected to provide the judge with the issues of the case, the relevant facts, the discovery they expect to perform, and a timetable for completing discovery in time for trial.

§ 2.32 State Court Dispute Resolution without Trial

Because of the lack of speed and the inefficiency of the public court system in resolving civil matters, some states have initiated new systems to attempt to cure or reduce these problems. The courts must perform a delicate balancing act in achieving efficiency of the court systems without sacrificing the constitutional rights of parties to civil disputes. The Constitution guarantees citizens access to the court system and provides for such procedural protections as the right to trial by jury and the right to confront and cross-examine witnesses.

To increase efficiency, some judges have imposed time limits on examination or cross-examination of witnesses, or have limited each party to a set number of hours to present its case. Judges who impose these types of limitations are being challenged with arguments that such limitations violate a party's constitutional right to a fair trial.

Other methods being used by the state public court systems to minimize the court delays in processing civil disputes include the use of attorneys to act as **pro tem judges** (attorneys temporarily appointed to act as judges), and the use of magistrates as judges (full-time civil servants who are not actual judges). Trial judges may also request that parties agree to the use of juries containing fewer than the required number of members. Although some states have provided by statute for smaller juries in certain types of disputes, it is common for judges in systems that normally require twelve jurors to request that the parties agree instead to the use of nine, or even six, jurors.

§ 2.33 Case Management in the State Courts

In order to reduce the backlog of civil cases in the state systems, some courts have taken the position that one reason for the backlog is the fact that the attorneys often control the progress of cases to suit their own needs. Although a case may be filed in order to comply with the applicable statute of limitations, the plaintiff's attorney may delay serving the lawsuit; or, the attorney may serve the suit, which is then answered by the defendant, but then take no other steps to move the case to trial.

Under the traditional court system, the party ready to proceed files a Memorandum to Set, or an At Issue Memorandum, with the court, stating that all necessary discovery has been completed, and the case is ready for trial.

Under the newer systems, the court takes control of the lawsuit from the attorneys at the outset of the litigation. Filing of the suit triggers court deadlines for completion of discovery, and sets a time for readiness for trial. These dates must be met by all parties. The case either settles shortly after filing, or the parties are forced to prepare it for trial. In some state systems, as in the federal system, a single judge is assigned to the case for all purposes, and that judge controls the case, up to and including trial. The exception to the rule that one judge handles all aspects of the case usually involves the settlement conference, which demands frank disclosure by the attorneys of the strengths and weaknesses of their cases. This settlement conference is usually heard by a different judge, a magistrate, or a pro tem judge.

These expedited proceedings skew the balance of litigation even further than normal, based on the economic strength of the parties. Attorneys who typically handle cases on a contingency basis are restricted to handling fewer cases, when they are not able to control the pace of the litigation. Attorneys handling matters on an hourly basis are less likely to be affected negatively, because they can pass on the cost of compliance to their clients, in the form of higher hourly rates or more billed hours needed to comply with the new and more complex court procedures.

2.34 Conclusion

The formal court process provides certain protection for parties to a lawsuit. It also removes the parties in the dispute resolution process from direct contact by essentially requiring the use of attorneys.

Chapter 3 of this book describes in detail the alternative methods of dispute resolution—arbitration, mediation, and private trials—and provides an explanation of some of the reasons why these systems may be more desirable for resolution of construction disputes than conventional litigation.

CHAPTER 3

ALTERNATIVES TO TRADITIONAL LITIGATION

§ 3.1 Introduction

When you have reached the point where you cannot resolve a construction dispute, you will need to bring in third parties to help resolve the dispute. The most common first step you will take, in your attempt to resolve the dispute, will be to consult the contract in order to identify the dispute resolution procedures that govern the dispute.

You may wish to speak with an attorney, to determine the significance of the contract provisions regarding dispute resolution. If no procedures are set out, the attorney should explain the options available to resolve the dispute. These options will usually include arbitration, litigation, and mediation.

Many attorneys who practice construction law prefer alternative dispute resolution (ADR) methods to litigation as a means to resolve construction disputes. In the fall of 1991, approximately 20 of the largest law firms in California signed a Law Firm Policy Statement on Alternatives to Litigation, developed by New York's Center for Public Resources, Inc. In the statement, these law firms pledged that they will advise clients to use alternative dispute resolution whenever possible. The Center for Public Resources, Inc., a nonprofit consortium of in-house and private counsel, promotes ADR. The Center recently studied 142 companies that used ADR in 1990 and discovered that they saved more than $100 million in litigation expenses.

§ 3.2 Design Professionals as Dispute Resolvers

It is common for a clause in construction contracts to identify the architect or engineer working on the project as the first person to resolve construction disputes. For disputes between the prime/general contractor and the subcontractors, the design professional may function adequately to resolve disputes. However, in disputes involving the owner and other parties, the neutrality of the architect or engineer may be in question. The design professional is hired by, and is an agent of, the owner. The prime/general contractor and subcontractors may feel they cannot trust the design professional to be impartial and fair, given his or her interest in maintaining a good relationship with the owner. Although different trade organizations have prepared and made available form documents and contracts for construction projects, the documents of the American Institute of Architects (AIA) are the most commonly used. It is important to recognize that a form contract of any organization usually favors the group that the organization represents. For this reason, owners, contractors, or engineers should be cautious in using the AIA documents.

The first page of AIA Documents A101, "Standard Form of Agreement Between Owner with Contractor" (1987) (hereafter, "A101") and A201, "General Conditions of the Contract for Construction" (1987) (hereafter, "A201") include the statement: "This document has been approved and endorsed by the Associated General Contractors [AGC] of America." It is likely that AGC approval of these AIA documents was the result of negotiations between the AIA and the AGC.

Another concern when using AIA documents is that the AIA documents are intended to be used as a package; each document contains numerous cross-references to other documents. Failure to use all related AIA documents on a project, failure to use the most current version of a document, or substantial modification of AIA forms may lead to ambiguities and conflicts within the documents, which may cause problems in the event of a dispute. If the documents are ambiguous, investigation of other sources as to the parties' intent may be necessary.

Because the AIA documents are the forms most commonly used in the construction industry, AIA clauses will be used as examples in discussing dispute resolution and arbitration procedures. References will be made to other forms where appropriate.

Many contracts call for the design professional to assess the quality of the work performed, and to determine whether the work complies with the applicable codes and standards. For example, paragraph 4.4 of A201 appoints the architect as the initial reviewer of claims on the construction project (see **Appendix B**).

Paragraph 4.4 and its subparts are not a part of A101, which the contractor actually executes. These provisions are contained in A201, which is usually attached to A101; these two documents are designed to be used together. If a party to a contract only reviews A101, the document requiring his signature, he will not be aware of this clause. Even upon reviewing A201, it may be difficult for a party to locate this or other significant provisions, unless the document is inspected carefully.

The method by which the dispute resolution clause is incorporated into A101 highlights the risk posed by using certain AIA documents and not others. A101 contains no arbitration clause, but incorporates A201 at Article 9 of A101. Failure to use A201 with A101 will result in the parties' entering into an agreement with no arbitration clause.

A set of procedures for the architect's treatment of a claim is described in paragraphs 4.4.1 through 4.4.4 of A201. In essence, it is a prearbitration procedure in which "evidence" is presented and a timetable is established for actions by the architect and the claimant.

An example of the fact that AIA documents favor architects can be found in the clause, in A201, that allows the architect the option of notifying the surety of a claim, but fails to require the architect to do so.

§ 3.3 Arbitration

Those in the construction industry have a history of dissatisfaction with the court system as a dispute resolution mechanism. Simpler and less technical disputes may take a long time to be resolved by the overcrowded court system, and are usually costly to resolve. The reasons for delays of all lawsuits generally apply, to an even higher degree, to construction disputes. The complexity of the technical issues requires more extended discovery proceedings. The multiplicity of parties involved in pretrial discovery tends to extend delays before trial and requires longer hearings and more witnesses during trial.

Use of an arbitrator who is an expert in the field, or who is trusted by the parties, may be the determining factor as to whether arbitration will resolve the dispute. Suppose a roofing contractor and a roofing supplier have a dispute about the grade of roofing tile delivered to a job site. The roofing contractor, claiming that the tile is not the grade specified in the plans for the building, refuses to accept and install the tile. The parties may be able to have an employee of another roofing contractor examine the tile and give his opinion as to whether the tile meets the specifications. Because of that arbitrator's expertise in the field, the parties accept his opinion. The dispute has been resolved without resort to costly attorneys or to the court system.

Another type of arbitration may involve a more subjective dispute, such as the number of ironworkers required to safely staff a particular commercial construction project. A general contractor's foreman, if he is familiar both with the type of building and with the needs and capacities of the workers, may be able to hear evidence and make a decision that is acceptable to both sides. In this instance, because of the trust and respect of the parties for that individual's opinion based on his experience and knowledge, the arbitration has been successful in resolving the case.

Unfortunately, many construction disputes are not easily resolved by arbitration. Some construction disputes involve legal issues such as interpretation of contract clauses, admissibility of certain types of evidence, or the standard of care that applies for architects or engineers. These issues, as well as the number of parties involved in construction disputes, often make resolution by arbitration difficult.

Another potential problem with using arbitration in construction cases is the fact that, in some states, the filing of a mechanic's lien can bar the right to arbitration. For instance, if a subcontractor wishes to make a legal claim against an owner, his primary cause of action is likely to be under the mechanic's lien laws, because he will have signed no contract with the owner. In those states, the subcontractor may elect to forgo the cause of action for foreclosure of the mechanic's lien and arbitrate his claim instead. To try to deal with this problem, the current version of the AIA form subcontract, AIA Document A401 (1987), states that the arbitration clause will not be deemed a limitation of rights or remedies under mechanic's lien laws.

An attempt to enforce a mechanic's lien right by filing a lawsuit may bar a right to arbitration. In a recent case, a court ruled that a party who filed an action to enforce a mechanic's lien waived any right to arbitration when he did not, at the same time, request that the action be stayed pending arbitration.[1]

§ 3.4 —Predispute Contract Review

Before signing a contract, there are relevant factors to examine regarding the desirability of arbitration as a dispute resolution mechanism. In arbitration, a neutral person serves as a judge of the merits of each side of the dispute, hears argument and evidence, and renders an award or makes a decision. Some states

[1] R. Baker v. Motel 6, Inc., 180 Cal. App. 3d 928, 225 Cal. Rptr. 849 (1986).

favor arbitration, and clauses originating in those states will reflect their attitude toward arbitration. California will uphold an arbitration clause unless it clearly cannot be applied to the dispute.[2] Illinois, on the other hand, requires an absolutely clear arbitration clause before it will be enforced.[3]

If you have no leverage, the other contracting party may simply state that the contract is to be accepted as is or not at all. In other words, you may not have the option of deciding whether you want to have an arbitration clause. Where an arbitration clause is included in such an imposed contract, once a dispute has arisen, a court may decide that arbitration was foisted on a weaker contracting party in order to prevent that party from having any forum in which to have its dispute heard. If so, a court may refuse to enforce the arbitration clause, particularly if the weaker party is an unsophisticated consumer.[4] A contract in which one party forces the other to adhere to certain unreasonable terms is often called an **adhesion contract**. Even where the weaker party is an experienced subcontractor, the theory is that such a contract will not be enforced because it is an adhesion contract.[5]

Some arbitration clauses require only one party, and not the other party, to arbitrate. Although it seems grossly unfair, a court recently upheld this type of one-way arbitration clause.[6]

A potential benefit to creating a custom arbitration clause, as opposed to using a standard clause, is that the parties may tailor the arbitration process to their needs. Specific options that might be addressed by a custom clause would be: (1) to restrict arbitration to disputes involving more than a certain dollar amount, (2) to provide for consolidation of all claims on a particular project, (3) to provide a right of full discovery, (4) to eliminate the use of attorneys in the arbitration process, (5) to specify whether the arbitration award should be binding or nonbinding, and (6) to allow the prevailing party to collect attorney's fees and costs as part of the award.

Important concerns, if you are considering arbitration as a mechanism for dispute resolution, will include: (1) the existence of statutory requirements that must be met for an arbitration clause to be enforceable, (2) methods for creating the right to arbitrate, including creation of the right by contract, (3) types of disputes that may be arbitrated, (4) whether the agreement to arbitrate provides for consolidation of disputes, and (5) necessary preparation for the arbitration proceeding.

There is currently a difference of opinion regarding the desirability of arbitrating construction disputes. Some attorneys are convinced it is less expensive

---

[2] Bos Material Handling, Inc. v. Crown Controls Corp., 137 Cal. App. 3d 99, 186 Cal. Rptr. 740 (1982).

[3] Silver Cross Hospital v. S. N. Nielsen Co., 8 Ill. App. 3d 1000, 291 N.E. 2d 247 (1972).

[4] Spence v. Omnibus Industries, 44 Cal. App. 3d 970, 119 Cal. Rptr. 171 (1975).

[5] Player v. Geo. M. Brewster & Son, Inc., 18 Cal. App. 3d 526, 96 Cal. Rptr. 149 (1971).

[6] Kalman Floor Co. v. Joseph L. Muscarelle, Inc., 196 N.J. Super. 16, 481 A.2d 553 (N.J. Sup. A.D. 1984), *aff'd*, 98 N.J. 266, 486 A.2d 334 (1985).

and more effective than conventional litigation; others prefer the familiarity of litigation.

Most construction contracts contain an arbitration clause. For any contract with an arbitration clause, one of the first questions to be asked is whether the agreement to arbitrate is enforceable. Most states will enforce such an agreement. In those states that do not enforce arbitration agreements for construction contracts, a party not wishing to arbitrate may be successful in defeating an attempt by another party to arbitrate. In states that permit arbitration of construction disputes, a party who wants arbitration under the contract must assert that right or he will find that he may have waived his right to have the dispute arbitrated.[7]

At common law (case-developed law rather than law created by statute), either party may revoke the agreement to arbitrate prior to an award being made pursuant to arbitration. Some states still follow the common-law rules.

§ 3.5 —Arbitration under State Law

Many states have passed laws requiring enforcement of agreements to arbitrate. Although some state laws are based on the common law, a large number of these state arbitration statutes are based, either in whole or in part, on the Uniform Arbitration Act, which was created to provide a uniform method for resolving business disputes. The purposes of arbitration have been stated to be: (1) to discourage litigation, (2) to allow the parties to resolve their disputes in a specialized forum that is likely to be familiar with the needs of the parties and with the customs and usages of the construction industry, rather than in a court of general legal or equitable jurisdiction, and (3) to allow for speedy resolution of disputes, permitting the construction work to be completed without undue delay.[8]

The Uniform Arbitration Act alters the common-law procedures drastically by: (1) making agreements to arbitrate future disputes irrevocable, (2) empowering courts to force arbitration on a reluctant party, (3) giving courts the authority to suspend litigation proceedings until arbitration procedures are followed, (4) setting out procedures for a court to appoint arbitrators or provide methods for selection of arbitrators, if the parties are unwilling to agree on a method to select arbitrators, (5) providing that arbitration decisions are only subject to limited judicial review, and (6) providing for judicial enforcement of an arbitrator's award.

---

[7] For cases of waiver of the right to arbitrate, *see* Burton-Dixie Corp. v. Timothy McCarthy Construction Co., 436 F.2d 405 (5th Cir. 1971); Post Tensioned Engineering Corp. v. Fairways Plaza Associates, 412 So. 2d 871 (Fla. Dist. Ct. App. 1982) (waiver not found); Christiansen v. Dewor Developments, 33 Cal. 3d 778, 661 P. 2d 1088, 191 Cal. Rptr. 8 (1983); E. C. Ernst, Inc. v. City of Tallahasee, 527 F. Supp. 1141 (N.D. Fla. 1981) (waiver not found).

[8] Pettinaro Construction Co. v. Harry C. Partridge et al., 408 A.2d 957 (Del. 1979).

Although most states have based their arbitration statutes on the Uniform Arbitration Act, there are variations. Florida's statute requires that Florida law be specified in the arbitration clause, or either party may void the arbitration clause.[9] Although a prior Texas statute had expressly excluded construction disputes from arbitration,[10] a 1979 amendment of the statute allowed parties to a construction contract to have both present and future disputes submitted to arbitration. Parties wishing to have their construction disputes arbitrated should carefully review applicable state laws prior to executing a contract with an arbitration clause, or prior to entering into an agreement to arbitrate an existing dispute.

Because construction projects often involve interstate commerce, a dispute may be controlled by both state and federal arbitration statutes.[11] An arbitration award on a project involving interstate commerce may be reviewed by either a state or federal court. The arbitration award therefore may not end the dispute, and the parties may invoke the legal process.

The issue of which law governs may be unclear even when the choice of law is based solely on the AIA documents. Under AIA Document B141, paragraph 9.1, governing law is the principal place of business of the architect. Under paragraph 13.1.1 of A201 (see **Appendix B**), the governing law is the place where the project is located. Unless the body of law that applies to both locations is identical, the arbitrator may have trouble determining which law applies. Clarifying the confusion of some state courts, the U.S. Supreme Court has now held that the Federal Arbitration Act may govern over state law relating to multiparty disputes.[12]

Different state laws may also affect the validity of an arbitration decision entered in one state and confirmed by another. In *Brinker v. Superior Court (Page Realty Corp.),*[13] a party (petitioner) received money damages in an arbitration that was heard in California. The petitioner then filed in New Jersey to have the arbitration award confirmed. After the New Jersey court confirmed the arbitration award, the petitioner obtained a writ of execution on the judgment. The defendant filed in California to vacate the arbitration award, on the basis that the arbitrator exceeded his authority. The California court held that the arbitration award could not be vacated because the California court was required to give "full faith and credit" to the New Jersey decision.

---

[9] Damora v. Stresscon International, Inc., 324 So. 2d 80 (Fla. 1975), *reh'g denied* (1976).

[10] Vernon's Ann. Civ. Stat. art. 224 (1966).

[11] Moses H. Cone Memorial Hospital v. Mercury Construction Corp., 460 U.S. 1, 103 S. Ct. 927, 74 L. Ed. 2d 765 (1983).

[12] Volt Information Services, Inc. v. Board of Trustees of Leland Stanford Junior University, 489 U.S. 468, 109 S. Ct. 1248, 103 L. Ed. 2d 488 (1989).

[13] _____ Cal. App. 3d _____, 1 Cal. Rptr. 2d 358 (1991).

§ 3.6 —Arbitration under Federal Law

Contracts dealing with interstate commerce or with maritime transactions may be subject to the Federal Arbitration Act.[14] For those situations in which the Act applies, federal rather than state law will apply.[15] Because federal policy favors arbitration, if the contract involves interstate commerce and includes an arbitration clause, the dispute is likely to be subject to arbitration under federal law.

Contractors working on government projects should also check to be sure that arbitration rights granted by subcontract do not take precedence over rights given to subcontractors under the Miller Act. The subcontractor must fulfill all the requirements of an arbitration provision under the subcontract before bringing a Miller Act suit to enforce the payment bond.[16]

§ 3.7 —Initiation of Arbitration

If you wish to use arbitration to resolve a dispute, you will usually notify the opposing party of your desire to arbitrate. Form contracts published by the American Institute of Architects (AIA), the National Society of Professional Engineers (NSPE), and the Associated General Contractors (AGC) use language for their arbitration clauses that resembles the wording used by the American Arbitration Association (AAA). The AAA provision states:

> Any controversy or claim arising out of or relating to this contract, or any breach thereof, shall be settled in accordance with the Construction Industry Arbitration Rules of the American Arbitration Association, and judgment upon the award rendered by the Arbitrator may be entered in any court having jurisdiction thereof.

The AIA has a history of encouraging arbitration. AIA Documents A201 (general conditions; **Appendix B**), A401 (general contractor–subcontractor; **Appendix G**), B141 (owner–architect agreement; **Appendix H**) all contain general arbitration clauses. One possible reason for the AIA's preference may be that the availability of arbitration as a method for resolving disputes may relieve architects of some of their responsibilities for resolving disputes on the project.

AIA Document A201, paragraph 4.5.1 (**Appendix B**) contains the language as to which disputes are subject to arbitration. Comparison with the AAA paragraph reveals that an exception is made for claims pertaining to "aesthetic effect." This may be a reflection of the fact that the architect is most likely to

---

[14] 9 U.S.C.A. § 2 et seq. (West, 1991).

[15] Paul Allison Inc. v. Miniken Storage of Omaha, Inc., 486 F. Supp. 1 (D.C. Neb. 1980).

[16] United States *ex rel.* Capolino Sons, Inc. v. Electronic Missile Facilities, Inc., 364 F.2d 705 (2d Cir. 1966), *cert. denied,* 385 U.S. 924, 87 S. Ct. 239 (1966).

be involved in disputes concerning the subjective approval of the owner. Under the form contract, the arbitrator has no power over decisions by the architect concerning aesthetic areas, except where they appear to be arbitrary or manifestly in bad faith. Some courts have been unhappy with this exception.[17]

A common source of owner-initiated disputes is the owner's belief that the project does not "look" or "feel" the way the owner believed it would. This problem is particularly common with residential projects, where the owner has a strong personal interest in the finished product.

The issue of compliance with specifications on a project, where the variation from specifications make little or no functional difference, may also create substantial problems, even on a large project where the owner has no personal interest in the finished product.

* * *

Case 3.1
(a) Prelitigation

A residential subdivision was built, and contained numerous defects. One defect involved the curbing that had been installed by the cement subcontractor. The subcontractor had not used a machine that would have perfectly curved the shape of the curb. In addition, the height of the curb varied between 4 and 7 inches, when the specifications called for curbing to be uniform at 6 inches.

Negotiation was unsuccessful in resolving the issue of the curbing. The prime contractor insisted it was entitled to have the curbing redone. The cost of such work was estimated to exceed the original contract amount by approximately 15 times because the already installed curbing and the gutters would have to be removed before new curbing and gutters could be installed.

Analysis. In this case, it is clear that the method of installation of the curbing should have been discovered by the prime contractor or owner at a much earlier time. The inadequacy of the machine used for installing the curb should have been apparent to the representatives of the prime contractor when the machine arrived at the job site and the curbing work began. Either the prime contractor's representatives were not paying attention or they were not present at the time when that portion of the work was performed.

The same criticism applies to the height differential. Had a check been made of the curb height after the first portions were installed, the height differential would have been discovered. At that point in time, the problem would have been minor and could have been cured for a nominal cost. After curbing had been

---

[17] Baker v. Keller Construction Co., 219 So. 2d 569 (La. Ct. App. 1969); *but see also* Mississippi Coast Coliseum Commission v. Stuart Construction Co., 417 So. 2d 541 (Miss. 1982).

installed throughout the subdivision, the cost of fixing the problem escalated dramatically.

(b) Post-Filing of Litigation

After the other defective conditions were resolved, the issue of the curbing remained the sole dispute. The cement subcontractor tried to reason with the prime contractor, stating that the height differential and lack of perfect curving were not visible to the casual observer and made no difference functionally, and that the cost of redoing the curb and gutter far exceeded the aesthetic benefit.

Eventually, through the use of a mediator, the issue was resolved by the subcontractor's refunding the contract amount for the curbing work to the prime contractor. The prime contractor used the money to have a finish cement contractor do some minor patching and cutting on the most severe locations of height differential. The lack of true curve to the curbing was left as originally installed, because it was purely an aesthetic variation from the specifications and was not noticeable to the casual eye. The prime contractor was able to demonstrate to the developer that the curbs were functionally completed, although some variation in the height still existed.

Comment. The general rule is that a failure to comply with performance specifications is a breach of contract. However, in *Gurney Industries, Inc. v. St. Paul Fire and Marine Insurance Co.,*[18] the court determined that substitution of an equivalent product was not a material breach. Some cases have also applied the doctrine of "substantial performance" to find performance, which does not meet specifications, as adequate to justify payment of the unpaid balance of the contract price, with damages caused by the breach being deducted.[19]

For cases on the obligation of a party to mitigate damages, see: *C. A. Davis, Inc. v. City of Miami*[20] (defendant contractor argued against recovery of owner, based on owner allegedly having spent more than was necessary to correct contractor's work; court stated contractor could only dispute completion cost based on waste, extravagance, or lack of good faith by owner); *Great Lakes Gas Transmission Co. v. Grayco Constructors, Inc.*[21] (contractor contended that owner could pass its loss to customers through rate increases; argument unsuccessful); *S. J. Groves & Sons Co. v. Warner Co.*[22] (mitigation argument advanced by breaching contractor, who maintained prime should have mitigated damages by firing subcontractor, not accepted by court).

---

[18] 467 F.2d 588 (4th Cir. 1972).

[19] Votaw Precision Tool Co. v. Air Canada, 60 Cal. App. 3d 52, 131 Cal. Rptr. 335 (1976)].

[20] 400 So. 2d 536 (Fla. App. 1981).

[21] 506 F.2d 498 (6th Cir. 1974), *cert. denied,* 95 S. Ct. 1130 (1975).

[22] 576 F.2d 524 (3d Cir. 1978).

Analysis. The curbs' varying heights were not visible to the average person, and the function of the installed curbs was unimpaired. For a dispute to continue for years over this type of item is a strong indication that the parties to the dispute, or their attorneys, were not adopting a pragmatic approach. Although under certain circumstances aesthetic appeal can be important, height and shape of a curb are unlikely to fall into that category.

* * *

Even the AIA has allowed for some variations in the general rule for binding arbitration of all disputes that arise under a contract. In 1990, the AIA published its Document B511, which contains language that modifies AIA Document B141. At paragraph 7.1, the AIA provides a blank for placing a dollar limitation on disputes that will be arbitrated. Paragraph 4.5.1 also provides for waiver of the arbitrability of certain claims, and allows for arbitration of other types of claims upon written notice. The arbitration clause of most construction contracts usually specifies the types of disputes that will be subject to arbitration. If the clause refers to disputes arising out of the performance of the agreement, an owner may not be able to enforce a dispute arising from another area, such as the issue of payment.[23]

Paragraph 4.5.2 of AIA Document A201 sets out the procedures for initiation of the arbitration (see **Appendix B**). This clause sets out the time frames in which a demand must be filed, and allows for notice to the architect, who is the first "arbiter," or judge of the dispute, under the contract language.

Paragraph 4.5.3 of AIA Document A201 provides for continuance of the work while a dispute is being resolved (see **Appendix B**). Because of the interrelationship of work on a construction project, one of the difficult issues to be addressed when a conflict arises is the effect that the dispute will have on the progress of the project. At least one court has interpreted this type of clause to only require a claimant to continue working on other parts of the project, and not on the disputed portion.[24] Under most circumstances, the general contractor will try to convince the subcontractor to continue work and to submit a change order whose validity will be determined at a later point in time. This procedure may only delay the dispute, rather than resolve it.

The EJCDC (Engineers Joint Contract Documents Committee) considers arbitration desirable for smaller claims. Paragraph 7.6.1 of EJCDC Document No. 1910-1 ("Standard Form of Agreement Between Owner and Engineer for Professional Services") specifies the disputes that are subject to arbitration:

> 7.6.1 All claims, counterclaims, disputes and other matters in question between the parties hereto arising out of or relating to this Agreement or the breach

---

[23] Galt. v. Libby-Owens Ford Glass Co., 397 F.2d 439 (7th Cir.), *cert. denied,* 393 U.S. 925, 89 S. Ct. 258 (1968).

[24] Village of Cairo v. Bodine Contracting Co., 685 S.W.2d 253 (Mo. Ct. App. 1985).

> thereof will be decided by arbitration in accordance with the Construction Industry Arbitration Rules of the American Arbitration Association then obtaining, subject to the limitations and restrictions stated in Paragraphs 7.6.3 and 7.6.4 below. This Agreement so to arbitrate and any other agreement or consent to arbitrate entered into in accordance herewith as provided in this paragraph 7.6 will be specifically enforceable under the prevailing law of any court having jurisdiction.

Paragraph 7.6.3 makes arbitration binding and compulsory only if the claim is for less than $200,000, exclusive of interest and costs. In other words, the EJCDC seems to favor arbitration only for smaller claims, and apparently believes that large claims belong in the traditional court system of litigation.

Paragraph 7.6.4 of EJCDC Document No. 1910-1 excludes from joinder or consolidation at the arbitration any claim relating to a person or entity who is not a party to the agreement. In other words, the arbitration can only cover the parties who actually signed the agreement. This reduces the utility of arbitration. If the arbitration will not cover all disagreements on a project, the parties are less inclined to treat the arbitration seriously as a dispute resolution mechanism.

The EJCDC forms reflect that organization's discomfort with arbitration as a method for resolution of all construction disputes by providing that parties wishing to use arbitration as a dispute resolution mechanism must execute a separate document, Exhibit OG-B, which is incorporated into the contract.

§ 3.8 —Timing for Demand for Arbitration

Paragraph 4.5.4 of AIA Document A201 describes the procedures and timing for demand of arbitration (see **Appendix B**). Demand for arbitration under the AIA clause requires the person making the claim to first submit it to the architect for resolution. This requirement may present a problem: as mentioned previously, the architect may not be viewed by all parties as wholly impartial. The person making a claim is protected from inaction by the architect, because other events can cause the claim to progress, distinct from action by the architect.

EJCDC Document No. 1910-1, paragraph 7.6.2, states that notice of demand for arbitration must be filed in writing with the parties to the agreement and with the AAA within a "reasonable time" after the claim or dispute has arisen, and makes no reference to submission of the claim to the architect. However, EJCDC Document No. 1910-8, Exhibit GC-A, at paragraph 16.2, provides that demand for arbitration must be made within a fixed period of time—30 days after the engineer's decision.

Paragraphs 4.5.4.1 and 4.5.4.2 of AIA Document A201 provide for penalties to be imposed on a person making a claim who fails to initiate the demand for arbitration within the time allotted (see **Appendix B**). Failure of the person making the claim to demand arbitration within 30 days of the architect's decision may result in the architect's decision being binding on the claimant. Strangely, if the architect does not render a decision until after arbitration has begun, that decision may be used as evidence at the arbitration. This is contrary to the usual relationship between most arbitration decisions and litigation. For example, in a nonbinding judicial arbitration, if a party rejects the arbitration decision, the arbitration decision is not admissible at trial. Once the arbitration decision is rejected, for most purposes, it is as though the arbitration never occurred.

§ 3.9 —Consolidation or Joinder of Claims at Arbitration

The issue about consolidation or joinder of claims to be heard in arbitration is critical. With paragraph 4.5.5 of AIA Document A201, which describes the limitation on consolidation or joinder (see **Appendix B**), the AIA seeks to preclude consolidation of separate claims that involve an architect as a party, thus preventing joinder of an architect as a party to an arbitration, unless the architect consents. From an efficiency standpoint, it is logical to have numerous claims regarding a single project heard by one arbitrator. Courts have varied in their holdings as to whether a court has the power to consolidate arbitrations or to add a party to a particular arbitration.[25]

In some states, where the case law has not permitted consolidation of multiple arbitration claims arising from a single project, statutes have been enacted by legislatures to allow the courts to consolidate.[26] In the current version of AIA Document A401, at paragraph 6.2, consolidation or joinder of arbitrations between the subcontractor and the general contractor and other disputes may be allowed, as long as they do not involve disputes with the architect (see **Appendix G**).

As stated previously, the EJCDC also opposes joinder of claims at arbitration. Paragraph 7.6.4 of EJCDC Document No. 1910-1 states that arbitration under the contract will not include, by consolidation or joinder, any person or entity who is not a party to the agreement. Under this clause, those working on a project who have signed other types of contracts will not be a part of the arbitration.

The courts have not always upheld provisions seeking to preclude joinder or consolidation of matters in arbitration. In *Garden Grove Community Church*

---

[25] Grover-Diamond Associates v. American Arbitration Association, 297 Minn. 324, 211 N.W.2d 787 (1973) (permitted consolidation); Stop and Shop Cos. Inc. v. Gilbane Building Co., 364 Mass. 325, 304 N.E.2d 429 (1973) (not permitted); Episcopal Housing Corp. v. Federal Ins. Co., 273 S.C. 181, 255 S.E.2d 451 (1979) (permitted); Protective Life Ins. Corp. v. Lincoln Nat'l Life Insurance Corp., 873 F.2d 281 (11th Cir. 1989) (not permitted).

[26] Cal. Code of Civ. Proc. § 1281.3 (Deering 1991); Mass. Gen. Laws Ann. ch. 251, 2A (West 1988).

v. Pittsburgh–Des Moines Steel Co.,[27] a California state court held that the clause did not prevent the same arbitrator from conducting two separate arbitrations, one between the owner and contractor, and one between the owner and architect.

Architects or engineers may not wish to have consolidation or joinder of claims. As design professionals, they may fear that they could be exposed to greater liability if forced to arbitrate disputes with the owner and with the contractor in a single arbitration, because the owner and contractor may both blame the architect for problems that have resulted on the project. By handling the two disputes separately, they may be able to assign blame at each arbitration to those parties not present.

On the other hand, an owner will not want to be required to arbitrate separately the different issues that have arisen on one project. Separate arbitrations may be more costly and may produce inconsistent results. The court in *Garden Grove* held that the construction contract only required the owner to arbitrate with the contractor if the owner could join all parties in the arbitration.[28]

If an owner is going to be required to proceed to litigation on some claims, it might as well litigate all claims. In addition, issues may arise as to whether litigation should be postponed during the pendency of arbitration. One of the generally accepted rationales for arbitration—that it will result in an expedient and inexpensive resolution of a construction dispute—may be eliminated if consolidation of all disputes pertaining to a project is not allowed.

* * *

Case 3.2
(a) Prelitigation

Numerous homes were destroyed by a serious fire in a residential area. Three of the homes destroyed were under construction by Builders, Inc. The homes were 60 percent completed at the time of the fire. The principals of Builders, Inc. were the owners of the property as well as the builders of the homes. The property under construction was covered by builders' risk insurance.

After the fire, the construction lender notified Builders, Inc. that full payment was due on the construction loans because of the destruction of the homes in the fire. Although destruction of the structures on property is a common triggering event for payment of homeowner and construction loans, the loan agreement signed by Builders, Inc. included no such clause. After consulting with counsel, Builders asked Bankcorp, the lender, to advise Builders what provision of the loan agreement Bankcorp relied on to maintain it was entitled to

---

[27] 140 Cal. App. 3d 251, 191 Cal. Rptr. 15 (1983).

[28] Garden Grove Community Church v. Pittsburgh–Des Moines Steel Co., 140 Cal. App. 3d 251, 191 Cal. Rptr. 15 (1983).

full repayment of the loans. After Bankcorp obtained legal counsel, Bankcorp realized it could not continue to claim that it was entitled to full repayment of the loans.

There was a second lender involved in providing financing for construction of the homes. Both lenders delayed in providing payments to allow reconstruction of the structures to proceed. At the outset, Builders had to clean up the debris resulting from the fire, and inspections had to be made of the remaining concrete retaining walls. Each home had three retaining walls, and they had been exposed to over 2000 degrees Fahrenheit of heat in the fire. Some cracking had occurred in the walls, and reconstruction could not begin until full inspection and testing of the walls had been performed.

Eventually, the lenders agreed to set aside a certain amount of funds to be used in the reconstruction. Unfortunately, the builders, lenders, and insurance company could not agree on an amount that represented the reconstruction cost of the homes. Builders, Inc. was required to sue both lenders for breach of fiduciary duty, and to sue the insurance company for breach of contract and for breach of the duty of good faith and fair dealing.

Analysis. Before beginning any construction project, the parties should ascertain the insurance necessary to cover reasonably foreseeable risks. In this situation, it would have been helpful to ensure that the fire insurance was adequate, and that provisions existed within the construction loan agreement to cover delays encountered because of acts of God such as fire, earthquake, or other natural disaster.

The principals in this situation did not have the financial strength to weather substantial delays in construction. Construction projects usually tie up substantial amounts of cash and assets of the parties. If multiple projects are scheduled simultaneously, or in sequence, delays will have a domino effect. Assuring that you have a financial cushion will make it more likely that you can make it through the problems or delays that frequently occur on projects.

(b) Post-Filing of Litigation

As a defense to the litigation, the insurance company relied on a clause in the insurance policy requiring that disputes between the insured and the company about insurance coverage under the policy must be arbitrated. The provision stated that any such disputes would be arbitrated through the American Arbitration Association. The clause also required that the arbitration be binding.

The critical issue in this case was whether Builders, Inc. wished to include its dispute with Bankcorp in the arbitration of its dispute with the insurance company. Arbitration of the Bankcorp dispute was not mandatory; the loan agreement contained no arbitration clause. After much consideration, Builders decided that it would arbitrate the entire dispute with the American Arbitration Association, because it did not want to be tied up in litigation for years. The financial investment in the properties was substantial for the principals of Builders. If the process dragged on for years, they would probably lose the properties at foreclosure, because they could not afford to keep making the payments under the construction financing agreement terms.

In addition, the homesite area, which had been devastated by the fire, was no longer as desirable, because so much of the vegetation had been destroyed in the fire. Many of the older, distinctive homes that had been located nearby had also been destroyed in the fire. Builders, Inc. needed to settle the dispute and get its money out of the project.

Prior to arbitration, the parties agreed to the use of a mediator through the American Arbitration Association. After four sessions, the mediator was able to convince the insurance company that it had potential policy limit exposure, and Bankcorp agreed to extend the construction financing at no fee, with enough additional time allowed for the reconstruction to be completed.

Analysis. The issue of whether arbitration is desirable as a means to resolve construction disputes depends heavily on the facts and circumstances of each case. In this situation, the contractor and owners were overextended financially. They could not have survived the lengthy delay that could have resulted if they had not agreed to consolidate both disputes for arbitration. Once all parties had agreed to arbitration, the use of a mediator through the American Arbitration Association was a logical step and resulted in a cost-effective resolution of the disputes.

* * *

Some legislatures have enacted statutes allowing consolidation and joinder of claims at arbitration.[29] Most current cases now find in favor of consolidation.[30]

AIA Document A201, paragraph 4.5.6, provides a penalty for failure of a person making a claim to put other parties on notice of all claims (see **Appendix B**). Under this document, there is a strong incentive to ensure that all claims you wish to make are contained in the original request for arbitration: the arbitrator or arbitrators can refuse to permit amendment to add claims at a later date. This is contrary to the situation in litigation, where the courts' policy is usually liberal regarding the amendment of claims.

---

[29] Cal. Code Civ. Proc. § 1281.3 (Deering 1991); Mass. Gen. Laws Ann. ch. 251, § 2A (West 1988).

[30] Maxum Foundations, Inc. v. Salus Corp., 817 F.2d 1086 (4th Cir. 1987); Higley South, Inc. v. Park Shore Development Co., 494 So. 2d 227 (Fla. App. 1986); Episcopal Housing Corp. v. Fed. Ins. Co., 273 S.C. 181, 255 S.E.2d 451 (1979); *but see* Manchester Township Board of Educ. v. Carney, 199 N.J. Super. 266, 489 A.2d 682 (1985) (architect's action not consolidated because there was no arbitration clause in agreement between architect and school district).

§ 3.10 —Review of Arbitrator's Decision

According to AIA Document A201, paragraph 4.5.7, under most circumstances, the arbitrator's decision is final and will not be overruled (see **Appendix B**). Courts usually refuse to reconsider the decisions rendered by arbitrators. Failing a showing of gross error of law, or of a bias of the arbitrator that was not known by the parties at the time of selection of the arbitrator, the decision of the arbitrator will stand. Other bases for nullifying the decision are fraud of the arbitrator, corruption, or a finding that the arbitrator exceeded the authority he was allowed by the contract clause.

The EJCDC varies from the AIA position that arbitration awards are final unless there is a finding of gross error of law or undisclosed bias of the arbitrator. The EJCDC states that the arbitration award is final, as long as it meets the review standards under the Federal Arbitration Act.[31] Under EJCDC's general condition statement, EJCDC Document No. 1910-8, published in 1990, the standard of judicial review for arbitration awards is no longer used.

As a general rule, a prospective arbitrator should make disclosure to the parties if he is in a relationship with a party, or other person, that might affect his impartiality as an arbitrator. The AAA Construction Industry Arbitration Rules specifically refer to the obligation of a person appointed as a neutral arbitrator to disclose any circumstance that might affect his impartiality as an arbitrator.

In *Commonwealth Coatings Corp. v. Continental Casualty Co.,*[32] an arbitration agreement provided that a contract dispute between a general contractor and a subcontractor would be resolved by a panel of three arbitrators. The parties followed the usual procedure for selection of the arbitrators. Each of the two parties appointed an arbitrator of his or her own choosing, and the two arbitrators then selected a third arbitrator. In this case, however, the third, supposedly neutral, arbitrator was an engineering consultant who had the general contractor as a regular customer.

As a part of the relationship between the engineering consultant and the general contractor, the consultant had received fees from the general contractor over a five-year period. The engineering consultant had even performed services on this specific project. The connection between the engineering consultant and the general contractor was not revealed until after the arbitration award had been made. The subcontractor challenged the award on the basis of the bias of the supposedly neutral third arbitrator, and the case reached the U.S. Supreme Court.

---

[31] 9 U.S.C.A. § 2 et seq. (West 1991); EJCDC Doc. No. 1910-1, par. 7.6.6.

[32] 393 U.S. 145, 89 S. Ct. 337 (1968), *petition for reh'g denied* (1969).

The Supreme Court vacated the arbitration decision. The Court stated that, although arbitrators are not required to sever all ties to the business world, arbitrators should disclose to parties any dealings that could create even the appearance of bias. The Court did not find that there was any actual fraud or bias in deciding this dispute, but the Court stated that the rules of arbitration require that any arbitrator or court that is authorized to hear cases must avoid even the appearance of bias.

There are situations in which arbitration produces the desired result of timely and low-cost resolution of disputes. In one case, a large section of cement was poured, but then cracked. The cement contractor claimed that the cement cracked because of defective specifications provided by the architect. The architect contended that the contractor had guaranteed the performance, that the specifications were proper, and that the cement failure was caused by the defective materials supplied by the cement contractor. When no agreement could be reached between the parties as to who was responsible for the problem, the parties agreed to submit the dispute to arbitration. The arbitrator decided in favor of the cement contractor. Although the other parties "lost" the dispute, they were able to resolve it without resort to litigation procedures, which would have been expensive and time-consuming.

§ 3.11 —Arbitration Procedures under the AAA Construction Industry Rules

The arbitration clause in the construction contract may specify that a particular organization, such as the American Arbitration Association (AAA), will be used to arbitrate disputes arising out of the work. Parties may also choose to use AAA procedures even when these procedures are not specified in the contract clause. The AAA has standardized rules and procedures for deciding disputes, and has particular rules for arbitration of construction matters. The AAA Construction Industry Arbitration Rules are incorporated into AIA Document A401, and were most recently amended as of January 1991 (see Appendix I).

If you are using AAA arbitration procedures, it is important to ensure that you are operating under a current set of the rules. Under the AAA rules, if you wish to make a claim, you must provide written notice to the other party of your intent to seek arbitration. The notice should describe the dispute and the remedy or solution you will seek, and should contain a complete description of the claim. Once an arbitrator is appointed, the claim may not be amended subsequently without permission from the arbitrator. You should mail the notice to the AAA, along with a copy of the contract or agreement that contains the arbitration clause. The arbitration will be initiated with payment of a fee of $300 (mailed to the AAA with the Request for Arbitration), as long as the remainder of the fee, determined under the AAA fee schedule, is paid within 60 days of filing, or prior to the first hearing, whichever comes first.

The AAA has established expedited procedures that normally apply to any case where no claim or counterclaim exceeds $50,000. You may use the expedited procedures for claims in excess of $50,000, if all parties to the dispute agree to the use of those procedures.

In addition, the AAA now has "Guidelines for Expediting Larger, Complex Construction Arbitration" and recommends that these guidelines be used for disputes involving $250,000 or more. They include holding an administrative conference early in the arbitration, which is intended to help the AAA arbitrator understand the nature of the case. Also available under these guidelines is a preliminary hearing at which the arbitrator will set out a schedule for the exchange of information by the parties and set dates for the hearing.

The other party is required to send a response to the claim within ten days of receipt. If an opposing party fails to file a response, the claim will be deemed denied by the opposing party, for the purposes of the arbitration.

After the AAA receives the claim, a list of names of potential arbitrators will be compiled by the AAA and mailed to the parties. Under the expedited procedures, the list will consist of five proposed arbitrators, from which one arbitrator will be appointed. Each party may strike two names without cause or evidence of bias. The parties have ten days in which to strike the names of any potential arbitrators whom they find objectionable. A party is not required to provide a reason for striking an arbitrator from the list. The parties are asked to number the remaining names on the list of arbitrators, based on the parties' preference.

Failure by either party to strike names or to list preferences will result in an assumption that all arbitrators are acceptable to that party. If a party strikes all the names on the list, the AAA will ask both parties to individually select a name from another list and to submit it to the AAA. The AAA will then select a third panel member to serve. This procedure mirrors the arbitrator selection procedure described in many arbitration clauses contained in contracts, such as insurance policies used in many states.

If the arbitration agreement does not specify the number of arbitrators, the disputes will be heard by a single arbitrator. However, the AAA may direct that a greater number of arbitrators be used.

In the past, arbitrators of construction disputes were experts in the construction field, or they were practicing attorneys, rather than persons making their sole livelihood from acting as arbitrators. Because of the congestion of courts in many states, it is becoming more common for lawyers, judges, contractors, engineers, or architects, retired from their previous occupations, to act as arbitrators on a full-time basis. Those who work through AAA as arbitrators must meet certain prerequisites. For those hearing construction disputes, eight years' experience in the arbitrator's given field of expertise is usually required, although AAA has the ability to waive this requirement.

AAA arbitrators agree to hear the first day of the dispute without compensation. Beyond the first day, the arbitrators' compensation will be based on the amount of service involved and on the number of hearings. If the parties are unable to agree on a suitable fee for the arbitrator, an appropriate rate will be established by the AAA and communicated to the parties. Administrative fees are based on the amount in issue. For a dispute involving up to $10,000, the fee is $300. For amounts between $10,000 and $25,000, the fee is 3 percent of the amount in issue. The administrative fees are then scaled up to an arbitration fee of $15,000, plus 10 percent of the excess over $5 million, where the amount in dispute is between $5 million and $50 million. With most arbitrations, including AAA arbitrations, the parties share the expenses of the arbitration. These expenses include both the administrative fees of the AAA and compensation for the arbitrator's time beyond the first day.

The parties usually provide the AAA with information about scheduling, and the AAA selects the dates for the arbitration to be held. If the parties wish to obtain documents from each other or to take depositions of witnesses, the arbitration date will usually be postponed until discovery is complete. Once an arbitrator or panel of arbitrators has been selected, any discretion in selection of the arbitration date will be with the arbitrator or arbitrators, rather than with the administrative personnel of the AAA.

Many of the powers granted by the courts to parties in traditional litigation, such as the power to subpoena witnesses or to obtain documents, are not within the powers of AAA arbitration. To demand attendance of a witness at an arbitration or appearance of a witness for a deposition requires a court order.

An arbitration hearing is conducted much like a court trial. The person making the claim will present his or her case, then the party responding will present its case. The presentations are usually composed of opening statements, testimony, and documents. The arbitrator sits as a judge. The arbitrator may ask that closing statements, or argument, be presented in writing rather than orally. The arbitrator may allow all parties additional time to present other evidence or legal briefs.

The closing statements may be particularly significant when the arbitrator is a nonlawyer. The closing oral statements or written briefs provide parties with an opportunity to educate the arbitrator as to the law each party believes applies to the dispute. Judges or lawyers, who are presumed to have a working knowledge of the law pertaining to the area involved in the dispute, will not require the same amount of guidance as a nonlawyer arbitrator.

After each of the parties has presented all of the relevant evidence and briefs, the arbitrator will render his decision. Under the AAA rules, the decision must be entered within 30 days of the closing of the record on the claim. The arbitator's decision is binding (unless otherwise provided for by contract or by agreement). If the losing party is unwilling to comply with the arbitrator's ruling, the prevailing party can obtain a court order to enforce the judgment.

§ 3.12 —Arbitration by Mutual Stipulation

Even if the contract does not contain an arbitration clause, you and the other parties to a dispute may agree to resolve it by either binding or nonbinding arbitration. You and the other parties are free to select appropriate procedures, select an arbitrator, and decide whether the decision of the arbitrator will be final and binding on you or will merely serve to provide guidance to the parties as to the likely judicial result.

§ 3.13 Arbitration Compared to Litigation

A strong argument in favor of arbitration is the speed with which arbitration can resolve disputes, as compared to litigation. Another argument in its favor is the technical expertise some arbitrators can bring to the process. Contractors, architects, and engineers are commonly used by the AAA or similar alternative dispute resolution organizations. On the other hand, the arbitrators are unlikely to have a legal background, and they may be more disposed to base their decision on what they perceive to be "fairness," rather than legal rules. Although a party may have complied with all of his legal responsibilities, he still may lose a dispute at arbitration because the arbitrator felt that he acted unfairly and did not deserve an award.

Each dispute may have particular aspects that make arbitration or some other form of alternative dispute resolution more desirable than litigation. If a party is faced with a dispute and has a choice about whether to insist on arbitration or resort to more formal court action, he should consult an attorney.

As a general rule, arbitrators do not state the reasons for their decisions. Occasionally, arbitrators may wish to explain their decision's rationale, particularly in complex cases, but this practice is discouraged by the AAA. Finality of outcome of the dispute is one of the reasons for arbitration. If arbitrators explain the reasons for their decisions, parties might be more likely to try to overturn the arbitrators' decisions.

§ 3.14 Mediation

Parties are also free to select a mediator who will work with them, and/or their attorneys, to attempt to resolve a dispute. If parties are already involved in an AAA arbitration, mediation is available to them through the AAA without payment of any additional administrative fee. A neutral mediator appointed by the

AAA may assist the parties in reaching a settlement, but the mediator does not have the authority to make a binding decision or award. If the parties to an AAA mediation cannot agree on their own procedures, AAA procedures will be used for mediations handled through the AAA. The AAA procedures are set out in the AAA Construction Industry Rules.

Mediation has become increasingly popular for resolution of construction disputes because of the number of parties involved in interrelated issues, and because of the high cost of a trial where complex technical problems, numerous claims, and large sums of money are involved. Advantages of mediation include the fact that it usually allows the parties to freely exchange all relevant information in an informal manner and to use practical problem-solving approaches and solutions that reflect the particular needs and interests of the parties. Disadvantages of mediation may be that it fails to impose a solution on the parties, and that the informal methods may permit parties to conceal relevant facts or to make assertions when the evidence to support those assertions is lacking.

From the parties' perspective, one of the strengths of mediation is that the process empowers the parties to act on their own behalf. By diminishing the role of the attorneys, each party can rely on its own senses, perspective, and intuition in selecting negotiation strategies or in creating solutions for the dispute.

Where the outcome of a dispute rests heavily on determination of a specific legal or factual issue that is contested between the parties, litigation may be a more desirable alternative. Hostility between the parties may poison the cooperation necessary for mediation to be successful. Determining whether each party wishes to use mediation to resolve a dispute includes deciding how much protection each party needs from the other. Another important factor to consider is the extent to which each party's case relies for its success on nonlegal perceptions of fairness, and the common-sense equities of the particular situation, as opposed to a rule of procedure or law.

A mediator will seek to protect the parties by advising them that the role of a mediator is not to represent either party, and that, unlike an attorney, he is not an advocate. The mediator will inform the parties of the risks of the mediation process and the limitations of a private "hearing," which is not governed by rules and procedures.

The mediator should also advise the parties of the fact that the procedure is strictly confidential. Various state and federal rules of evidence deal with the confidentiality of settlement discussions. Statements or documents produced at mediations typically are not admissible at trial.[33] Despite the fact that mediation is usually considered to be a settlement discussion, or a compromise negotiation, the mediator should still have the parties explicitly agree to that fact and should have them sign a document affirming their agreement. This will guarantee the confidentiality of the proceedings, if the case is not disposed of by mediation.

---

[33] Cal. Evidence Code § 1152 (Deering 1991); Fed. R. Evid. 408 (West 1991).

The mediator usually makes an independent determination of whether the dispute, given all relevant factors, is appropriate for mediation. If the mediator concludes that one or the other of the parties is unable to adequately protect himself or herself, the mediator should terminate the mediation. Likewise, if the hostility between the parties reaches a high level, the mediator should terminate the proceedings.

If you agree to mediation, the mediator will probably advise you and the other parties as to how the court system would be likely to resolve your dispute. Although the focus of mediation is on pragmatic concerns, its backdrop is the other dispute resolution methods available to the parties. If a party has a position that would be strong if presented in a courtroom, that party's incentive for accepting a mediated settlement is less.

One of the important decisions for each party to make is whether the party wishes to have his attorney present and representing him at the mediation, or whether the party will act on his own behalf. The parties can agree to self-representation and still have their attorneys present, or available by telephone, to provide consultation or advice if a question arises about a particular point.

The conduct of the mediation itself varies. The mediator may advise the parties about their legal rights and then elect not to become directly involved in the discussions between the parties. The mediator may facilitate discussions, or may divide the parties and speak with each side individually.

Agreement on procedures and issues is reached between the mediator and the parties to the mediation before beginning the mediation. Typically, the parties will relate the facts of the case to the mediator and explain their respective positions. Guidelines on what types of issues will be addressed by the mediation will be discussed. The parties will usually decide before the mediation whether they will agree to be bound by any oral agreements made during the mediation. Most mediations provide that the parties are not bound by oral agreements until the parties sign a settlement agreement.

Some mediators will prepare a memorandum of agreement regarding the settlement; others will prepare the settlement agreement itself. If a party represents himself at the mediation, the mediator may encourage the parties to have independent counsel review any final agreement reached at mediation. The purpose of this step is to ensure that all parties have knowingly accepted any settlement negotiated at a mediation. This review by independent counsel is an important step to consider if a party has not consulted his or her attorney during the mediation.

The amount in issue in the dispute may determine each party's willingness to use alternative methods such as mediation or arbitration. If a dispute involves several million dollars, the mediator's hourly rates of $250 to $600 per hour, when divided among all the parties, will be cost-effective. With disputes involving smaller amounts, the parties may be less inclined to use alternative methods that may not avoid the necessity of a trial.

EJCDC Exhibit OG-B, which is used with EJCDC Document No. 1910-27-A, calls for mediation to precede arbitration or litigation. Instructions to users of EJCDC Contract Document OG-B state that the Association of Soil and Foundation Engineers (ASFE; formerly called the American Association of Soil and Foundation Engineers), which approved the EJCDC document, suggests using the same person to preside over the mediation and the arbitration.

* * *

Case 3.3
(a) Prelitigation

A hospital hired Smithco, a large contractor, to perform construction work on expansion of a hospital. Smithco agreed to act as both general contractor and construction manager on the project. Before all the plans and specifications were assembled, Smithco informed the hospital that the work could be completed for a guaranteed maximum price (GMP). In fact, Smithco told the hospital that it might be able to perform the work for *less* than the GMP, and agreed to share any reduction in the GMP with the hospital.

Because the plans and drawings were incomplete, essentially the work each day proceeded by improvisation. Pipeco, the mechanical subcontractor, and the HVAC contractor were asked by Smithco to redesign and scale down their work; Smithco told those subcontractors that it was impossible for Smithco to perform the contract within the GMP as the project was originally designed.

The project went through a number of architects, many of whom were located halfway across the country. The inspector hired by the owner required standards to be met that were far in excess of the specifications. The doctors at the hospital had ongoing design input into the Surgery Center, which was to be located on the first floor of the five-story building. The floor layout for the first floor required that work be performed out of sequence by the mechanical subcontractors, the drywall contractor, and others.

Through private negotiation, all claims were resolved except for those of the mechanical subcontractor, the electrical subcontractor, the drywall subcontractor, and the general contractor (Smithco).

The single largest problem was created by Smithco's preproject sales technique. It had sold the project to the owner for an amount less than the actual cost of the work. Perhaps Smithco believed that, once the project was begun, the shortfall it would suffer if forced to perform for the contract price could be offset by change orders, based on claims that the plans provided by the architect contained work beyond the scope of the original agreement. The general contractor probably believed that, after the subcontractors were committed to the project, they could be persuaded to submit their own change orders, or to accept losses for those portions of the work that could not be compensated for by change orders.

One of the mechanical subcontractors was forced by the owner's inspector to comply with particular standards for hanging water pipe. The standards were not justified by the plans or by the type of material to be carried in the pipe. It was planned that the pipe would carry, at most, 80 pounds of material. The inspector tested the pipe hangers at 200 pounds of pressure. The testing to this standard caused many hangers to fail, and more than half the pipe hangers had to be redone by Pipeco.

Pipeco tried to reason with the inspector and attempted to get Smithco to intercede on its behalf. Smithco refused to intervene and told Pipeco to do the work as instructed by the inspector. It was not practical for Pipeco to walk off the job or to stop work. The subcontractor completed the work, hoping that, ultimately, compensation for the extra time and materials would be received.

Analysis. The use of a guaranteed maximum price (GMP) as a marketing tactic for contractors raises an array of questions. In this particular situation, Smithco, the prime contractor did not have plans and drawings, and the project had only reached a conceptual stage. In addition, the parent company was trying to sell Smithco, providing Smithco with an incentive to have a certain number of projects on the books, regardless of the eventual profit from each project.

The GMP is used as a sales technique by prime contractors because owners are attracted by the apparent ceiling on the cost of the construction. After signing the contracts, owners are battered with change orders and extras. In addition, as in this case, the prime contractor goes back to the specialty contractors and tries to have them agree to perform the specialty work for a lesser amount.

On this particular project, Pipeco, the wet-side mechanical subcontractor, disagreed even on the original contract price for Pipeco's work. Rather than abandon the project because Pipeco did not have a signed contract, Pipeco worked for six months without a signed agreement. This was a substantial mistake. Pipeco had no way of knowing from week to week whether Pipeco was going to be paid because of the lack of a signed contract. Eventually, the contract Pipeco did sign was executed under protest and did not reflect the amount Pipeco expected to be paid for the work. Smithco assured Pipeco that Smithco would see that Pipeco was "taken care of."

Although Pipeco had an attorney who had regularly performed work for Pipeco and who was knowledgeable about construction matters, Pipeco's principals failed to consult with the attorney until matters had reached the stage of litigation; they considered legal consultation "too expensive." Before the case was over, Pipeco had lost hundreds of thousands of dollars that might have been saved had Pipeco approached the project in a more businesslike manner and consulted with an attorney at an earlier stage.

The input by the doctors associated with the hospital added significant delays. Even if it was appropriate for the doctors to have input into the design of the Surgery Center portion of the building, this input should have been received before the project left the conceptual stage. Construction was delayed while doctors wandered around the structure deciding where the machines were going to be located. Pipeco, in particular, had to install different types of piping, such as oxygen and gas conduits, in several locations, to deal with different possible floor plan options that were being considered for the Surgery Center.

Individual architects were changed a number of times during the course of the project. Some of the architects were located in different parts of the country. This contributed to delays in implementing design changes and in responding to requests for information by specialty contractors. Establishing standard procedures for promptly communicating changes and necessary information, and adhering to these procedures, would have substantially reduced the delays encountered on this project. If the architects were not not going to be available on a regular basis, some other individual employed by the owner or architects should have been charged with the responsibility of handling these types of scheduling matters.

The issue of the inspector's actions regarding the testing procedures is problematical. An inspector may be privately retained, as he was in this case, or he may be a public employee. As discussed in other sections of this book, government entities are immune from suits by the public under most circumstances. However, because the inspector had been privately retained by the hospital, any claims for the inspector's negligence or failure to properly perform duties would be included in the claims made by the parties against the hospital.

(b) Post-Filing of Litigation

As the project was being completed, approximately fifteen of the major subcontractors remained unpaid. They filed mechanic's liens on the property, as did the prime contractor. Extensions of time were stipulated, to enable the parties to negotiate settlements. The parties had numerous meetings over a period of approximately two years, and some of the subcontractors settled with the prime contractor and the owner.

The county in which the actions were filed had recently instituted a **fast-track system** for handling litigation. Designed to expedite the handling of lawsuits by the courts, this system requires the courts to take control of the litigation and push it forward, even if the parties wish to hold off from proceeding to trial. Shortly after filing suits to foreclose the mechanic's liens, the parties requested exception from the fast-track procedures so they could try to resolve the situation themselves. Although the court refused to allow an exception, the court agreed to continue the status conferences to allow the parties negotiating time. Eventually, the court consolidated all the actions. At the time of consolidation, there were twelve claims, including that of the prime contractor against the owner.

A number of the parties hired claims consultants. These consultants were nonattorney experts who were versed in construction and adept at the preparation of construction claims. One consultant charged its subcontractor client

$40,000 as an initial fee for preparing a claim package consisting of narrative descriptions of the problems that occurred on the project, charts of work flow, and calculations of costs caused by delay, lack of coordination, and other problems encountered by that subcontractor. The consultant was to receive one-third of the eventual recovery from the prime contractor and owner, less the $40,000 paid at the outset.

Approximately eighteen months after construction was essentially complete, five subcontractors were still unpaid, in spite of all the money spent in attempts at nonlitigation methods for resolution of the problems.

Trial of all the disputes arising from this one hospital addition project was expected to last approximately eight weeks. After all the attempts at private settlement and negotiation had failed, and approximately four months before the first scheduled trial date, the parties hired a mediator with substantial experience in mediating construction disputes. The mediator required each party to limit itself to a presentation to the mediator of no more than twenty minutes. The mediator then met with each of the parties individually, explaining his views of the strengths and weaknesses of each party's position.

A large stumbling block to settlement was the general contractor's refusal to accept any responsibility for the cost overruns, delays, and other problems that occurred on the project. Smithco claimed that all the problems arose because of incompetent work by the subcontractors, poor plans by the architect, lack of the architects' response to requests for information, massive numbers of architectural bulletins, lack of responsiveness of the owner to requests for information or change orders, and indecisiveness on the part of the doctors participating in the design of the Surgery Center.

After approximately nine hours of mediation, a settlement was reached between the hospital and the three subcontractors, without the agreement of the general contractor. The hospital agreed to pay certain sums directly to the subcontractors, and to indemnify the subcontractors from any claims by Smithco, the general contractor. Because there were no typewriters or word processors available, an architect (no doubt because he had the most legible handwriting) wrote out the settlement agreements in longhand. Somewhere around midnight that evening, Smithco and the hospital were able to work out a settlement with the help of the mediator.

Although the mediator charged $600 per hour for his time, the mediation was extremely cost-effective for the parties involved, when compared to an eight-week trial, which would not have taken place until six months to a year had passed. The rights of the parties were taken into account by the mediator, but the overwhelming issues at the mediation were: (1) how much money was available from the owner to settle the claims, and (2) how much the general contractor and each of the subcontractors would accept, to avoid the high cost of a trial.

Analysis. The implementation of fast-track procedures to expedite trial delays can cause new problems of its own. For a case of this complexity, and with the number of parties involved, the California goal of resolving the case within twelve months of filing cannot be met. The judge in this case dealt with the complexity and nature of the case by continuing the **status conference**, which is the proceeding in which the attorneys explain what discovery needs to be performed, and then set a date for trial.

There were three status conferences. Finally, the matter was referred to a judge for a early settlement conference. There were actually two sessions with the settlement judge. He ended the settlement conferences by advising the three subcontractors who still had claims that he believed the owner had $1.8 million available for settlement, and that if the attorneys for the three subcontractors could go out in the hall and agree on a split, the case might be able to settle. At that time, the figure of $1.8 million was less than half of the amount of the demands. The attorneys for the subcontractors decided that the settlement conference was not going to resolve the case and turned their efforts to scheduling a private mediation to resolve the case.

A number of the parties in this case hired "claims consultants" to prepare claims analysis. Although the consultants acting for the owner were probably paid on an hourly basis, the claims consultants for the others may have been compensated on a contingency basis. At least one, the consultant for Pipeco, received over $200,000 for his work. This is probably far more than it would have cost Pipeco to have consulted with Pipeco's attorneys when it was first involved in the project, without even adding in the cost of taking the case through trial.

The position of Smithco, the general contractor, in refusing to acknowledge any responsibility for damage or delay in this project was untenable. Smithco had sold the project to the owner with the concept of the GMP (guaranteed maximum price), the general contractor's project managers were switched numerous times, and Smithco refused to press the owner to pay for the subcontractors' extras and change orders. Smithco was probably the largest contributing factor to problems on the project. Even when Smithco's position is viewed solely as a negotiating strategy, Smithco's failure to accept any responsibility signified lack of good faith in participating in the negotiation and mediation process. The only way the mediator was able to resolve the case was by arranging a settlement directly between the owner and the specialty contractors—without Smithco. After this agreement was reached, Smithco finally became reasonable. Faced with the prospect of going to trial against a united front consisting of the owner and all the specialty contractors, Smithco agreed to settle.

* * *

Although a party to a construction project is expected to comply with the specifications, claims of impossibility of compliance may be argued by the non-complying party. In *Foster Wheeler Corp. v. United States,*[34] Foster Wheeler

---

[34] 206 Ct. Cl. 533, 513 F.2d 588 (1975).

agreed to provide the government with boilers and to perform certain tests to demonstrate that the boilers could withstand a specified level of shock. After design work, and creation of mathematical models, Foster Wheeler decided it could not perform. Although the court determined that the promised performance was not literally impossible, it was deemed sufficiently impracticable to excuse Foster Wheeler's performance.

Although generally public officials (such as the inspector in **Case 3.3**) are immune from liability, liability may be imposed on public agencies or officials under certain facts.[35]

If the changes required on a project so far exceed the work agreed to be performed by a contractor, compensation beyond the contract price or GMP may be allowed.[36]

§ 3.15 Mini-Trials

The private dispute resolution organizations that provide arbitrators and mediators often also provide full courtroom facilities and trial procedures for parties upon payment of suitable fees. The parties can elect to have the trial binding or nonbinding. If the parties have not agreed to have the decision of the judge of the mini-trial be final, a party dissatisfied with the result may be required to go through the additional expense and delay of a trial in the public court system.

§ 3.16 Conclusion

Many systems and procedures exist that provide alternatives to filing and pursuing conventional lawsuits to resolve construction disputes. Arbitration as an alternative method may involve attorneys, or the principals may represent themselves. Mediation may pragmatically focus on the dollars involved, rather than on legal rules or the admissibility of evidence.

---

[35] Sunlake Apartment Residents v. Tonti Development Corp., 522 So. 2d 1298 (La. App. 1988) (attempt to hold local government official liable for approving defective building plans). Regarding liability for negligent inspections, see also *Note, Local Government Liability in Virginia for Negligent Inspection of Buildings, Structures and Equipment,* 18 U. Rich. L. Rev. 809 (1984). Regarding negligent issuance of building permits, see *Comment, Municipal Tort Liability for Erroneous Issuance of Building Permit,* 58 Wash. L. Rev. 537 (1983); Holman v. State, 53 Cal. App. 3d 317, 124 Cal. Rptr. 773 (1975); Morris v. County of Marin, 18 Cal. 3d 901, 559 P.2d 606, 136 Cal. Rptr. 251 (1977) (injured worker allowed to sue county, which had issued building permit to worker's employer, who, in violation of state law, had failed to file workers' compensation certificate).

[36] C. Norman Petersen Co. v. Container Corp. of America, 172 Cal. App. 3d 628, 218 Cal. Rptr. 592 (1985) (cardinal change in work eliminated contract price as a limit on recovery); Rudd v. Anderson, 153 Ind. App. 11, 285 N.E.2d 836 (1972) (reasonable value of work rather than contract price governed, because of number of changes to work).

Arbitration clauses are commonly contained in construction contracts. When an arbitration clause exists in a contract, if one party to the contract demands arbitration, usually the others must submit to the process.

In certain circumstances, an element of choice may be involved in deciding whether to arbitrate construction disputes. This choice could occur before the signing of a construction contract that contains an arbitration clause. It could also occur where a dispute has already arisen but it is unclear whether the arbitration clause will be enforced, or where the opposing party has not demanded arbitration.

Determining which procedure is most suitable to resolve a construction dispute may be dictated by the complexity of the dispute, the amount of dollars involved, and the question of whether the parties have, or wish to have, a continuing business relationship with the other parties to the dispute.

The choice of a dispute resolution system should be made carefully. Construction projects generally involve substantial amounts of capital. Selecting the wrong system may signify whether an organization will emerge from a project with a profit, will bear a substantial loss, or may even be forced into bankruptcy.

CHAPTER 4

THE CONSTRUCTION PROCESS

§ 4.1 The Traditional Construction System: The Parties and Their Roles

Although many businesses in the past fifty years have been consolidating, and large conglomerates are now dominating industries, the construction business has remained highly decentralized. It is composed primarily of perhaps twenty very large firms and a large number of small and medium-size firms. The majority of construction firms have remained locally oriented, with most serving a single metropolitan area.

Competition is the dominant characteristic of the construction business. Most firms are comparatively small; the average number of employees for each construction firm in the United States is fewer than ten. Many firms are family-owned.

Two out of every three construction companies specialize in a particular trade—plumbing, electrical, masonry, carpentry, plastering, excavation, or mechanical. Specialization has altered the nature of the legal relationship between owners, who commission projects, and the individuals who provide the labor.

The traditional construction project involves at least four principals: the owner, the design professionals, the builders, and the financing entities. The owner acquires the land, hires one or more design professionals (an architect and/or engineers) to design a structure, and selects a prime contractor to build the structure. (To avoid repeating "architect and engineer" throughout the text, the more general term "design professional" will be used.) The prime contractor then subcontracts with specialty trades to handle each facet of the construction. This traditional construction framework has formed the basis of existing construction law, despite the fact that alternative systems are also being developed and used.

For many projects, the owner must retain a design professional before he even acquires the property. The design professional, and possibly other consultants, will help the owner determine which land to acquire, prepare a rough design of the structure, and help the owner to obtain the necessary preliminary approvals required for construction. The plans or drawings at this stage will be sketchy and will usually not provide the basis for a finished project. Lenders for a project often become involved at an early stage, because most projects require financing.

When construction begins, the design professionals may have some control over progress and inspection of the work, to ensure that it conforms to the intent of the owner and is consistent with the design, plans, drawings, and applicable laws. Lenders control payment of funds for construction as the project proceeds. On a small minority of projects, if suppliers or subcontractors are not paid by the prime contractor, the lender may be required to stop payment to the prime contractor for the work that is progressing on the project. (The term "prime contractor" is used here rather than the more common term "general contractor," to identify the contractor that has a direct contract with the owner and is generally supervising the work.)

Changes in the design of the project, or in its construction, may be made during the construction phase. Changes may originate with the owner, the design professional, the prime contractor, or subcontractors on the project.

In this traditional system, the prime contractor meets the obligation to build the project by performing some of the labor, as well as by soliciting and receiving bids from subcontractors and coordinating their performance. These subcontractors perform specialized aspects of the construction. Certain types of projects will even have sub-subcontractors, who perform more highly specialized work.

Suppliers of materials used on a project affect the progress and success of the project. The timeliness of delivery and the quality of the supplies are important factors in ensuring prompt and defect-free completion of work on the project.

Many projects are built using this traditional approach to construction. Most of the legal rules that apply to construction law problems assume the traditional approach is being used.

§ 4.2 Prime Contractors as Coordinators Rather Than as Providers of Labor

The percentage of projects performed in the traditional manner is decreasing. In the traditional structure, the general contractor provided much, if not most, of the labor for the entire project. Currently, prime contractors tend to subcontract over 95 percent of the actual labor. The design professionals (architects, engineers, and others who participate in the design of the project) have a direct contractual relationship with the owner.

Under the more modern system, the prime contractor functions more as a coordinator, seldom performing any of the actual labor on the project. In this system, the prime contractor organizes the work, provides scheduling, and coordinates the work of the specialty trades.

The use of numerous specialty trades to perform the work requires that the prime contractor have multiple contracts with the subcontractors. The master contract and the subcontracts must be coordinated, and complex and differing levels of relationship must be maintained among all the parties involved on a project. A typical project may require dozens of contracts between owner and prime contractor, prime contractor and subcontractors, subcontractors and suppliers, owner and sureties, owner and architect, and owner and insurers.

Owners are turning, more often, to a *construction manager,* to ensure proper budgeting, scheduling, and quality of construction.

§ 4.3 Owners

Wide variations can be found in the level of experience and sophistication of owners. At one end of the owner spectrum, there may be a representative of a public agency who deals daily with the award of contracts and with contract negotiation. At the other end of the spectrum is an individual who wants remodeling work done at a residence and has had no experience whatsoever with the construction process.

Owners with less experience will need more guidance and supervision from design professionals, contractors, or others involved with a project. Such owners may be more likely to use form contracts, such as the AIA documents, in order to avoid incurring legal fees for preparation of personalized contracts.

The attitude of courts faced with making decisions in construction disputes involving owners will depend, in part, on the courts' perception of the amount of protection that should be accorded to the owner. The more sophisticated the owner, the less a court will see a need to protect the owner or to give the owner the benefit of the doubt in the event of ambiguities in the contract.

In *Travelers Indemnity Co. v. Ewing, Cole, Erdman and Eubank,*[1] the owner, a public entity, had purchased from a contractor components that had been built to specifications of the architect. The contractor was required to declare bankruptcy, and the bankruptcy trustee took possession of the units. The owner argued that the architect should have advised the owner, earlier in the construction process, of the risk that the contractor might declare bankruptcy and the possibility of the owner's losing possession of the components. If the owner had been aware of these risks, precautions would have been taken against such a loss.

The *Travelers* court refused to impose this type of liability on the architect. The court referred to the relative sophistication of the owner and stated that: (1) the owner was a government agency, (2) it was represented by counsel throughout the negotiations, and (3) it would be an inefficient allocation of professional responsibilities to hold architects liable for not alerting the owner's attorneys to the legal ramifications of the bankruptcy of a contractor. A contrary ruling would have sent a message to architects that they are required to provide legal advice to owners, even when the owner is knowledgeable and represented by counsel.

In *Travelers,* the owner's representative was experienced—in effect, a "professional" owner. The court was not convinced that there was any reason to shift responsibility for legal advice from the owner's attorneys.

If an owner is involved in a dispute and can prove to the court that he is an inexperienced party in the construction process and has been taken advantage of by others, he is more likely to win the case. In other words, if it appears that the owner *deserves* to win, chances of success are substantially greater. Although the legal system has many rules and doctrines, it is a mistake to underestimate the extent to which a court will try to compensate deserving parties, or to find parties not liable where imposing liability would be unfair.

---

[1] 711 F.2d 14 (3d Cir. 1983), *cert. denied,* 464 U.S. 1041, 104 S. Ct. 707 (1983).

§ 4.4 —Private Owners

Private construction amounts to nearly 40 percent of all construction. Residential construction is approximately one half of private construction, or approximately 20 percent of all construction. Commercial construction has increased in the past twenty years in the categories of office buildings, factories, plants, shopping malls, and regional marketing centers.

The construction business is at the mercy of changes in the economic climate. The amount spent on residential construction expanded drastically during the early 1970s. It grew by more than 150 percent between 1970 and 1977. However, from 1977 until 1980, residential construction grew by only 3 percent, a decline in real-dollar terms because of the rate of inflation at that time.

A private owner may be an individual, a corporation, or a partnership. A more sophisticated owner is likely to have established procedures for handling construction of a project. Such an owner may have developed his own contract forms and other documents to provide structure and consistency to his construction projects.

Many owners, however, will rely on the prime contractor, the architect, or an attorney to provide proper contract forms and procedures for a project.

The level of complexity and the amount of capital involved in construction projects dictate that failure of any of the parties to the process to perform properly may have dire consequences. Workers or members of the public can be injured if proper construction practices are not followed. Property owners next to or near the project may be damaged. When claims are filed, the public pays for these claims in the form of higher insurance premiums.

* * *

Case 4.1
(a) Prelitigation

Joe, a contractor, agreed to replace a deteriorated retaining wall for $34,000. The wall was a part of a driveway at a multiunit residence located in an area near a river. The agreement was made during a winter of heavy rains. During the progress of the work, Joe demanded that Diane, the owner, sign change orders totaling approximately $4,000. The change orders were for work such as clearing of trees, and for movement of soil. This work should have been within the scope of the original contract. Diane signed the change orders, because Joe refused to proceed further with any work unless Diane did so.

Joe failed to follow the plans and specifications prepared by the engineer. The retaining wall was not completed before rains further damaged the condition of the driveway. In addition, the retaining wall was improperly located, resulting in the wall's being shorter than building code requirements.

As problems became apparent, Diane secured additional engineers to investigate the drainage of the wall and to check the compaction of the soil under and around the retaining wall. Diane withheld final payment of $20,000 from Joe, pending a determination of the amount of damage caused by Joe's defective and uncompleted work. In addition, Diane was forced to have $4,000 worth of temporary work done to prevent a mudslide, when heavy rains fell after the original promised date of completion by Joe.

Diane retained an attorney, who advised her to continue to withhold final payment to the contractor. The county building department had signed the permit for completion of the work, a fact heavily relied on by the contractor in his demands for final payment. It turned out that the permit had been granted in error. The permit was rescinded by the county because the work had not, in fact, been completed in accord with building code requirements.

Analysis. In essence, the owner of this residence was acting as her own general or prime contractor. She arranged for Joe to do the work, and she hired the engineers. She also took on herself some of the risk involved. Although owners of homes occasionally decide they can do the supervision themselves, when they total up the costs at the end, they sometimes conclude that they should have hired someone else to take on the role of supervisor.

(b) Post-Filing of Litigation

The engineer and the new contractor retained to complete the work on the driveway and the wall maintained that the contractor's work was defective. Joe filed a mechanic's lien and a suit for breach of contract, to enforce his rights under the mechanic's lien. After pretrial discovery, depositions, and exchange of documents, Joe agreed to accept $5,000 as payment in full. It appeared likely that Joe's claim of $4,000 as extras fell within the scope of the original contract. Estimates to repair and redo his defective work had ranged from $25,000 to $35,000. In the interest of saving legal fees that would be incurred in taking a lawsuit to trial, Diane decided to accept in settlement the $15,000 she still held of the $20,000 she had withheld from Joe. Diane used this settlement to pay a new contractor to complete the work properly.

Analysis. Usually, it is not a successful argument, at trial or arbitration, that a party signed an agreement or contract under duress and should not be held to the terms of the agreement. In this case, perhaps Diane should have fired Joe and retained a new contractor when the issue of the change orders first arose. At the least, she could have brought in another contractor to provide her with an opinion on whether those claimed extras were part of the scope of the original work.

The original approval of the work by the county complicated Diane's case, and a deposition had to be taken of the county building inspector. He testified as to the reasons he originally granted the final approval for the permit, and as to why he subsequently rescinded the approval.

Of great value in this case was the fact that Diane did not provide additional payments to Joe. Because she held back further payments, she had financial leverage when a dispute arose regarding the quality and completeness of Joe's work. Joe had to take affirmative steps to receive final payment; Diane did not have to try and recover money from Joe. Diane was able to use the withheld funds to have the remedial work performed and to retain an attorney. Had she paid Joe in full, Diane would have had to sue Joe. To fund the litigation and completion of the work, Diane would have had to borrow money.

* * *

The contractor in **Case 4.1** has authority to claim that he is entitled to additional compensation if the changes are not within the scope of the original contract.[2]

The amount of retention in this case was the $20,000 held back by the owner. It provided the owner with money for claims against the contractor, without the owner's being required to file a lawsuit. The amount of retention does not, however, constitute a limit on the amount an owner can collect for damages against a contractor.[3]

The owner in **Case 4.1** cannot collect attorney's fees. They are normally not available in construction cases, unless specifically provided for by agreement or by statute. However, they may be awarded under extraordinary circumstances, such as the commission of an intentional tort by a third party.[4] Other than a claim for negligence, the owner in **Case 4.1** might have a claim in strict liability against the contractor, based on the theory that the wall the contractor built is defective. The rule applied in many states regarding liability for unsafe or defective products is articulated in Section 402A of the *Restatement of Torts* (1965):

> That one who sells any product in a defective condition unreasonably dangerous to the user or consumer or to his property is subject to liability for physical harm thereby caused to the ultimate user or consumer, or to his property, if (a) the seller is engaged in the business of selling such a product, and (b) it is expected to and does reach the user or consumer without substantial change in the condition in which it is sold.

---

[2] Rudd v. Anderson, 153 Ind. App. 11, 285 N.E.2d 836 (1972).

[3] Gurney Industries, Inc. v. St. Paul Fire and Marine Insurance Co., 467 F.2d 588 (4th Cir. 1972).

[4] Waldringer Corp. v. Ashbrook-Simon-Hartley, Inc., 564 F. Supp. 970 (C.D. Ill. 1983), *modified sub nom.* Waldinger Corp. v. CRS Group Engineers, Inc., 775 F.2d 781 (7th Cir. 1985) (a specialty contractor was held entitled to recover from an engineer who had rigged the specifications for the project and who refused to consider a substitution).

> This rule applies although (a) the seller has exercised all possible care in the preparation and sale of his product, and (b) the user or consumer has not bought the product from or entered into any contractual relation with the seller.[5]

§ 4.5 —Public Owners

Surprisingly, although public construction has experienced explosive growth since the end of World War II, it accounts for the same proportion of total construction dollars spent as it did thirty years ago. Public construction is made up of a wider variety of projects than private construction. In 1980, one-third of the budgeted public construction funds in the United States was being used for buildings; 30 percent was devoted to construction of water and sewer systems; 23 percent, to highway construction; 10 percent, to dams, irrigation, water diversion, and resource conservation projects; 4 percent, to military construction.

One important distinction between projects involving public, as compared to private, owners is the lack of ability of most public owners to freely select contractors, design professionals, or consultants. Public owners must follow selection procedures set out in the regulations or laws that control the public entity. Generally, they must award construction contracts to the lowest responsible bidder.

Social engineering may also be one of the goals in the award of public contracts. Affirmative Action policies are designed to provide preference to minorities or women. Many public contracts require that a set percentage of subcontractors be minority contractors: Minority Business Enterprises (MBEs), Disadvantaged Business Enterprises (DBEs), or Women's Business Enterprises (WBEs). The effect of injecting these public policy considerations into bidding for public projects is that dealing with public owners is likely to be more difficult than dealing with private owners. The public agencies concerned with awarding contracts may have goals to satisfy other than simple profitability or maximum savings.

Public entities, as owners, are concerned with controlling the resolution of disputes. Until recently, public agencies resisted the use of arbitration to resolve construction disputes. As arbitration has become more widely accepted, public agencies have moved toward favoring arbitration. Some laws, in fact, require

[5] Cases applying these standards include: Vandermark v. Ford Motor Co., 61 Cal. 2d 256, 391 P.2d 168, 37 Cal. Rptr. 896 (1964); Bexiga v. Havir Mfg. Corp., 60 N.J. 402, 290 A.2d 281 (1972). In some cases, purchasers of services, rather than "products," have attempted to hold the sellers of services liable under a strict liability standard. Typically, these cases have not been successful; *see* Schipper v. Levitt & Sons, Inc., 44 N.J. 70, 207 A.2d 314 (1965). ***But see also*** Kriegler v. Eichler Homes, Inc., 269 Cal. App. 2d 224, 74 Cal. Rptr. 749 (1969); Wright v. Creative Corp., 30 Colo. App. 575, 498 P.2d 1179 (1972), *reh'g denied* (June 6, 1972). For an article that explores this subject in detail, *see Comment, Strict Liability in the Building Industry,* 33 Emory L.J. 175 (1984).

certain public agencies to arbitrate disputes rather than to pursue litigation. Different procedures and rules usually govern arbitration or mediation of disputes when public entities are owners.

Another significant difference between work for public owners and private owners is the lack of lien remedy available to contractors on projects with public entities as owners. On private work, unpaid contractors, design professionals, or suppliers may be entitled to file a lien on the property they have improved. If necessary, the unpaid party can foreclose on the lien, force sale of the property, and be paid from the proceeds. This is generally not an available remedy in public work.

In some states, an alternative to the lien remedy exists for unpaid subcontractors or others working on a private project. They can issue a **stop notice** to lenders providing financing for the project. Upon receipt of the stop notice, the lender is required to stop paying the general contractor until the issuer of the stop notice is paid.

There is no lien right on public projects. However, on a public project, if the general contractor has posted a bond, unpaid subcontractors may be able to recover payment for their services or materials from the bond.

§ 4.6 Design Professionals

Architects, engineers, and other types of consultants may be used by owners during the design phase of a project. The design professional will incorporate the owner's needs into a design, and may help the owner predict the cost of the project, often with the assistance of subconsultants. The design professional may also advise the owner on the selection of the prime contractor. The design professional is no longer likely to be responsible for supervising the performance on the project during the construction phase. Other functions of the design professional on more complex projects include providing help to the owner in locating the property and in securing financing. During construction, the design professional usually provides input on design content.

It was once considered unseemly for design professionals to compete actively against each other to obtain contracts for projects. The rationale for this policy was that, as professionals, engineers and architects were above the general commercial marketing methods used for other services or products. In the 1970s, association activities or restrictions on competition were attacked and held to be illegal restraints on trade under antitrust law. In *Goldfarb v. Virginia State Bar,*[6] either mandatory or suggested fee schedules for professional services, such as those for an attorney or design professional, were found to be illegal. Competition among professionals, based on price, nature, and quality of service provided, is now commonplace.

---

[6] 421 U.S. 773, 95 S. Ct. 2004, 44 L. Ed. 2d 572 (1975).

§ 4.7 —Architects

Most states require that architects be licensed by state authorities. States generally require that the architects applying for licenses must be citizens of the state and reside within the state. A common requirement is a minimum age, typically 21 to 25. Most states also require that a specified number of years must have been spent by the architect obtaining practical experience. This requirement usually is coupled with a required amount of professional schooling. States inquire into the architectural candidate's character and honesty. Architects are also required to pass a complex written examination, which reflects the concern of the states with protecting the public from the danger of allowing unqualified individuals to design buildings.

In the traditional building construction system, the architect is the owner's primary consultant. The architect communicates for the owner with the prime contractor, subcontractors, or any of the necessary approving public entities. Usually, the architect will monitor construction and progress of the project, handle requests for information from the contractors, approve progress payments or requests for change orders, and approves change order payments. As described in **Chapter 3**, under the AIA documents, presentation of a claim to the architect is the first step taken to resolve claims that arise regarding the project (other than those between the architect and the owner).

AIA contracts define the architect as the agent of the owner. Unless the architect is an employee of the owner, his function is not truly that of an agent during the design phase of the project. Exceptions to this statement would arise if an agency relationship were formally and expressly set out in the contract documents, or if the owner caused third parties to believe that the architect was his agent by giving the architect "apparent authority" to act for the owner.

As a general rule, the architect's role as agent for the owner begins when the design phase ends and construction begins. On the construction site, the architect speaks for and acts for the owner. Information conveyed to the architect is deemed to have been transmitted to the owner as well.

As owners seek to complete projects as economically as possible, they may avoid using architects in the traditional manner. If the owner is a large organization, many of the functions traditionally handled by an architect may be handled by employees of the owner. The owner may also choose to have his own employees, or one designated employee, supervise and monitor construction of the project, so that the architect is used only during the design phase. If standard construction forms based on the traditional system are used, deviation from the traditional mode of performance may give rise to confusion or conflicts in contract interpretation.

Tasks performed by the architect will vary greatly and will generate differing levels of potential liability for the architect. Although it is unwise to do so, some architects offer advice to owners on the financial viability of a project. This type of advice should be avoided by the architect if he wishes to recover his fee.[7] AIA Document B141 provides a list of services that may be performed by the architect. The same document also provides protection for the architect by putting responsibilities on the owner and by requiring the owner to indemnify the architect for claims made against the architect that arise from his work on the project. If the owner decides to dispense with many of the functions usually handled by the architect, AIA Document B141 may be unsuitable, because its use will require extensive alteration or revision. AIA Document B161 (an unstructured alternative form to B141) or B801 (designed for projects using construction management) may be more suitable forms.

In 1990, the AIA issued a new form, AIA Document B511, as an adjunct form to be used with B141. Paragraph 1.2 of B511 addresses services provided by the architect in remodeling, rehabilitation, and renovation projects. Paragraph 2.6 of B511 provides for owner administration of the construction contract. Paragraph 2.7 allows for fast-track design and construction. These provisions are intended to help an owner who is using B141 in a nonstandard manner. They are also intended to ensure that following a nonstandard arrangement does not impose additional responsibility on the architect. Paragraph 1.2 of B511 provides that the owner will indemnify the architect for claims arising out of the work on the project.

Where an architect is retained only for design services and has no site-supervision responsibilities, the architect may not be held liable for the problems associated with the construction performance. However, should a problem arise, it is likely that some of the parties will assert that the problem was caused, in whole or in part, by faulty design by the architect. Lenders and professional liability insurers may also be less comfortable with a project on which the architect does not monitor progress. An architect should be aware that certain states do not permit design-only projects. (See § **4.20**.)

§ 4.8 —Engineers

Engineers fulfill a role similar to that of architects. The engineer licensing process is parallel to that of architects. In the AIA documents, the engineer is not usually identified as the party to whom claims should be brought for resolution. However, there is no reason why an owner could not modify the contract to so identify the engineer. This may be particularly desirable if no architect is being used on the project.

As described earlier, the EJCDC (Engineers Joint Contract Documents Committee) form documents are also used in construction. A 1992 edition of the EJCDC owner–engineer agreement, scheduled for publication in the last quarter of 1992, will depart from use of a single document and will provide for

[7] Ballinger v. Howell Manufacturing Co., 407 Pa. 319, 180 A.2d 555 (1962).

expanded use of exhibits for inclusion with the contract, if desired by the contracting parties.

§ 4.9 —Construction Managers

Some owners will use a full-time representative on a job site. That individual may be an employee of the owner's company, an employee of the architect or engineer, or a construction manager (CM), who is a separate entity.

Owners have been concerned about a perceived lack of attention by design professionals to cost control, quality control, and adherence to construction schedules. CMs are individuals with management skills who act as overall supervisors of projects. As owners continue to attempt to build higher-quality projects at a lower price, use of CMs has increased. Under certain circumstances, public agencies have allowed contracts to be awarded without the use of competitive bidding, where it was known that a CM would be used.[8]

Under some owner contracts with architects, the architect agrees to provide CM services. The legal effect of this type of agreement may be to effectively convert the contract to a design-build contract.[9] (See also **§ 4.20**.) The usual arrangement is that a CM is hired in addition to the architect, and there is no contractual relationship between the CM and the design professional. The CM advises the owner and the design professional regarding design, and the CM usually handles the on-site services for the project.

One problem with the use of a CM is identifying the limits of the CM's authority to act for the owner. On some projects, the CM hires specialty contractors. Unless the relationships are clearly defined, there may be confusion as to whether those contractors were hired by the CM acting for the owner, or whether the contractors are in contractual privity only with the CM. This may be an important distinction. In some states, such as California, a contractor who is in a direct contract with the owner is not required to send a preliminary notice of lien to the owner or lender prior to filing mechanic's liens.

Under other circumstances, the CM may agree to perform its work for a guaranteed maximum price (GMP) and to provide its own labor for the project. When functioning in this manner, the CM appears to be more of a prime contractor than a manager acting on behalf of the owner.

Sometimes, the prime contractor will agree to perform the work as a prime contractor, and, for payment of separate compensation, to perform services as a CM on the project. Under such a contract, compensation to the prime/CM for its services as CM may be based on the cost of construction plus a fixed percentage.

---

[8] Willman v. Children's Hospital of Pittsburgh, 74 Pa. Comm. 303, 459 A.2d 855 (1983).

[9] First National Bank of Akron v. Cann, 503 F. Supp. 419 (1980), *aff'd,* 669 F.2d 415 (6th Cir. 1982).

Because the body of construction law has relied primarily on the use of the traditional construction structure, participation in a project that uses a CM may result in confusion regarding relative responsibilities. In one case, there were actually *two* construction managers on a job, with one denominated the "Overall CM."[10]

The Construction Management Association of America (CMAA) has developed a series of contract documents to define the role of the construction manager in regard to the performance of construction. The use of these contract forms has caused concern among both engineers and architects, who see potential liability arising from the lack of involvement of these groups in the construction process because of the CM designation. One possible solution is to have the owner indemnify the design professional for costs arising from liability claims that develop during the construction phase. Another solution is for EJCDC to prepare its own agreement for use between the owner and engineer, in which EJCDC defines the duties of the CM. As CMs are used more frequently, forms will be developed and some of these concerns may subside.

Use of innovations such as CMs may lead to confusion regarding the rights and responsibilities of the parties under a construction contract, unless these issues are clearly dealt with in the contract documents. If the dispute goes to litigation, the rules of law developed for use with the traditional structure may allocate liability in a way that the parties did not foresee.

§ 4.10 —Compensation of Design Professionals

Various methods are used to compensate architects for their services. The amount of the fee paid by an owner is largely a matter of contract or agreement between the parties. In the event of a dispute, the court may be asked to interpret whether the services performed by the architect are within the scope of the contract and are covered by the basic design fee, or whether they constitute extra services. A general rule of contract interpretation is that courts do not look beyond the written words of the contract. If a portion of the contract is unclear or ambiguous, the courts will construe this portion against the party who wrote the contract, rather than against the party who merely signed or agreed to it. If the architect drafted or provided the contract and the description of the basic design services is unclear, it is likely that the court will determine that the services performed are covered by the basic design fee and that no additional compensation is due.

One method for calculating the architect's fee is the **stated percentage of construction costs** method: multiplication of the construction costs by a designated percentage produces the fee amount. This fee usually covers only basic

[10] DiSalvatore v. United States, 456 F. Supp. 1079 (E.D. Pa. 1978), 472 F. Supp. 816 (E.D. Pa. 1979), *appeal dismissed,* 622 F.2d 578 (1979).

design services. Additional services or reimbursable expenses would be added to this type of fee arrangement to arrive at the architect's total compensation. The percentage figure used may depend on the amount of work the architect must perform, as well as the degree of risk of failure of the project. If the project is not completed, the architect will be uncompensated for his time on the project because his fee depended on completion.

Some architects use a flexible percentage, or a sliding scale percentage (the fee decreases as costs rise). This fee arrangement continues to be used, although it may not be an accurate measurement of time spent on any one project. In the overall picture, it may average out to reasonable approximation of the time spent by the architect.

Another fee method used for compensation of architects is based on multiplying the direct personnel expense. Typical multiplier figures range between 2 and 4, the average being 2.5 to 2.7. According to AIA Document B141, paragraph 10.1.1, direct personnel expenses include both the direct salaries of the architect's personnel engaged on the project and a portion of the cost of the mandatory and customary contributions and benefits related to their salaries.

Yet another fee arrangement for architects is a professional fee plus expenses, also called a **cost contract**. Under this method, a fixed fee is calculated and agreed on, but necessary costs are reimbursed above and beyond this fee. Failure by the architect to make it clear that a preliminary statement of likely costs is only an estimate may result in that preliminary cost statement acting as a ceiling on the amount the architect may recover.[11]

Some architects use a fixed fee contract. Under this arrangement, the architect is to be paid a stated amount, regardless of the amount of time he spends on the project or the final cost of the project. This system works most effectively when the project is routine and well-organized. Most form contracts state that the fixed fee agreement only covers basic design services. Nevertheless, to avoid later disputes, it is good practice for the architect and the owner to specify precisely the services that are covered by the fixed fee.

If the architect performs work outside the scope of the original fee arrangement, whether at the owner's request or with the owner's knowledge, or because such work appears necessary, performance of that work usually entitles the architect to additional payment. In the absence of any provision in the original agreement specifying the rate of compensation for extra work, the architect's services would be compensated at the reasonable value of the services. Reasonable or market value for a product or service is usually defined as the price to which a willing buyer and willing seller would agree, absent any compulsion or duress. This method would include consideration of the complexity of the work, the level of skill of the architect, the number of hours worked, the nature of the work performed, and any other relevant factors.

---

[11] Martin Bloom Associates v. Manzie, 389 F. Supp. 848 (D. Nev. 1975) (architect forfeited right to fee).

§ 4.11 Contractors

The generic use of the word *contractor* simply refers to one who performs labor for another, but who is not an employee. A contractor provides labor or services pursuant to an agreement, or contract. Modern development has somewhat altered the original system, in which a general contractor was in a contract with the owner, and the general contractor hired subcontractors.

§ 4.12 —Prime/General Contractor

A prime contractor ("the prime") is one who bids for the master contract for construction. He usually has overall responsibility for the construction of the work. On some occasions, the prime may also act as construction manager for the owner. The prime obtains bids from other contractors (subcontractors), who perform specialty trades.

§ 4.13 —Subcontractors

The specialty trades, or subcontractors, have contracts with the prime contractor. Usually, none of these contracts governs the relationships of the subcontractors with each other, with the owner, or with the design professionals.

When the prime contractor is negotiating with an owner or bidding on a project, he is also negotiating with, or receiving bids from, subcontractors and suppliers to perform specialized portions of the project, or to supply goods to be used on the job. A few generations ago, the intricate subcontracting of portions of construction did not exist.

Under the present system, contractual rights and obligations of the various parties may be unclear if they are not specifically addressed in the contract documents. On some occasions, they can actually be in conflict. Coordination by the prime contractor of both contract terms and subcontractors' performance is essential to maintaining the balance that will avoid conflicts and, ultimately, litigation.

Because of the two separate sets of negotiations, competitively bid projects present problems. The subcontractor submits a bid to the prime contractor. This bid is neither accepted nor rejected; it is incorporated into the bid package submitted to the owner. If the prime's overall bid is accepted, the prime notifies the subcontractors of his formal acceptance of the subcontract bid.

The subcontractor, however, is free to modify or rescind his bid up to the moment the prime expressly accepts the subcontractor's bid. If the subcontractor raises his bid after the prime has submitted his bid to the owner, the prime faces the possibility of having to perform the contract for a price greater than he had anticipated. However, the prime is not bound to use the subcontractor on whose bid the prime relied. After the prime's bid is accepted, the prime is free to ask other subcontractors for new bids and to accept the lowest bid then available.

Just as the prime is not bound to accept the subcontractor's bid, absent statutory or administrative requirements, the subcontractor may withdraw its bid to the prime contractor at any time prior to the prime contractor's acceptance of the subcontractor's bid. Formerly, few prime contractors were successful in using the courts to enforce subcontractors' bids. Generally, courts have refused to enforce subcontractors' bids, because of the lack of reciprocity in the prime's refusal to give a firm acceptance of subcontractor offers prior to the award of the prime contract. In the past, this has made it difficult for primes to prove the existence of a binding contract.[12]

Recent cases have allowed primes to recover under the theory that the prime had relied on the subcontractor's bid and had taken steps based on that reliance. This theory is called **promissory estoppel**. The most well-known case to apply this doctrine is *Drennan v. Star Paving.*[13] In *Drennan,* the subcontractor (Star Paving) had withdrawn its bid because it had discovered a mistake in its computations, which formed the basis of the telephone bid it had made. The California Supreme Court refused to hold the subcontractor liable under contract law for withdrawing the bid offer after the award of the prime contract to the general contractor (Drennan). However, the court found Star Paving liable under a theory of promissory estoppel. The holding of *Drennan,* however, is qualified: the prime contractor must accept the subcontractor's offer without delay after award of the prime contract, and the prime contractor must not reopen bargaining on the bid of that subcontractor or the bids of any other subcontractors.

Once the prime is awarded the bid, the prime may press the subcontractors for more favorable bids. This procedure is known as **bid shopping**. In some cases, the awarding owner will encourage bid shopping, because it can result in the owner's receiving the project completed for a lower amount, furthering the goal of competitive bidding.

Bid shopping alters the bidding process in a number of ways. Bids of both primes and subcontractors are less likely to be accurate or reliable. Primes are encouraged to underbid, on the theory that the bids of subcontractors later may be lowered if the prime shops bids. Subcontractors are encouraged to bid higher than their true fair bid, to allow room for downward scaling of the subcontractor bid in response to bid shopping by the prime after award of the prime contract.

The practice of bid shopping prevents some subcontractors from participating in the process. If a small number of subcontractors control a specialty, they may be successful in maintaining the subcontract bids at an artificially high level. If bids are artificially maintained at a high level as a result of collusion, it is anticompetitive activity that can violate antitrust laws. However, these types of

---

[12] James Baird Co. v. Gimbel Brothers, 64 F.2d 344 (2d Cir. 1933).

[13] 51 Cal. 2d 409, 333 P.2d 757 (1958).

violations are exceedingly difficult to prove. In addition, the current federal government's enforcement of antitrust laws has been exceedingly lax.

In the worst case analysis of this bidding problem, subcontractors will submit irresponsibly low bids—either initially, or upon later pressure from the prime—and then sacrifice quality or fail to take safety precautions on the job, in their attempt to make a profit.

The Association of General Contractors (AGC) condemns bid shopping in its Code of Ethical Conduct. Bid shopping, however, continues to exist. One theory, at least as to public contracts, is that it provides a desirable form of competition and results in construction at a lower price to the public, which bears the cost of public projects.

In order to prevent some of the abuses resulting from bid shopping in public construction, some states have enacted statutes that require a prime contractor to file all solicited subcontractor bids with the appropriate awarding agency before preparing the prime contractor's bids. The agency may reject master bids that are not responsive.

The agency sends the prime contractor a list of the subcontractor bids that are acceptable to the agency. The prime chooses from the list of subcontractors and prepares his bid in accord with the sub-bids he has selected. If awarded the prime contract, the prime contractor may not substitute other subcontractors, except with the approval of the awarding agency.

Other state legislatures have passed statutes that require prime contractors to list all subcontractors whose bids were used in calculating the prime bid. The statutes also limit the circumstances in which the prime contractor who is awarded the contract may substitute subcontractors or alter the provisions of the subcontracts.[14]

§ 4.14 Lenders

Lenders are increasingly taking an active role in construction—inspecting the progress of the work, ensuring that the checks written for construction financing are being used as represented by the prime contractor, or by the owner, and verifying that the progress of the project is on schedule. Some lenders use a system of draws: subcontractors and suppliers are paid directly, to avoid use of funds for unauthorized purposes by prime contractors or owners. Another way to prevent abuse is for the lender to issue checks for work or supplies to joint payees. The two payees are usually the prime contractor and the specific subcontractor or supplier.

Some states provide subcontractors or suppliers with stop notice rights on private projects. If a subcontractor or supplier is unpaid for work performed

[14] Cal. Pub. Contr. Code § 4104 (West); Mass. Gen. Laws Ann. ch. 149, § 44E; N.J. Stat. Ann. § 52:32-2.

on a job, the subcontractor or supplier may demand, by means of a stop notice, that the lender withhold payment to the prime contractor. The lender may have liability to those parties if it fails to withhold payment of funds as demanded.

Construction lenders often are concerned with whether the owner of the property has a substantial stake in the success of the project. One way to satisfy this goal is to require the owner to put up enough of the construction funds to dissuade the owner from abandoning the project when problems arise. Problems such as increased costs of construction, claims by contractors, and other delays on the project all increase the risk for the construction lender.

§ 4.15 Sureties

The bonding demanded of contractors by owners provides security for acceptance of the bid, for completion of the project, and for the payment of those performing on the project. Although this subject will be explored in depth in **Chapter 8**, a general discussion is appropriate here. The common types of bonds used in construction are licensing bonds, payment bonds, performance bonds, and bid bonds. Licensing bonds are usually of minimal value unless a project is very small. The contractor, or other principal, must post a nominal bond to obtain his license. Upon default of the contractor, a claimant may make a claim on the bond. The amount available under contractors' licensing bonds usually ranges between $5,000 and $15,000.

The general purpose of payment, performance, or bid bonds is to ensure that, if the principal under the bond fails to perform a contractual obligation, the bonding entity will take responsibility. The surety usually has the option of either stepping in to pay the cost of performing the uncompleted work, or hiring a new contractor to complete the contract left unfinished by the principal under the bond. Bonding companies should, but often do not, monitor the progress of the work closely. Bonding agreements frequently require that the bonding company be notified in the event that a principal is faltering in its performance on the project.

§ 4.16 Insurers

The necessity for insurance has become a fact of life for nearly every aspect of our lives, both business and personal. In the construction business, contractors must carry workers' compensation insurance to cover the employer's liability for injuries to workers. The workers' compensation system was developed to provide expeditious help for workers who were injured in the course of their work. Workers' compensation is subject to a legal system that is completely different from the state or federal civil court system. It has its own judges, rules, and formulas for compensation. The system was intended to pay for the cost of medical bills and disability for a worker injured on the job, regardless of whether the worker was negligent in causing the injury. When a worker is injured on the job, his exclusive remedy against his employer is usually through this system. There are few exceptions to this rule.

The other most common type of insurance for contractors is general liability insurance. This insurance covers risks such as loss or damage to the project as a result of fire or other disaster. It also covers liability for injuries to third parties.

Most general liability insurance policies for contractors do not cover claims by owners for defective construction. These policies do not serve as warranties that the project will be completed in a proper manner. They serve as insurance in the event of an injury of a third person, and they cover liability for damage that is incidental to performance of construction on the project. The language of the policies is tailored so that coverage is only provided for "consequential" damage caused by representatives of the insured. The policies are intended to cover only incidental risks, not the risk that the contractor will not perform properly.

As an example, after construction is complete, tenants in a commercial building sustain damage to interior improvements; because the roofing was done improperly, water leaked into the space of the tenants. The primary source of loss or damage to the owner is the fact that the roof was built incorrectly: the roof will need to be repaired, or rebuilt. Liability for the cost of repair of the roof (or replacement, depending on the severity of the problem) would not be covered under the roofer's general liability policy. Damage to the possessions of the tenants would be covered, because it is damage caused by the defective work, or **consequential** damage.

It is also common for the contracts between prime contractors and subcontractors to require that the subcontractor show proof of general liability insurance. If an injury, or damage to adjacent property, occurs, the prime will not be liable and will not have to use its own insurance coverage. Instead, the prime can rely on the coverage provided by the subcontractor's insurance. In some cases, the construction contract may require that the subcontractor name the prime contractor as an additional named insured under the subcontractor's policy. In the event of litigation, the insurer for the subcontractor will then be directly obligated under the insurance contract to defend the prime contractor as a named insured.

§ 4.17 Other Alternatives to the Traditional System

As described in § **4.2**, the traditional system has been altered with the increased use of specialty contractors, prime contractors who do not provide labor for projects, and construction managers (CMs). Currently, there are other, more radical systems also being used in construction. These include the use of multiple prime contractors, fast-tracking, and design-build.

§ 4.18 —Multiple Prime Contractors

The use of multiple prime contractors, or of separate contracts for a project, has developed to provide owners with more ability to control the specialty contractors. Owners' reliance on the prime contractors' management skills is eliminated. One of the other goals of this method is to lower the cost of the specialty contracts. By breaking the specialty aspects of the work into smaller units, more contractors may bid. A greater number of contractors bidding can lead to more competition and lower bids. Under this system, the prime contractor does not increase its bid to compensate for project managerial responsibilities.

In the traditional system, subcontractors expect the prime contractor to coordinate the subcontractors and to resolve disputes or conflicts. With multiple primes, the various contractors must look to whoever has the job of supervision of the work. This may be the owner, a CM, or a design professional. The shift in the role of the prime contractor, from primary source of labor for the project to manager of the project, results in legal ambiguities regarding the responsibility of parties when problems arise. Where the multiple prime system is used, even greater legal problems and confusion may arise.[15]

§ 4.19 —Fast-Tracking

Fast-tracking, also called **phased construction**, is an alternative system in which design of the project may not be completed prior to commencement of construction. Typically, a construction manager is used when a project is fast-tracked, although a CM may also be used in conjunction with the traditional method of construction. Parts of a project may be fast-tracked, with other parts performed in the traditional method. Because construction with fast-tracking may begin before the design is complete, construction should, theoretically, be completed at an earlier date than if the inception of construction awaits the completion of the design.

An additional reason for using fast-tracking is that the developer may be able to obtain construction financing earlier. The costs of acquiring the land, and other fixed costs associated with owning the land, continue during the design phase. Interest on the land loans accrues. If the construction funds are received sooner, and construction begins and ends sooner, the total cost of the project should be reduced.

---

[15] Broadway Maintenance Corp. v. Rutgers, 90 N.J. 253, 447 A.2d 906 (1982).

One disadvantage to fast-tracking is that bids for the cost of construction are more speculative, because the design and specifications for the project are not available. With fast-tracking, it is common for the prime contractor to provide the owner with a contract price that is the contractor's cost plus overhead and profit. The owner will probably demand a guaranteed maximum price (GMP).

A potential disadvantage to fast-tracking is that a greater number of claims may occur during construction, because of the incompleteness of the design. It may then be argued that the project effectively has no guaranteed maximum price (GMP), and there is simply a contract for cost plus overhead and profit.[16] Portions of the work may not be covered by the original total cost of the design.[17]

With fast-tracking, the importance of coordination between contractors is even greater than with other nontraditional methods. The scheduling for performance of each of the contractors is usually tighter, and the progress of each contractor depends on the work schedules of the preceding contractors. Claims related to contractors' delays, often caused by the failure of other contractors to do their work on time, can result in higher construction costs.

§ 4.20 —Design-Build

Under certain circumstances, a contractor or an architect may agree to handle all facets of a project. This agreement will usually take the form of presenting an owner or a developer with a package in which all aspects of the work are covered; the owner is not required to hire other parties to perform the construction, or to monitor or coordinate the job.

Generally, this "design-build" proposal will occur at a very early stage. It may be made before the owner has even decided to proceed with the project. A contractor who learns of the owner's desire to build or remodel will do some research and present a proposal to the owner. Such a proposal may provide that there is a guaranteed maximum price (GMP). The contractor promises that, except under narrowly drawn, specific circumstances, the owner will not be required to spend any more than a certain sum. In some cases, the contractor may even promise that if the project is completed for a lesser amount, the owner will share in the cost reduction.

Architects may also propose or agree to perform a design-build project. For example, a company that wishes to relocate to another city may agree to accept a design-build package from an architect who has previously built projects in

---

[16] Armour & Co. v. Scott, 360 F. Supp. 319 (W.D. Pa. 1972), *aff'd,* 480 F.2d 611 (3d Cir. 1973); C. Norman Petersen Co. v. Container Corp. of America, 172 Cal. App. 3d 628, 218 Cal. Rptr. 592 (1985).

[17] Grinnell Fire Protection Systems Co. v. W. C. Ealy & Associates, Inc., 552 S.W.2d 747 (Tenn. App. 1977).

that city and has secured land for the project under a long-term lease. By accepting the architect's design-build package, a relatively inexperienced owner may be able to complete a project at a lower cost and avoid active involvement in the construction process.

§ 4.21 Conclusion

Traditional contracting practices are still frequently used for construction, and the security and certainty associated with knowing each party's role in the construction process is of significant value. The increased use of alternative methods may complicate the construction process, and may create more confusion when disputes occur. As alternative modes of construction are explored, increased legal uncertainty and greater likelihood of claims are inevitable.

Because existing construction law is based primarily on use of the traditional construction process, the courts' application of those rules may not produce equitable or consistent results where alternative methods are used.

Standard forms may be an economical method of formalizing the agreements of the parties for some simple construction projects that follow the traditional method of construction. However, use of standard forms, such as those generated by the AIA, EJCDC, CMAA, or AGC, may be less appropriate if the plan for construction deviates drastically from the traditional system.

As with many solutions, the problems created by the use of alternatives may be greater than the problems the alternatives were designed to solve. However, as more nontraditional construction projects are built, disputes regarding these projects will be litigated. Court opinions will be rendered, resolving disputes arising from these projects. The law that will evolve will realistically reflect the rights and responsibilities of the parties on these projects. As a result, some of the legal risks currently associated with use of alternative construction methods will be reduced.

CHAPTER 5

CONTRACT AS A MEANS OF REGULATING DUTIES AND BEHAVIOR

§ 5.1 The Purpose and Function of Contracts

Contracts are binding agreements between parties. Contracts are the parties' direct expression of the terms and conditions to which they have agreed. The use of formal written contracts for transactions helps to bring predictability and stability to the commercial marketplace. For the parties, the greater level of reassurance and predictability derived from the use of express contracts contributes to the success of business ventures and makes it more likely that entrepreneurs will initiate commercial ventures that are a benefit to the economy as a whole.

The right of parties to form contracts is not entirely unrestricted. Some restrictions are based on statutes and public policy, which are used in interpreting parties' rights and duties. However, freedom from impairment of the ability to form contracts is one of the basic provisions of the federal and state constitutions.[1]

This right of parties to execute contracts and to perform them with whomever they choose assumes that parties in the commercial marketplace are equal. Within the construction process, it is obvious that not all parties are in equal bargaining positions, and that some parties have far greater leverage than others. Where one party appears to have been exploited, or circumstances show that it would be unfair to impose liability on a particular party, courts may be reluctant to literally enforce the written terms of contracts.

---

[1] U.S. Const. art. I, § 10, cl. 1; Cal. Const. art. I, § 9.

Parties to the construction process will have contracts with owners, architects, contractors, lenders, insurers, and suppliers of goods. As a rule, most of the participants will have a contract with only one primary party in the construction framework. In the traditional setting for construction projects, the owner will contract with one or more design professionals and with a prime contractor. The prime contractor will have contracts with the specialty subcontractors and suppliers. As a rule, the subcontractors will have no contracts governing their relationships with each other, the suppliers, the design professionals, or the owner. The design professional usually does not have a contract with the prime contractor or with other design professionals who may be used on a project. However, the contract forms used with subcontractors often will incorporate, by reference, the terms of the master contract.

Many contracts are fully and properly performed without any reluctance or resistance on the part of the parties to the agreements. In spite of this fact, it is important not to downplay the importance of the ability of the parties to a formal contract to enforce the terms and conditions of the contract in court, should that be necessary.

§ 5.2 Elements for Contract

A valid contract requires the presence of certain basic elements. There must be an offer (or counter-offer) and an acceptance. There must be a true agreement between the parties regarding the terms of the contract, often called a **meeting of the minds**. **Consideration** (something of value), such as goods, services, or money, must usually be provided by each party. Contracts generally fall into two different categories: express and implied.

There is also a requirement that the terms of the contract be ascertainable and definite. This requirement is often satisfied by a written document, but it can also be established by other means, such as the custom or practice in a particular trade, or a history of transactions between the parties. Whether price and quantity are specified is often critical in determining whether a contract is valid and enforceable.

§ 5.3 —Express Contracts

An **express** contract is one in which the parties signal their intent to agree to be bound by specific terms and conditions, either orally or in writing. The vast majority of construction contracts are express. However, many individual terms and conditions to construction contracts may be implied, based on a pattern of conduct between the parties, common practice within the construction industry, or patterns or practices in commerce generally.

§ 5.4 —Implied-in-Fact Contracts

Contracts may be created even where there are no expressly discussed terms and conditions and there is no written document. Such contracts are called **implied contracts**. Implied contracts may be either implied-in-fact or implied-in-law. In an **implied-in-fact** contract, the existence of the contract is determined based on actions of a party, rather than on his words.

For example, in an express contract, a contractor provides a bid to an owner to build a roof for $5,000. To accept the bid, the owner would either sign the bottom of the bid form, or the owner would tell the contractor that he or she had agreed to the terms stated in the bid. The two parties would then have a contract.

In implied-in-fact contracts, acceptance of an offer could be based on actions or on a pattern of conduct. A contractor makes a bid for construction of a roof. Instead of signing the proposal, or agreeing to the terms orally, the owner provides the contractor with the deposit amount specified in the written offer. This act of providing a deposit in the amount specified in the contract is likely to constitute an acceptance of all the terms of the contract, making this an implied contract.

Often, implied-in-fact contracts are found in circumstances where acts are not customarily performed gratuitously. A contractor offers to paint a house for an owner. The owner agrees. Although the parties did not specifically discuss the fact that the owner would pay for the service, it is generally recognized that people do not paint houses without expecting to be paid. By agreeing to have the contractor paint the house, the owner has impliedly agreed he would pay for the work. Although in this situation price has not been discussed, if a court found a valid contract existed, the court would award the contractor the reasonable value of the painting service.

In *United States v. Young Lumber Co.*,[2] an employer of a subcontractor tried to argue that it had no obligation to pay the supplier of a piece of equipment on the job because it had not formed any express agreement to do so with the supplier. The court found there was an implied contract, which could be inferred from the conduct of the parties. The court stated that, where one party performs for another a useful service, of a character for which there is usually a charge, and the service is rendered with no objection of the party receiving the service, the law implies a promise on the part of the recipient to pay the reasonable value of the service.

Another example of an implied-in-fact contract is based on a pattern of conduct between parties. A supplier of sheet metal has had a written agreement with a contractor that a certain type of sheet metal is to be shipped to the contractor, who then pays a certain price. After thirteen written contracts, deliveries of the goods, and subsequent payment, the supplier ships the sheet metal at the request of the contractor, although there is no written contract or purchase

[2] 376 F. Supp. 1290 (D.S.C. 1974).

order that specifies the price. The supplier's shipment of certain goods, and the contractor's payment of a certain price for the goods, on the previous thirteen occasions is probably a sufficient pattern of prior conduct to imply that the contractor has agreed to pay the same price on the fourteenth occasion.

§ 5.5 —Implied-in-Law Contracts

An **implied-in-law** obligation is more often found as a part of an agreement. This means that, although the parties did not expressly agree to a certain term or condition in the contract, the law implies that the condition exists in the contract. For example, most states have provided by statute that, if an agreement contains a clause that one party is entitled to recover attorney's fees if that party prevails in an action that has to be brought for failure to perform under the contract, the other party (if he prevails) will be entitled to an award of attorney's fees as well. This is sometimes called the **equal dignities rule**.

Many form contracts used by contractors contain an attorney's fees clause. In a state that has codified the equal dignities rule, a court deciding a case involving an attorney's fees clause would use the statute to imply that the clause applies both ways, in the interests of fairness. Although the contract does not expressly state that either party (whoever prevails) will get his or her attorney's fees, the law implies that it does, and a court would rule accordingly.

The lack of an express contract does not signify that an agreement will not be enforced by a court. It may be more difficult to prove the existence of the agreement, but, if the contract is proven satisfactorily by documentary evidence or by the testimony of relevant witnesses, an implied contract may be held to be as valid as one that is express.

§ 5.6 —Statute of Frauds

Contrary to popular belief, not all valid and enforceable agreements must be in writing. The first Statute of Frauds was enacted in 1677 in England. The United States inherited this statute, which provides that certain agreements must be expressed in writing to be enforceable. The statute is called the Statute of Frauds because it is designed to prevent the perpetration by individuals of false or deceitful acts. In other words, the statute is intended to prevent persons from falsely asserting that a contract or conditions of a contract exist, or denying the existence of a contract or of particular terms in a contract.

Most state statutes of frauds required that an agreement be stated in writing if it cannot be performed, by its stated terms, within one year; if it pertains to transactions in real property; or if it leaves property by will. Contracts that, by their terms, cannot be performed within one year are likely to be contracts of a more serious nature, and therefore should be in writing and not dependent on memory. The emphasis, inherited from the English, on the value and importance of property is the reason that land transactions fall within the rule. Contracts to leave property by will represent a similarly serious transaction.

Under the Uniform Commercial Code, Section 2-201, and the corresponding state commercial codes (e.g., California Commercial Code, Section 2201), contracts for the sale of goods in excess of $500 must be in writing. This dollar threshold is to eliminate business transactions that are of a minimal nature.

Unless an agreement is required by the applicable statute of frauds to be stated in writing, an oral contract is as valid as a written contract. However, proving the existence of the oral contract, and of the contract terms, may be more difficult than proving the validity of a written contract. Necessary terms will need to be proven by the testimony of the parties, the testimony of the witnesses, or the existence of relevant secondary documents. Because of the difficulty of proving an unwritten contract, it is always preferable to have a contract set out in writing, whether or not this is required by applicable law. It is easier to prove the necessary elements of a contract with a single written instrument than it is to attempt to prove a contract based on the fading recollection of the parties, the memory of witnesses, or inference based on secondary documents.

§ 5.7 —Meeting of the Minds

In order for negotiated contracts to be valid, each of the parties must have agreed to the terms and conditions. This is sometimes referred to as a **meeting of the minds**. This phrase is used to refer to situations where both parties understand the terms and conditions of the contract and knowingly assent to the contract provisions. Ideally, the parties have met and discussed the terms and conditions, explored possible problems that might come up in the course of performance of the contract, and mutually drafted a written document that incorporates all the essential terms and conditions of the contract. Unfortunately, this is often not the case. Instead, standard contract forms are used and they contain provisions that are not discussed, understood, or effectively assented to by all the parties. If a dispute arises under these circumstances, the dispute will be far more difficult to resolve fairly.

§ 5.8 —Doctrine of Mistake in Formation of Contract

A contract founded on a basic mistake of fact may be void and unenforceable, because the parties have failed to agree to an effective contract. Each party thought the agreement referred to a different set of terms and conditions. One significant consideration taken into account by a court being asked to apply the doctrine of mistake is whether the party who made the mistake is the party seeking to be released from the contract. A party relying on another's mistake or ambiguity is more likely to be released than the party who created the ambiguity.

For example, a prime contractor describes the proposed construction of a project to a subcontractor as consisting of a mixed-use building that will have stores and commercial tenants on the first two floors and condominium apartments above. The prime tells the subcontractor that the proposed building site is across from the post office in a particular town. Unknown to the prime contractor, there is more than one post office in that town. The subcontractor goes to view the land across from a different post office than the one intended by the prime contractor, and the subcontractor provides a bid based on use of that parcel of land. The subcontractor's bid is unreasonably low because of differences between the two parcels of land.

If the subcontractor refuses to perform in accord with its bid because of the mistake regarding the building site, a court deciding whether the contract was void because of the mistake might hold that the parties never really had an agreement because there was no meeting of the minds. The court would analyze the reasonableness of the mistake regarding the building site, and would examine what steps, if any, before making its bid, the subcontractor took or could have taken to verify the true building site. The prime contractor will have a difficult time prevailing against the subcontractor on a contract theory if the subcontractor, in good faith, believed the location was across from the other post office, this mistake caused him to bid incorrectly, and the subcontractor was without fault in not discovering the mistake before submitting the bid.

If the difference in the two parcels of land is substantial, and if it would be unfair to require the contractor to perform the work according to a bid that is too low because it was based on the use of a different parcel, the court is less likely to force the subcontractor to perform according to the bid.

§ 5.9 Offer and Acceptance

Most contracts are created by one party's making an offer, and an acceptance of that offer by the person to whom it was made. Sometimes, the person receiving the offer makes a **counter-offer**, an offer in response to an offer. As a general rule, an offer may be accepted by the person to whom the offer was made, until the person making the offer expressly revokes it, or until the person receiving the offer has revoked the offer by making a counter-offer, or a reasonable amount of time for acceptance has elapsed.

For example, an architect presents a written proposal for a design package to an owner who needs a certain type of building. The owner reviews the documents that make up the package and signs the proposal, which has been presented as an offer. By signing the document, the owner is signaling his acceptance of the terms and conditions contained in the offer. The written proposal, and the other documents in the design package, together with the signature signifying acceptance of the offer by the owner, form a binding contract.

In many situations, a person receiving an offer switches to the role of a person making a counter-offer, who then must await the acceptance of the counter-offer. In the situation described above, instead of simply accepting the package of documents, the owner, after reviewing the terms of the written offer, decides that he agrees with much of the proposal, but wishes to change a number of the terms and conditions. He now presents the altered or modified written document to the architect as a counter-offer. The previously presented offer has been revoked by the owner's making the counter-offer; the first offer is no longer capable of being accepted by the owner. In some situations, an entire series of offers and counter-offers may be made before the parties reach an agreement.

In certain cases, parties may be negotiating for a contract, but have not reached the offer and acceptance stage and are not yet ready to finalize the contract. Early communications may be characterized as negotiations, or proposals for offers. One definition of a true offer is that it creates a **power of acceptance** in the **offeree** (the person receiving the offer). Use of words such as "promise" or "offer," within the language of a proposal, is an indication that the person sending the document intends to make an offer that is capable of being accepted by the person receiving it.

Also important in determining whether statements made by a party constitute an offer is ascertaining whether the essential terms of the agreement have been set forth or have been communicated between the parties. A proposal that specifies price, location, time for performance, quantity, quality, and other details is likely to be a valid offer, as opposed to a situation where the parties are still negotiating. Where a person is furnishing proposals to a number of other parties, rather than to only one other party, it is more likely that the proposals are invitations to offer, or negotiations, rather than valid offers.

A party may wish to provide to another party a document that discusses proposed performance of work; the document is not intended to be a firm offer capable of being accepted. This document is sometimes called a **letter of intent**. The party may even specifically state, within the body of the document, that the proposal is not to be construed as an offer that may be accepted. The purpose of the proposal is to outline the terms and conditions of a possible future contract. Because the proposal is not sufficiently definite as to all necessary terms and conditions, it does not constitute a valid offer. One purpose of this type of document is to keep the negotiation process going and to lead to final negotiations, a valid offer and acceptance, and a contract.

§ 5.10 Consideration

Consideration exists as a contract requirement to eliminate promises that are too negligible to enforce, or promises whose enforcement would be unfair. The promise by one party to perform his part of an extremely unequal bargain should not be enforced by a court if it appears that the other party is giving little or nothing of value.

As described earlier, even if a traditional contract is not found, either for lack of consideration or for lack of mutual assent, the concept of reliance may be used by a court to enforce certain promises where it appears one party heavily relied on the promise of another.[3] The concept of reliance is being used more frequently by courts to enforce promises where such enforcement appears fair. It is more straightforward to allow recovery on the theory that a party made an offer on which the other party reasonably relied, than it would be to try to analyze the communications of the parties to see whether the facts support an analysis of a valid offer, acceptance, and the existence of sufficient consideration on both sides.

Past consideration is usually insufficient to make a promise enforceable. For example, a well-driller agrees to provide potable water for a homeowner by digging a well for payment of $15,000. The well is drilled, but the water is not potable because of a high lead content. The driller installs a filtering system to remove the lead, and then requests payment of an additional $2,500 for the system. Because the driller already had received payment not only to drill the well but to provide potable water, he will probably not be successful in claiming that he is entitled to additional payment for the filtering system. There was no new promise by the homeowner to pay an additional amount for the filtering system.

Despite this general rule, some states will enforce a promise based on the conferring of past benefits, if the enforcement is necessary to prevent injustice.[4] For example, although a developer has built a house for a new owner in accord with a contract, the land has no water and is not usable as a residence. Because the true intent of the contract was to provide a residence in which the owner could live, a court might require the developer to provide a well that will supply water to the owner, without the developer's receiving additional payment.

§ 5.11 Certainty, Specificity, and Definiteness of Contract Terms

Even when a valid offer has been made and accepted, lack of specificity of essential terms to a contract may preclude the existence of a valid contract. The lack of specific contract terms may be an indication that the parties have not completed their negotiations. In the event that a dispute arises under the contract, a judge or arbitrator will not be able to ascertain whether the contract has been performed, because the agreement is too vague.

Factors to be examined, in the event of a dispute centering on a lack of certainty to the terms of a contract, include the importance of the terms or

[3] Bethlehem Fabricators, Inc. v. British Overseas Airways Corp., 434 F.2d 840 (2d Cir. 1970).

[4] Restatement (Second) of Contracts § 86 (1981).

conditions in dispute, whether and to what extent performance has been rendered, and whether the party claiming that the contract is not sufficiently specific to be binding has some external reason for seeking to be released from his obligation under the contract.

For example, suppose a roofer is claiming his contract to provide a roof for a school library is not valid because the grade of roofing shingle was not specified in the contract. The school district claims the highest grade of shingle was intended by the contract. The roofer claims a lower grade of shingle was to be used. Since the signing of the contract, the cost of all grades of roofing shingle have escalated dramatically. Under these circumstances, a court could determine that the grade of roofing tile is reasonably ascertainable from the type of roof, the type of roof tile used previously for the library, or other information. In other words, by referring to other relevant information, the contract terms are sufficiently explained and definite enough to ascertain the grade of shingle to be used, and the contract should be enforced. The court is particularly unlikely to release the roofer from the contract if it appears that the roofer is merely attempting to be released from a bad bargain resulting from the increase in the cost of the roof shingles.

§ 5.12 Adhesion Contracts

The type of construction contract typically used has changed over the years. At one time, the contracts were freely negotiated and individualized agreements. Now, printed standard construction contract forms are common. In many cases, either the form contract used does not contain the language and understanding of the parties, or one party has drafted a form and presented it to the other party on a "take it as is or leave it" basis. With either situation, it is difficult for a court to find that the written contract or agreement between the parties is a true representation of the parties' intent to be bound by the document.

Where a stronger party forces a weaker party to accept a contract because of unequal bargaining power, the contract is usually called a **contract of adhesion**. In this situation, one or more of the parties to the agreement is unable to refuse to accept some or all of the terms of a contract, or to specify alternative terms. The weaker party must "adhere" to the contract imposed by the stronger party. With some contracts, price may be the only term that can be negotiated by the buyer. Insurance contracts, whether purchased by individuals or businesses, rarely allow the buyer to exercise any discretion over the terms of the contract. In most instances, the buyer of insurance never even *sees* the terms of the contract until after the buyer has secured the insurance.

If all provisions of an insurance contract were strictly enforced according to their terms, unfairness might result. In the early and middle part of the twentieth century, there was a strong movement within the courts to go beyond the limits of the written document and carefully examine the intent of the parties, based on factors outside the written contract. In the past ten or fifteen years, there has been a return to stricter enforcement of the written terms of an agreement, unless it would be demonstrably unfair to do so. This return to stricter contract interpretation is probably based on the time involved in a court's evaluation of the complex subjective intent of the parties to a contract, and the difficulty of attempting to assess their relative bargaining power.

§ 5.13 Methods of Creating Contracts

Generally, construction contracts are created either by negotiation between the parties or by letting contracts for competitive bidding.

§ 5.14 —Negotiation

With negotiation, the parties have a dialogue in which they each express their needs. Typically, one party to the construction contract will be providing services or goods, and the other will be supplying money. If the parties mutually draft an individually tailored agreement, their contract should be an incorporation of each party's needs and wants, and the contract should represent an agreement for which each has bargained. If they have used a form contract, their intent regarding the terms and conditions of the contract may be obscured by their lack of understanding of the meaning of the clauses or legal terms used in the form contract. For those contracts that are fully performed without any dispute, the fact that each party has a different concept of what the language of the clauses means is irrelevant.

However, when a dispute arises in such a situation, a court or arbitrator will have to determine the meaning of the form contract, as applied to the particular set of facts. If the document itself is ambiguous, the court or arbitrator will have to go beyond the confines of the form to examine other manifestations of the parties' intent. An inequitable decision may occur because the negotiation of the parties did not produce a written contract that truly expressed the parties' intent.

§ 5.15 —Bidding

With competitive bidding, an owner typically releases specifications and drawings to qualified prime contractor bidders, and solicits bids. The bidders then prepare and submit bids to the owner, based on the information provided by the owner, investigation by the bidders of relevant information, and solicitation by the prime contractor bidders of subcontract bids for performance of portions of the work.

With public contracts, bidders must be qualified by the relevant government authority before they can bid. In many public contracts, certain percentages of the subcontractor bids must be allocated to Disadvantaged Business Enterprises (DBEs), Minority Business Enterprises (MBEs) or Women Business Enterprises (WBEs).

§ 5.16 Particular Contractual Clauses

Most contracts contain certain standard clauses. Construction contracts usually include these standard clauses, clauses that pertain specifically to construction, as well as others directed to the particular issues that arise during performance of construction contracts.

§ 5.17 —Time for Performance

It is standard in most contracts to provide that prompt performance of the promised acts under the contract is important to the parties. This concern is often phrased in contracts using language such as "time is of the essence to the performance of this contract." Such clauses show up in construction form documents, for example, at paragraph 8.2.1 of AIA Document A201 (see **Appendix B**) and paragraph 12.3 of NSPE 1910-8.

Because of the nature of a construction contract, timely completion of performance of the contract is important. If completion of a project is delayed, union wages may rise, newly enacted legislation may require that plans for buildings be redrafted, and the cost of supplies may increase. Contractors may have other projects to begin or to work on, and losses may result if they are forced to spend additional time completing a late project. The availability of construction funds, the interest paid on those funds, and the fact that a vacant and unfinished building does not earn any revenue, are all types of damage resulting from delayed completion of construction.

A major component of construction delay damages is known as "extended site conditions." The additional costs incurred by the contractor when construction is delayed include additional general condition costs and additional supervision costs. Contractor home office expenses (HOE) are often added to this category of delay damage.

Construction contracts are likely to factor anticipated bad weather into the calculation of the time for completion. Extraordinarily bad weather, or other circumstances, may extend construction performance into winter or into another season of the year that makes construction difficult or impossible, thus delaying completion of a project.

Because of the number of parties involved in the construction process, and the amount of resources committed by each party to a project, delays on a major project can bankrupt a small contractor or cause crippling losses to a medium-size or large contractor.

The more clearly the parties have specified the time allowed for performance of the contract, and the more accurately they have calculated contingent factors into the number of days allowed for construction, the less likely they will encounter delays in completion that will result in disputes or claims for damages.

§ 5.18 —Liquidated Damages Provisions

To try to prevent delays in completion, and to avoid litigation, the parties often stipulate in the contract a dollar figure that is expressly agreed to represent the damage an owner is likely to suffer if there is a delay in completion. These **liquidated damages** are a substitute for and an alternative to the judicial award of actual damages. Liquidated damages clauses are usually enforceable under certain circumstances: (1) if actual damages would have been impracticable or extremely difficult to estimate at the time of breach of the contract; and (2) if the amount of the liquidated damages is reasonable under the circumstances.[5] The parties are free to agree to liquidated damages clauses as long as they meet the above standards. Reasonable damages amounts are usually defined as those damages that do not appear to be an unfair penalty on the party on whom they are imposed.

The use of liquidated damages clauses is encouraged by the courts because their use avoids unnecessary litigation and promotes certainty in contracts. As long as a party knows the exact amount for which he will be liable in the event he is late in completing his construction contract, he may be encouraged to ensure timely completion. Without a liquidated damages clause, a contractor may allow construction delays to occur, taking his chances in later argument about the reasonable amount of actual damages to be awarded by a court or arbitrator.

§ 5.19 —No Damages for Delay

The frequency of contractor claims against owners for delay damages has made it common for owners to attempt to insulate themselves from these claims. One method for achieving this goal is for the owner to specify within the contract that it has the right to delay performance by the contractor, and that such an act by the owner is not a breach of the contract. Another method is for the owner to claim that the contract clause that provides for time extension of performance is the contractor's only remedy, and that no additional payment is due for delay.

---

[5] *E.g.*, Cal. Civ. Code § 1671 (Deering 1991).

Generally, courts have held that the use of the time-extension provision does not necessarily prevent a claim of damages for delay.[6]

The most common method of owner protection is insertion of contract clauses that attempt to place the risk of owner delay on the contractor. These clauses specify that there will be no payment to the contractor by the owner for delay damages. The clauses are usually enforced, except where there are special circumstances. In a recent New York case, a court held that no-damage clauses bar recovery of delay damages unless the owner has committed fraud or malice, or is guilty of bad faith, or has shown a reckless indifference to the rights of others.[7]

The holding of this case was subsequently modified in New York by a set of consolidated cases. These cases stated that delay damages could be recovered, even where a no-damage clause was in the contract, if the delay was not contemplated by the parties, if the delay was so unreasonable as to constitute an intentional abandonment of the contract by the owner, or if the delay resulted from the owner's breach of a fundamental obligation of the contract.[8]

A New Jersey court expressed its support for these clauses by stating that they are a part of the economic package on which the parties agree, and that the contractor who accepts these risks will reflect the acceptance of the financial responsibility for the risks in his price.[9]

A federal court case, *John E. Green Plumbing & Heating Co., Inc. v. Turner Construction Co.,*[10] held that this type of clause is strictly construed. It bars delay damages, which are defined in this case as time lost when work cannot be performed because materials are not delivered or necessary preliminary work is not completed. The *Green* court held that such a clause does not prevent a claim based on an alleged hindrance of performance of the work. Examples given are failure to coordinate the work, or failure to supply temporary heat to the job site.

Pennsylvania has taken a more forceful attitude regarding no damages for delay clauses. That state has held that the clauses will not be enforced if there is positive interference or a failure to act in some essential matter necessary to the prosecution of the performance of work.[11] An Ohio case articulated the same

---

[6] *See, e.g.,* Bender-Miller C. V. Thomwood Farms, Inc., 211 Va. 585, 179 S.E.2d 636 (1971) (remedy stated in contract is not exclusive remedy unless so explicitly stated by parties).

[7] Kalisch-Jarcho, Inc. v. City of New York, 58 N.Y.2d 377, 448 N.E.2d 413, 461 N.Y.S.2d 746 (1983).

[8] Corrinno Civetta Construction v. City of New York, 67 N.Y.2d 297, 493 N.E.2d 905, 502 N.Y.S.2d 681 (1986).

[9] Broadway Maintenance Corp. v. Rutgers, 90 N.J. 253, 447 A.2d 906 (1982).

[10] 742 F.2d 965 (6th Cir. 1984).

[11] Coatesville Contractors & Engineers, Inc. v. Borough of Ridley Park, Pennsylvania, 509 Pa. 553, 506 A.2d 862 (1986).

rule as Pennsylvania, but found that the facts supported enforcing the clause, barring additional recovery by the contractor.[12]

From the perspective of an owner, particularly if the work includes foundation work where subsurface conditions, or other facets of the project, entail substantial uncertainty, a no-damages-for-delay clause is important. Whether the clause will actually be enforced depends largely on whether the owner is culpable or has contributed to the delay, and the judicial attitude in a particular state regarding this type of clause.

From the perspective of a contractor signing a contract that contains this clause, the risk of being precluded from making a subsequent claim for damages should be considered when the contractor calculates the bid for the work.

§ 5.20 —*Force Majeure*

Force majeure contract clauses provide an excuse to parties in construction for delays caused by extreme and unpredictable weather conditions, earthquake, fire, or other natural disaster. AIA Document A201, paragraph 4.3.8.2, refers to *force majeure* due to weather conditions (see **Appendix B**). Most contracts refer to other natural disasters, such as floods or other "acts of god" in similar clauses.

§ 5.21 Indemnity

Indemnity is the concept that one party will be held responsible for the debt or obligation of another. Noncontractual indemnity is usually implemented when it appears that an innocent party should not have to pay a third party for the loss caused by another, more culpable party. This is sometimes called **implied** or **equitable indemnity**. A large component of our system of law is based on the theory that each person bears responsibility for his or her actions, and courts will therefore imply indemnity when the parties have failed to provide for it by express contract.

§ 5.22 —Public Policy Concerns

Physical or moral responsibility for losses can range from innocent negligence to intentional wrongdoing. In some cases, such as product defect law, liability may be imposed even in the complete absence of negligence or intentional acts, based on the social policy of having those with the greater ability to absorb loss pay for injuries (e.g., manufacturers of products that cause injury).

---

[12] Carrabine Construction Co. v. Chrysler Realty Corp., 25 Ohio St. 222, 495 N.E.2d 952 (1986).

Aside from the cases of strict liability without fault, the courts prefer to place liability on those who appear more culpable. A party who caused a loss by acting negligently is less culpable or blameworthy than a party who caused a loss by acting intentionally. For example, a subcontractor's worker is likely to be able to recover from an owner who has deliberately violated safety regulations. If the prime contractor can be proven to have been aware of the violations, both the contractor and the owner might be held jointly liable to the innocent injured worker.

In some circumstances, attempts are made to identify a similar distinction between the relative levels of culpability between parties, based on the theory that the negligence of some parties is **passive** (they failed to act) and that of others is **active** (they negligently or willfully caused damage).

* * *

Case 5.1
(a) Prelitigation

A prime contractor was building a three-story office building in a metropolitan area. The project supervisor for the prime, Ray, had a number of sons who were in the construction business. He suggested to one son, Jim, that the son apply for work with the steel subcontractor. Another son, Bob, was employed by one of the other subcontractors on the project.

After the steel framing was complete for the building, laminated wooden beams called "glu-lams" were being raised to the third story to provide the structural support for the ceiling of the second story and the floor of the third story. Jim was sitting astride one of the steel beams at the level of the third floor, and his brother Bob was operating the forklift that was lifting the glu-lams up to the third floor.

As one of the glu-lams was being raised, it became caught against one of the metal holsters into which the glu-lam was going to be placed. The holsters had been installed by the steelworkers. Jim leaned over to try and free the glu-lam. As he leaned over, and pushed on the beam, it suddenly sprang loose and knocked him off his position on the beam. He fell to the ground, and his back was severely injured. It was doubtful that he would ever be able to work in the construction trades in an active capacity again.

Prior to the accident, a question had arisen about installing a wire that would encircle each floor of the steel structure to keep workers from falling. The protective wire was not installed, because no agreement could be reached on who would pay for the cost of its installation.

Under the terms of the contract, the steel subcontractor had specifically assumed responsibility for compliance with all state OSHA (Occupational Safety and Health Administration) requirements. One of the OSHA requirements that had been clearly violated was the requirement that laborers working at a height in excess of fifteen feet were supposed to be wearing safety belts. A safety belt has a metal ring, and the metal ring was to be connected to a cable looped around a vertical girder, horizontal girder, or other stationary object near the location where the laborer was working.

Although safety belts were available in the construction trailer at the job site, the superintendent for the steel subcontractor did not insist that his crews use them. The men complained that the safety belts were uncomfortable and slowed down their progress. Ray, the prime contractor's supervisor for the project, had spoken to his son Jim about wearing a safety belt when Jim was working at height. Jim's response to his father had been that he had to keep his job. Ray spoke to Bill, the supervisor for the steel subcontractor, about the lack of use of safety belts. Bill, tired of being told how to do each and every task by Ray, walked away from him, and did not require the use of safety belts by those laborers who were working more than fifteen feet off the ground. Ray had last spoken to his son Jim about the need to wear a safety belt just days before the accident.

After the accident, Ray ordered everyone to stop work on the project until the state OSHA organization was brought in to investigate the accident. OSHA inspected the work procedures, found numerous violations, and cited and fined the steel subcontractor for these violations. Included among these violations was the failure to require all steelworkers working at height to wear safety belts.

Analysis. Because of the nature of the construction business, it is common for different members of a family to be involved in construction, and even to work on the same project. In this case, the father, Ray, was looking for work for his sons. Because he knew the steel subcontractor was looking for workers, he told his sons to apply for work with the steel subcontractor. Ray should have been more concerned about the safety of work practices on the project because his own sons' safety was involved.

A consistent theme of construction accident cases is that those on a project are aware of safety rules and procedures, but openly disregard them. Common examples are failures to enforce rules regarding use of hard hats or steel-toed shoes. This disregard of safety rules often occurs because performing the project in strict compliance with safety procedures results in slower progress for the project. Many of the laborers will not cooperate by complying with safety regulations. In this case, the laborers did not wish to use the safety belts because they were a hindrance to physical coordination, and because use of safety belts would have slowed down the work. Every time a worker moved more than ten feet, he would have had to wrap the wire around a new beam and reattach the safety belt.

At trial, the steel subcontractor would have argued that even though the contract specified that the steel subcontractor had responsibility for safety according to the contracts, common sense dictated that the prime contractor, who was responsible for overall progress, scheduling, and organization of the project also bore liability for the accident.

When an accident occurs as a consequence of failures to observe safety rules and procedures, primary liability for the injuries is likely to fall on the person charged with ensuring compliance with safety rules. In most cases, the responsible person would be the prime contractor or construction manager. However, under his contract, the steel subcontractor had assumed the responsibility for compliance with safety requirements.

Review of contract documents before signing should include a discussion and allocation of responsibility for safety compliance. Although the logical party to bear this responsibility will vary from project to project, it is an issue that should be addressed before contracts are executed.

(b) Post-Filing of Litigation

Counsel for the plaintiff (Jim, the injured worker) took depositions. Counsel established that Jim could no longer work as a steelworker, and that he would have to receive vocational training to learn a more sedentary occupation. If he had sufficient intelligence and managerial skills, he might be able to start his own business as a contractor. A certain amount of capital would be required to begin such a business.

The prime contractor settled with the plaintiff for $450,000. The steel subcontractor refused to offer any contribution to the settlement, believing that he had no liability.

Because Jim had worked for the steel subcontractor, Jim could not sue the steel subcontractor directly. His exclusive claim against his employer was under the workers' compensation system. However, the prime contractor had a cross-complaint for indemnity against the steel subcontractor.

To expedite settlement, the prime contractor demanded that the steel subcontractor pay only half of the settlement paid by the prime to the plaintiff, even though the prime contractor believed he had a viable argument that he was entitled to full indemnity by the subcontractor. However, no settlement was reached between the prime contractor and the steel subcontractor at this point.

The parties to this contract had used AIA Documents A101 (agreement between the owner and contractor), A201 (general conditions of the contract), and A401 (agreement between the contractor and subcontractor). The 1978 version of A401, current at the time of this case, had an indemnity provision that stated, at paragraph 11.11.1, that indemnity would be provided ". . . regardless of whether it is caused in part by a party indemnified hereunder."

The prime contractor believed it was entitled to complete indemnity from the steel subcontractor based on the indemnity clause. It interpreted the clause to call for complete indemnity, regardless of any negligence of the prime contractor.

In one California case, *Jiminez v. Western Pacific Construction Co.,*[13] a prime contractor sought indemnity from a subcontractor for damages the prime had

---

[13] 185 Cal. App. 3d 102, 229 Cal. Rptr. 575 (1986), *reh'g denied* (Nov. 25, 1986).

paid to an injured worker. The court found that the indemnity clause was not properly drafted to constitute a specific indemnity clause, because it would allow indemnity even where the party seeking indemnity is negligent. The court held that the clause in this contract was a **general clause**. Under California law, if the clause is general, and the party seeking indemnity has actual or constructive knowledge of the dangerous condition, no indemnity will be allowed.

In this case, the father, who was a representative of the prime contractor, had actual knowledge of the danger that caused or contributed to his son's accident (the lack of wire encircling the structure, and the non-use of safety belts). Although he subsequently denied he had the power to force the subcontractor to use safety belts, or take any other action, after the accident he ordered all work stopped until an investigation was conducted and appropriate safety measures were instituted.

The case was set for trial, but was continued numerous times by the court because of a lack of available courtrooms in which to hold the trial. Finally, the case was sent to a judge who concluded that the indemnity clause was not ambiguous. The court held that the clause called for complete indemnity by the steel subcontractor of the settlement amounts paid by the prime contractor, as well as interest from the time of the settlement. He called for judgment in the amount of $450,000 plus interest, which came to approximately $620,000. After filing for appeal, the steel subcontractor eventually negotiated a settlement with the prime contractor for $500,000.

Analysis. Because of the large differential between the workers' compensation recovery available to an injured worker from his employer, and the amount available from a third party, it is common for workers injured in construction accidents to sue third parties on the project, who can then sue the employer for indemnity. In this way, plaintiffs attempt to avoid the limited recovery available in the workers' compensation system. Originally, the workers' compensation system was intended to make recovery easier for the worker, because the worker could recover for his injuries regardless of his own negligence. However, the recoveries available through the workers' compensation system have not kept up with the larger recoveries currently available in the civil justice system.

In this case, the prime contractor settled with the injured worker because the prime believed he could recover a substantial reimbursement in the form of partial or complete indemnity from the steel subcontractor under the subcontract. However, in discovery, the principals of both the prime contractor and the steel subcontractor testified that their belief was that the indemnity clause signified that each party covered only his own negligence, and not that of the other party.

The insurance adjuster for the steel subcontractor, who believed the indemnity clause precluded recovery for the other's negligence, did a great disservice to his insured. When the adjuster took the position that he would not pay any money toward the prime contractor's proposed settlement, he alienated the prime contractor and made a compromise settlement impossible. Further offers just prior to trial were rejected. The prime contractor's demand had become full and total indemnity, plus interest from the time he had settled with the injured worker.

If the indemnity clause had been construed as a general clause under California law, the active negligence of Ray in failing to enforce the use of safety belts on the project would have precluded *any* indemnity, not just partial indemnity, associated with the proportionate responsibility of the prime contractor.

* * *

Cases in different jurisdictions have disputed the appropriateness of allowing noncontractual indemnity by employers of third parties for workers' injuries; see *Dole v. Dow Chemical Co.*[14] (indemnity allowed); *Galimi v. Jetco, Inc.*[15] (indemnity precluded). Currently, a small majority of jurisdictions preclude this type of indemnity.[16]

§ 5.23 —Contractual Indemnity

It is a given that each party to a construction contract will seek to use contractual clauses to shift to another party the risk of a loss, or of liability for damages, rather than having to pay those costs himself. In many cases, determining ultimate responsibility becomes a battle based on the contract clauses contained in each of the parties' contracts. Each indemnity clause has to be examined, to determine how all the indemnity clauses interact, before the liability is finally placed on a party.

The problems of adhesion to an unfair contract imposed by a stronger party on a weaker party are particularly significant in the area of indemnity. An agreement by which a hospital seeks indemnity from patients for its own acts may be held invalid.[17] On the other hand, in *Rutter v. Arlington Park Jockey Club,*[18] a contract clause in which a racetrack sought to be exculpated by horse owners from liability for loss or damage to horses stored at the racetrack, even if the losses were caused by negligence of the racetrack, was upheld when horses were killed at the racetrack in a barn fire. The *Rutter* court pointed to the fact that the individual owners were in a better position to obtain

---

[14] 30 N.Y.2d 143, 282 N.E.2d 288 (1972).

[15] 514 F.2d 949 (2d Cir. 1975).

[16] *See* A. A. Equipment, Inc. v. Farmoil, Inc., 31 Conn. Supp. 322, 330 A.2d 99 (1974). California has barred noncontractual indemnity by statute, at Cal. Lab. Code § 3864, but has allowed express contractual indemnity for worker injuries. For other cases on indemnity, *see* Becker v. Black & Veatch Consulting Engineers, 509 F.2d 42 (8th Cir. 1974); Warburton v. Phoenix Steel Corp., 321 A.2d 345 (Del. Super. 1974), *aff'd,* 334 A.2d 225 (Del. 1975).

[17] Tunkl v. Regents of the University of California, 60 Cal. 2d 92, 383 P.2d 441, 32 Cal. Rptr. 33 (1963).

[18] 510 F.2d 1065 (7th Cir. 1975).

insurance against losses for the horses, because it was impossible to know how many horses would be at the track, or the respective value of each of the horses, at any given time. The *Rutter* court deferred to the business judgment of the parties in placing the burden of insuring against the loss on the horse owners. Even where this type of indemnity or exculpatory clause is enforced, courts usually demand that the clause have a high degree of specificity and that it clearly set out the terms and conditions of the agreement to indemnity or exculpate.[19]

Parties should not assume that the indemnity clause in their contract will be enforced. An arbitrator or court will review the facts and equitable considerations before doing so. The decision of whether to enforce a contractual indemnity clause requires a court to balance competing societal interests—on the one hand, implementing the public policy that parties should bear responsibility for their own acts, and, on the other hand, respecting the constitutional rights of the parties to allocate risks by the contract as they see fit.[20]

§ 5.24 Differing Site Conditions Clause

One of the most common sources of an unexpected increase in construction costs is unexpected or undisclosed site conditions that cause the contractor to expend greater amounts of labor or materials to complete the work. In some circumstances, delays in completion of performance also may result from the contractor's having encountered unexpected site conditions.

§ 5.25 —Purpose

A contract clause that allows the contractor to collect amounts beyond the specified contract amount for unexpected site conditions is frequently used in construction contracts. The contractor can remove some of the uncertainty about the site conditions by performing subsurface exploration such as making test borings. However, these tests can be expensive and may be unreliable in predicting the actual job conditions that will be encountered. A contractor seeking to protect himself can also research public records regarding the site or the structures on the site. Nevertheless, construction contracts often contain clauses allocating the risk of unexpected site conditions, because all uncertainty regarding site conditions cannot be removed by precontract investigation.

---

[19] Cincinnati Gas & Electric Co. v. Westinghouse Electric, 465 F.2d 1064 (6th Cir. 1972) (indemnity not allowed because language insufficiently specific).

[20] For other cases on indemnity, *see* Becker v. Black & Veatch Consulting Engineers, 509 F.2d 42 (8th Cir. 1974); Warburton v. Phoenix Steel Corp., 321 A.2d 345 (Del. Super. 1974), *aff'd,* 334 A.2d 225 (Del. 1975).

§ 5.26 —Examples

Contractual provisions regarding differing site conditions attempt to allocate the risk of increased cost or delay in completion, in the event unexpected conditions are encountered on the project. It is common for owners to make some site information available to contractors prior to solicitation of bids. This information may be accompanied by disclaimers as to the accuracy of the information.[21] Courts generally enforce such disclaimers.

For example, in *Costanza Construction Corp. v. Rochester,*[22] Costanza was the successful low bidder on a public improvement contract. Costanza considered rock removal to be a small part of the job, and bid $25 per cubic yard to excavate rock, an amount below the actual cost. Job drawings prepared by Passero, a codefendant, showed only about 20 cubic yards of rock in the area to be excavated. The specifications for the project estimated about 100 cubic yards. In fact, Costanza removed over 600 cubic yards. Costanza was paid at the agreed-on unit price for the rock removal, and brought a claim based on breach of contract, negligence, breach of warranty, and equitable adjustment, all of which were dismissed. The contract with the city included a disclaimer for any responsibility for the accuracy or completeness of the information on the drawings concerning existing conditions, including rock, and obligated the contractor to satisfy himself as to site conditions and the work required. The court found no merit to Costanza's argument that the actual amount of rock encountered constituted a cardinal change that altered the very essence of the contract. Costanza argued that his time was too limited to make an in-depth inspection, and that to do so would have been prohibitively expensive. The court nevertheless found that the bidding specifications obligated Costanza to make an in-depth inspection.

Some contracts will contain a clause allowing the contractor an equitable adjustment, in the event unexpected conditions are encountered. A claim that the owner committed fraud in its disclosures might allow the contractor to obtain additional compensation, but a contractor may find it difficult to claim damages for reasonable reliance on site information provided by owners, where the site information contains a disclaimer regarding the accuracy of the information and a statement that the contractor should conduct independent investigation or testing to verify the accuracy of the information provided.

One way for a contractor to protect itself against the risk of encountering unexpected conditions and incurring additional expense is to add a contingency factor to the bid amount. To avoid artificial inflation of bids to cover such contingencies, many owners provide a system for handling unexpected conditions. In federal procurement, this type of clause was formerly called

---

[21] *See* United States v. Spearin, 248 U.S. 132, 39 S. Ct. 59, 63 L. Ed. 166 (1918).

[22] 147 A.D.2d 929, 537 N.Y.S.2d 394 (1989).

"Changed Conditions." This clause was subsequently changed to "Differing Site Conditions." The same problem is addressed in paragraph 4.3.6 of the current AIA Document A201 as "Claims for Concealed or Unknown Conditions." The purpose of these contract clauses is to allocate the risk of loss. In the absence of such clauses, the entire risk for the cost of unforeseen conditions rests on the contractor.

However, as compared to contractors, owners are typically in a position of superior knowledge regarding site conditions. If an owner deliberately conceals a site condition, or a court finds that the owner failed to disclose a material fact regarding a condition (constructive fraud), this may avoid placement of the risk on the contractor for the cost incurred because of a site condition that the owner failed to disclose.

The California Association of General Contractors (AGCC) has provided for differing site conditions in Article 4 of its prime contract and, at Article 3, requires the owner to supply the contractor with all information of which it is aware concerning surface and subsurface conditions, as a way to help the contractor avoid encountering unforeseen conditions.

If an owner is seeking to place the risk for differing conditions on the contractor by disclaiming responsibility for the accuracy of information, he should allow bidders sufficient time to inspect the site and to run appropriate tests. The owner should draft disclaimer language with care, placing the disclaimer language near the language regarding site information in the contract.

§ 5.27 Contract Clauses Requiring Insurance

A common method for protecting against third persons' losses arising from a construction project is to include in the contract a clause requiring that all parties performing work on the project must obtain general liability insurance. Certificates of insurance, or other proof of insurance, should be provided by each contractor before that contractor performs any work. If a third party is injured because of the negligence of one of the contractors on the project, the general liability insurance, also called "comprehensive general liability" (CGL) should cover this type of loss.

In the event of an injury to a worker on the job, workers' compensation insurance will pay for medical care and for any disability that results from the injury. Although the system of workers' compensation was developed to make recovery easier for workers, many people feel it results in second-class compensation for workers because no recovery is allowed for "pain and suffering." Pain and suffering are important and significant components of personal injury awards in the civil system. Personal injury plaintiffs are entitled to recover for the actual expenses they have incurred and will incur as a result of the injury, but pain, suffering, and emotional distress are often the most substantial portion of the damages claimed in an injury case in the civil system.

§ 5.28 Attorney's Fees Provisions

Although the United States inherited the English common-law system, the United States modified the system in some important respects. In England, the losing party in a civil suit must pay the winning party's attorney's fees. In the United States, attorney's fees are not recoverable under federal and state law, unless specifically provided for within the contract (in a case based on breach of contract) or by statute.

The argument in favor of the "English" rule is that it discourages a plaintiff from filing a frivolous suit; he will consider carefully the risk of losing the suit and of being held responsible for the defendant's attorney's fees.

The argument in favor of the "American" rule is that it provides freer access to the courts for disadvantaged people, and that few lawsuits, even those with merit, would be brought if the losing party had to pay the winning party's attorney's fees.

Most construction contracts contain a clause providing for recovery of attorney's fees in the event that suit must be brought for breach of any of the obligations under the contract. In addition, most states have provided by statute that, if an agreement contains a clause that one party is entitled to recover attorney's fees if an action has to be brought for failure to perform under the contract, the other party (if he prevails) will be entitled to an award of attorney's fees as well. This is sometimes called the **equal dignities rule**, applied to attorney's fees.

In a state that has codified the equal dignities rule as it pertains to attorney's fees, a court deciding a breach of contract case that includes a claim for recovery of attorney's fees clause would use the statute to imply that the clause applies both ways, in the interests of fairness. Although the contract does not expressly state that either party who prevails will recover his attorney's fees, the law implies that it does, and a court would rule accordingly.

On smaller projects, it is especially critical to have an attorney's fees clause in the contract. Although the amount in issue may be minor, a party can pursue legal action with the knowledge that the amounts incurred for attorney's fees will be a recoverable item if the matter goes to trial or arbitration. The transaction costs of participating in the court system are proportionally higher for small cases. It is difficult to justify spending $40,000 in legal fees where $20,000 is in issue. Where $1 million is involved, the $40,000 seems acceptable as a part of the cost of recovering the $1 million. For those cases (the majority) settled without trial or arbitration, the parties must include legal fees in their calculation of their settlement demand, if they wish to be compensated for the attorney's fees they have incurred.

§ 5.29 Integration of the Contract

Most written contracts contain a paragraph stating that any and all oral representations, or "agreements not contained within this writing," are invalid. The purpose of this clause, called an integration clause, is to eliminate reliance of a party on oral representations, and to ensure that all the parties have a clear understanding of the parameters of the contract. The use of an integration clause makes it harder for a party to argue that a representative of another party promised that a certain provision would be included in the contract, that extra payment would be made for certain services, or that a particular clause in the contract would not be enforced.

The use of an integration clause does not guarantee that the court will enforce the entire agreement as written. Its use does make it more difficult for parties to argue that they believed in and relied on oral assurances or on agreements set out in other documents and not incorporated into the written contract.

Requesting all parties to initial the integration clause provides additional security for a party wishing to avoid later arguments that the other party did not agree to the integration clause, that other conditions or terms existed, or that oral representations were relied on in entering into the contract.

§ 5.30 Arbitration Provision

As discussed in detail in **Chapter 3**, many construction contracts contain a provision requiring that disputes regarding the contract be arbitrated. In some cases, the clauses make arbitration the sole forum for resolution of the dispute. In other clauses, arbitration may be nonbinding or may be subject to review by a different authority after arbitration.

If you have a choice about whether to use an arbitration clause, or whether to execute a contract that contains an arbitration clause, see the discussion at §§ **3.4** through **3.13**.

§ 5.31 What Law or Laws Control the Contract Dispute?

The primary factors used in determining the law that applies to a contract dispute are: (1) where the contract was made (i.e., executed or negotiated), (2) the citizenship of the parties to the contract, and (3) the place of performance of the terms of the contract. In any suit where the federal government is a party, federal law will be applied.

Many construction contracts contain clauses that specify the law that will govern any lawsuits brought for breach of the contract. Some contracts contain a **forum selection clause**. This clause specifies that disputes regarding the contract must be brought in a court of a certain jurisdiction, usually that of the state where the preparer of the agreement resides or where it has its primary offices. The forum selection clause also usually states the law that will apply, usually the state or federal law of that jurisdiction.

States vary on the extent to which forum selection clauses will be enforced. The general rule is that freedom of contract allows parties to specify the laws or court system they wish to have apply. However, not all such clauses will be enforced. As an example, a construction contract is made by residents of the state of Illinois. The contract is to be performed in Illinois, and was executed in that state. The contract language specifies that any disputes must be litigated in Kuwait.

If asked to decide whether to enforce this clause, a court would probably determine that there is no plausible reason for disputes regarding this contract to be heard in Kuwait. Kuwait has no interest in hearing this dispute, and there is no reason to require the parties to litigate it in Kuwait. There would be no reason to have Kuwaiti law apply, which is the law a court in Kuwait would be most likely to apply.

Under certain limited circumstances, a U.S. court would agree to apply foreign law in deciding a dispute, but this would usually be in a situation where there is some substantial connection between the foreign location and the dispute. In *McGhee v. Aramco,*[23] an employee working in the oil industry in Saudi Arabia sued in federal court in California for wrongful termination, and for various torts (noncontractual claims). Because the tort claims arose from actions in Saudi Arabia, the federal court in California applied Saudi Arabian labor law to these claims. On the contract claims, however, California state law was applied by the federal court because of citizenship of the parties, and the place where the contract was negotiated.

§ 5.32 Requirement of Bonding

Public construction contracts, as well as many private construction contracts, often require that bidders obtain performance and payment bonds. This is a form of assurance for the owner that, if a prime contractor fails to perform its work, or subcontractors fail to perform, or any contractor fails to pay suppliers as required under the contract, the work will be completed by the surety, or the surety will compensate the owner for the cost of completion.

Absent these types of bonds, an owner who has a dispute with a contractor could proceed against the bond required for licensing under state law. The contractor must post this bond, typically in the amount of $5,000 to $15,000, when obtaining a license. If the contractor performs defective work, the owner can attempt to obtain damages for the defective work from the bond.

[23] 871 F.2d 1412 (9th Cir. 1989).

§ 5.33 Coordination of Work with Other Parties

Although individual subcontractors or specialty contractors are rarely in contract with the other subcontractors on the project, some subcontracts contain a clause requiring that each subcontractor coordinate his work with that of the others. The dependence of each subcontractor on the performance of the other contractors makes this type of clause desirable.

In the absence of such an express clause, if a subcontractor has a claim against another contractor, an action could be brought in contract in those states that recognize third-party beneficiary claims. Under these claims, a party outside a contract is entitled to recover damages from one of the parties to the contract because the performance of the contract is intended to benefit the third party.

In this situation, the argument is that each subcontractor is an intended beneficiary of the promised proper performance of the others, under their subcontracts with the prime contractor. A breach by one of the subcontractors damages the others. That damage is sufficiently foreseeable for a court to impose liability on the nonperforming contractor under a third-party beneficiary theory.

Where these remedies are unavailable, an aggrieved subcontractor could sue other subcontractors under a negligence theory, if he is damaged by the acts of the other subcontractors in failing to coordinate with him, or by their preventing him from properly performing under his subcontract.

* * *

Case 5.2
(a) Prelitigation

A large commercial prime contractor undertook construction of a 32-unit condominium complex. The project was located on a grade, which required a fair amount of cutting and filling. There were three separate structures, each located on a different level. The prime had never previously worked on a project involving residential construction, and was unfamiliar with the methods used in that type of construction.

The framing contractor had worked for many years, as a carpenter, on jobs assigned through the union, but he had never before acted as an independent contractor.

During construction, the representative for the prime contractor was changed several times. Requests by each of the subcontractors were handled in a disorganized manner. The plumbing subcontractor was allowed to perform its part of the work ahead of the other trades, simply because the plumbing subcontractor was more vocal than the other subcontractors.

As a result, after the exterior framing was completed, the plumbing subcontractor built plumbing stacks and began performing other parts of his work.

This out-of-continuity performance resulted in increased cost and delays being incurred by the framer and by other contractors.

The building site's grade was so severe that it was not possible to raise lumber to the third story of each of the three structures with a forklift. The prime attempted to use a crane to bring the lumber in from the air, but the crane operator refused to perform because there were live power lines adjacent to the building. The prime was unable to have the power lines turned off for a long enough period to allow the crane to safely load the lumber. The only solution available for bringing the lumber up to the third-floor area was for the men to hand-carry the lumber to that level. This process took approximately ten times as long as it would have taken to raise the lumber with a crane.

Although weekly meetings took place on the job site, the representative for the prime was insufficiently assertive to control all the subcontractors, especially the plumbing subcontractor. At one point, to show his displeasure, the plumbing subcontractor had all of his employees walk off the job.

The constant out-of-sequence work particularly affected the framing subcontractor. He was accustomed to performing framing on residential projects in what he described as a "production" type of construction method: he would be able to use a single crew of carpenters from the union hall, moving them continuously from task to task throughout the project. Because of the lack of sequential control by the prime contractor, the framing contractor was required to release carpenters, and then go out and hire new carpenters when the flow of work picked up. The framer was also required to redo work because other subcontractors performed their work out of sequence.

The lumber ordered by the prime contractor for the framing had significant defects. Large amounts of the wood were rejected by the framer, and new lumber had to be ordered. There were variations in the thickness of the planks. Much of the lumber was delivered to the job site with warping, a result of its having been stored in the open and exposed to rain and moisture.

There was no location near each building where the lumber could be stored. The single storage location for the lumber was insufficient to store all the lumber necessary for the project. As much lumber as possible was stored in the single location, and additional lumber was ordered as space became available. The distance from the site's lumber storage spot to the buildings where the lumber materials were to be used required additional man-hours for transport of the lumber.

The drywall contractor performed much of his work prior to completion of the final plumbing work. Much of the drywall had to be cut and patched after installation, again requiring additional labor and causing delays.

The framing contractor and others attempted to keep the project on track by presenting a schedule for work to be performed by each of the respective trades on a weekly basis. Unfortunately, the supervisor for the prime contractor was also supervising two other projects at the same time, in different locations.

Because of travel time to the other locations, the project supervisor for the prime was only at this project approximately 20 percent of the time.

This constant absence from the job delayed processing of change orders and requests for information. Work would either stop entirely, or crews of workers would have to move to other tasks, while waiting for responses from the prime and the owner.

The owner ran out of money, and the completed units were not selling. The owner started offering units in the project to unpaid subcontractors, in lieu of payment. Eventually, the project was completed.

Because the framing contractor had not paid all of his withholding taxes, both the federal and state governments placed tax liens on the framing contractor's contract proceeds. No entity, including the prime contractor, could make payments to the framing contractor without satisfying the liens. The framer tried to negotiate with the Internal Revenue Service by offering to accept a condominium unit in lieu of payment from the developer, but the IRS refused to accept the offer because it could not determine the precise value of the unit, due to fluctuations in the real estate market.

The framer eventually went bankrupt. The federal and state liens went unpaid, accruing interest to an amount where no settlement was possible between the framer and the prime, because of the size of the liens. The prime also asserted numerous causes of action against the framer, claiming that many of the delays associated with the project's completion were the fault of the framer. All of the claims by those involved in the construction project were resolved, except for that of the framer against the prime and owner.

Analysis. The fact that the prime contractor on this project had never performed a residential project should have signaled to the others involved in the project that problems could arise. The tight nature of scheduling for maximum efficiency on residential projects contrasted heavily with the type of construction scheduling to which the prime contractor was accustomed. Uneven quantity of materials and the lack of consistent on-site supervision by the prime contractor resulted in delays, defective work, and numerous lawsuits by subcontractors, union trust funds, and, eventually, the homeowner's association.

The framing contractor went bankrupt because of the out-of-sequence work and down-time resulting from the poor scheduling by the prime contractor. The prime contractor's representative was at the project once or twice a week, instead of every day. Questions, requests for information or detail drawings, or change orders took weeks instead of days to be processed. The plumbing contractor, a large organization that did a lot of business on commercial projects with the prime contractor, was able to intimidate and disrupt the work of the other trades.

Having a project supervisor who was on the site every day, who planned the progress of the project in an orderly fashion, and who was responsive to the demands and needs of all the subcontractors would have reduced the problems that occurred on the project. Cooperation, joint effort, and goals were destroyed on this project by the lack of discipline and organization.

The inadequate storage space for lumber, and the fact that there was no cost-effective means to transport the lumber to the top stories of the structures, further exemplified the problems resulting from poor planning.

(b) Post-Filing of Litigation

Because of the bankruptcy of the framer, the framer's claim lay dormant for a number of years. The counsel for the defendant owner had filed an answer. The framer had given counsel for the defendant prime contractor an open extension of time to respond to the lawsuit. Nearly ten years after the suit was originally filed by the framer, counsel for the owner and contractor moved for dismissal of the action, for lack of prosecution.

The court denied the motion, but strongly suggested that the case be brought to trial or dismissed within the following six months.

Because of the lapse of time since depositions had initially been taken in the action, counsel for both sides stipulated to the primary witnesses who were still available to have their depositions retaken. The witnesses had poor recollection because ten years had passed since the construction project had been completed. However, the framing contractor's principal had had his professional career destroyed by this project, and many of the events regarding the project were strongly imprinted on his memory.

The project supervisor for the prime contractor had much weaker recollection. He had been working on two other projects at the same time as this project. He had also supervised approximately 40 projects in the intervening years.

As the time for trial drew near, counsel for the parties attempted to negotiate a settlement. A significant problem was the Internal Revenue Service lien. With the interest mounting over the years, this lien was now in excess of $100,000. After some investigation, counsel for the framer learned that the IRS had not renewed its liens, and, instead of over $100,000 being owed to the federal government, only $2,000 was owed. After learning of this, counsel for the framer and the prime negotiated a settlement of $42,000 with the prime contractor.

Analysis. The delay of over ten years from the time the project was performed until the time the case was actually litigated is unusual. Most lawsuits may have a delay of two to three years before all nonlitigation alternatives are exhausted and the case is being prepared for trial. In this case, the long delay made many witnesses unavailable. The principals of the commercial prime contractor were still available, but their memory of the events was weakened by the passage of time and by the number of other projects they had worked on in the intervening years.

The case brought by the framing contractor lay dormant, in large part, because of disinterest by the framer and by his attorney, who had taken the case on a contingency basis. The Internal Revenue Service had refused to negotiate on the tax liens, and the state had also held liens on any proceeds that might be paid by the prime contractor to the framer.

A new attorney was brought in to prepare the case for trial. In the course of trying to determine the actual amount of the federal and state tax liens, he learned that the IRS had not renewed its lien, and that settlement in a much lower range was possible. Because he did not want to alert the prime contractor to this fact, he negotiated a settlement amount with the prime contractor without disclosing the lapse of the IRS liens to the attorneys for the prime contractor.

In this case, the framer should have refused to go forward with the additional work without agreement about additional payment. Additional payment is appropriate where changes are numerous[24] or drastic.[25] Even where no authority is obtained for changes, if the contractor is required to perform emergency work, compensation for changes may be allowed, according to AIA Document A201, paragraph 10.2.1.

Most jurisdictions provide that the prime contractor is obliged to ensure that subcontractors perform in a timely manner, and that their work is not delayed.[26] Some cases hold that the subcontractor may not collect delay damages from an owner or prime contractor.[27]

§ 5.34 Pay-When-Paid Clauses

A **pay-when-paid** clause usually provides that a prime contractor need not pay a subcontractor until the prime contractor is paid by the owner or developer. If there is a pay-when-paid clause in the contract, ensuring that everyone is aware of its existence will reduce the risk of subsequent misunderstandings.

Many states have limited the applicability of these clauses. Some states permit enforcement only when the lack of payment to the prime contractor is a result of, or related to, the work of the subcontractor from whom payment is

[24] Wunderlich Contracting Co. v. United States, 240 F.2d 201 (10th Cir. 1957) (6,000 changes to the original contract).

[25] Saddler v. United States, 152 Ct. Cl. 557, 287 F.2d 411 (1961) (doubling excavation on small contract).

[26] For a survey of cases, *see* Annot., 16 A.L.R. 3d 1252, 1254 (1967).

[27] Loughman Cabinet Co. v. C. Iber & Sons, Inc., 46 Ill. App. 3d 873, 361 N.E.2d 379 (1977) (general contractor granted duly authorized extension of 309 days to perform; subcontractor had no cause of action for delay against general because delay was beyond the control of general); but, in J.J. Brown Co. v. J.L. Simmons Co., 2 Ill. App. 2d 132, 118 N.E.2d 781 (1954), a subcontractor was able to recover damages based on a theory of *quantum meruit* (market value of services performed) for delay from the prime contractor when the first subcontractor's performance was delayed by that of another subcontractor.

withheld. The purpose of this interpretation of the clauses is to prevent a prime contractor from failing to pay a subcontractor when the prime is in a dispute with the owner and the dispute is not connected to the work of the subcontractor.

If a pay-when-paid clause is being used in one of your contract forms, or such a clause is contained in a contract you have signed, you should read the clause carefully. At one time, New York law held that construction contracts that tied the subcontractor's right to payment from the prime to payment of the prime by the owner were valid.[28] However, in *Thomas J. Dyer Co. v. Bishop International Engineering Co.,*[29] the leading case on pay-when-paid clauses, the court stated that most parties expect to be compensated for their labor. Upholding a pay-when-paid clause causes the laborer to suffer a forfeiture. The court found that the language of such a clause must be clear and specific to support the argument that the subcontractor assumed the risk that it will not be paid if the prime contractor is not paid by the owner. The *Dyer* court described the purpose of such a clause as: to allow the prime contractor a reasonable period of time to obtain the funds from the owner. The *Dyer* court held that the clause did not require that the subcontractor wait an indefinite period of time for an event that may never occur (payment of the general contractor by the owner). Recent cases use the *Dyer* interpretation of pay-when-paid clauses.[30]

§ 5.35 Disputes Clauses

Many construction contracts, such as the AIA clause referred to in § **4.11**, require that construction continue during arbitration of a dispute. Other clauses, framed more generally, provide that a contractor or subcontractor may not stop work or walk off a job, even if the owner or other person in authority orders the contractor to perform work the contractor feels is beyond the scope of the original plans or specifications.

It is not uncommon, in the absence of a disputes clause, for a contractor or subcontractor to refuse to perform what it deems to be "extra work," unless the person demanding the work provides a written change order. It is obvious that

---

[28] Mascioni v. I. B. Miller, Inc., 261 N.Y. 1, 184 N.E. 473 (1933).

[29] 303 F.2d 655 (6th Cir. 1962).

[30] Nicholas Acoustics & Specialty Co. v. H&M Construction Co., 695 F.2d 839 (5th Cir. 1983) (holding that the pay-when-paid clause provides the prime contractor with the time to ensure that the subcontractor has completed its work); Culligan Corp. v. Transamerica Ins. Co., 580 F.2d 251 (7th Cir. 1978); Midland Engineering Co. v. John A. Hall Construction Co., 398 F. Supp. 981 (N.D. Ind. 1975); Yamanishi v. Bleily & Collishaw, Inc., 29 Cal. App. 3d 457, 105 Cal. Rptr. 580 (1972); Schuler-Haas Electric Co. v. Aetna Casualty and Surety Co., 40 N.Y.2d 883, 357 N.E.2d 1003, 389 N.Y.S.2d 348 (1976).

there will then be disputes between the owner and the contractors regarding whether particular work is within the scope of the original contract, or whether this constitutes extra work.

Where the contract contains a disputes clause, the work must proceed, and the issue of payment will be resolved through (1) arbitration, (2) a claims procedure in place on the project, or (3) whatever dispute resolution means are used after construction is completed. The contractor's failure to continue with the work is likely to be considered a breach of the contract by the owner, who could then justify terminating the contractor and retaining a new contractor to complete the work.

§ 5.36 Form Contracts for Construction Projects

There are many good reasons to use a standard form agreement for construction work. Time and energy are saved and legal costs are reduced. In theory, forms that are widely used, such as those of the AIA or the AGC, have familiar language and contain clauses that encourage a party to feel comfortable with known rights and obligations.

Two major risks arise from the use of form agreements: (1) these parties did not draft the agreement, and its language may not really represent their understanding of the contract, and (2) the form may have been updated and altered significantly since the party last used or reviewed it; it is *not* the contract the party thought he was making.

As new standard forms are issued, commentary appears in publications to explain the changes made in the forms. Some writers in the construction field publish articles that discuss proposed changes to standard construction forms in trade journals, such as those issued by the Association of General Contractors, or in legal journals, such as the *Public Contract Newsletter,* a publication of the Section of Public Contract Law of the American Bar Association. Other articles published in journals analyze the changes made to form contracts after they have been altered. It is unclear how much access the average person in the construction business has to this type of information. Many of the attorneys advising construction clients may not even be aware of changes to the form agreements until long after the changes have been made in the forms.

Major trade organizations such as the American Institute of Architects (AIA), the Association of General Contractors (AGC), and the Engineers Joint Contract Documents Committee (EJCDC) all issue standard forms that are available for use by their members. When nonmembers are involved in contracts that use these forms, these nonmembers may also recycle the forms for use on future projects, making use of the forms even more widespread.

Use of standard forms may reduce short-term cost and allow the parties to assume that the form contract represents a known quantity. However, a party runs a substantial risk if he uses the forms without careful reading and review.

The standard forms should be tailored for each project and reviewed by an attorney prior to signing.

§ 5.37 Compensation for Breach of Contract

When you are suing a party for breach of contract, two types of remedies are usually available: either an award that represents the **benefit of the bargain**, or an award that represents **restitution**. The first remedy is considered "forward-looking," and the second, "backward-looking." The two types of remedy are mutually exclusive; both will not be awarded by a court for the same claim.

Compensation under the benefit-of-the-bargain theory means that you are compensated for the profit or benefit you would have derived from the transaction, had it been fully performed. Although, in some instances, determining the profit you would have derived from the completion of the contract may be too uncertain or speculative, lost profit is recoverable where satisfactory proof of the lost profit is provided. The remedy is forward-looking because the court seeks to place you in the position you would have been in if the other party had fully performed under the contract.

The rationale for attempting to provide damaged parties with the benefit of the bargain is that, if parties know they can receive this remedy, they will be confident, when entering into contracts, that they will not lose their anticipated profits if they are required to sue for breach of contract. The nondefaulting party will get from the court exactly what he had bargained for from the other party to the contract, if a court grants this remedy. Application of the remedy serves to discourage people from deliberately breaching contracts that are no longer profitable or are difficult to perform. The defaulting party will still have to provide the nondefaulting party with the anticipated profit under the agreement.

The remedy of restitution is more frequently awarded by the courts. In many situations, establishing the amount of lost profits with reasonable certainty is too difficult. In addition, if the lost profits are not sufficiently connected to the breach of contract to be reasonably foreseeable damages, they will not be awarded. The test of foreseeability is whether the other party, at the time he signed the contract, could have foreseen that the damages that occurred would result, in the event of a failure to perform.

As a matter of policy, unforeseeable damages are not generally allowed. For example, one "consequence" of breach of a construction contract by an owner might be that it forces a prime contractor, who has been in business for twenty years, to go bankrupt. All the good will built up over that twenty-year period is lost. The problem with awarding money for this type of damage is that *any* financial problem of sufficient magnitude can bankrupt a business. The courts define this as a **consequential damage**, which is often held to be too remote and unforeseeable to be compensable damage under the contract.

To a certain extent, parties need to be able to calculate their risks upon entering into contracts. To allow recovery for this type of attenuated or remote damage would discourage parties from entering contracts, or cause them to price their performance at such a high level that it would impede normal commerce.

§ 5.38 —Money Damages

The courts, using their legal powers (as opposed to equitable powers; see § **2.7**) can award a party money damages. The court of law, derived from the English court system, can provide only money to compensate successful litigants. The most common remedy for breach of contract is a money award.

The court does not actually secure the money for a party. The court only provides the party with a written judgment. The prevailing party can then take that judgment to execute on property or assets of the losing party, to satisfy the judgment.

In England there were two separate and distinct court systems—courts of equity and courts of law—and each had its own set of rules, procedures, and remedies. In the United States, the single legal system can act either through its powers of equity or its powers of law.

§ 5.39 —Injunctions

When a court acts as a court of equity (the court can provide nonmonetary relief), it can provide parties with injunctions (orders that a party do or not do a particular act) or order **specific performance**. A court judgment for specific performance means a party so ordered must perform his part of a contract, such as to transfer land or to deliver goods. It is not a common remedy for construction disputes.

There is a general prohibition against forcing one person to work for another, based on the constitutional proscription against slavery. In a dispute regarding a personal service contract, a court cannot order one party to perform for another under such a contract, but it can order that the breaching party *not* work for anyone else for a period of time.

Failure of a party to obey an injunction can result in a court's ordering that the noncomplying party be jailed for contempt. In construction contract disputes, the most common use of an injunction would be to stop an awarding agency, or other owner, from awarding a contract that has been let for competitive bids. Injunctions are an extreme remedy; they constitute a demand that the court take action without a full evidentiary hearing. For these reasons, the burden on the plaintiff seeking injunctive relief is high. In order to obtain it, the plaintiff usually must provide: (1) evidence to the court that the legal remedy of money damages is insufficient to compensate the plaintiff for the harm, or threatened harm, (2) proof that there is a strong likelihood of the plaintiff's prevailing at trial, and (3) usually, the posting of a bond that will be forfeited if the plaintiff does not prevail at trial.

Sometimes, the low bidder objects to the award of the contract to a higher bidder; or, a higher bidder may claim that it should have been awarded the contract because a lower bidder is not qualified. In these situations, the party seeking the injunction might ask the court not to permit the award of the contract to the disputed bidder. The plaintiff bidder might ask the court to enjoin the owner from awarding the contract to anyone other than the plaintiff bidder. In some situations, the plaintiff bidder asks the court to order the award of the contract to the plaintiff bidder.

Injunctions may also be used by a construction entity to prevent a public agency from enforcing regulations. In *Consulting Engineers v. Department of Transportation,*[31] an association of consulting engineers brought an action to prohibit the New York State Department of Transportation (DOT), and its commissioner, from implementing new practices outlined in the revised regulations controlling the consultant selection process. The DOT claimed the suit was moot and that petitioners were seeking an advisory opinion, or, in the alternative, that the practices of the department were permissible under State Finance Law, Section 136-a.

One of the causes of action against the DOT involved a requirement for the submission of additional information on selected projects. The engineers' association contended that the additional information converted the negotiation process to competitive price bidding, which was prohibited by State Finance Law Section 136-a.

The DOT had a two-step selection process that was used occasionally for exceptionally large or complicated projects. In step one, the DOT ranked the firms according to their qualifications. The most qualified firms were asked to submit technical proposals. The DOT selected the most highly qualified firm and commenced negotiation as to fee. In addition to the proposals, the firms were asked to submit a sealed envelope containing a staffing or manning table that showed the work-hour estimate by grade level. The DOT asserted that, until the ranking process, which is based on the technical proposals, the sealed envelopes were not opened. It stated that the staffing tables were used by the DOT to verify the accuracy of its own staff's estimate and to assist in negotiating the precise scope of the project with the consultant firm, should that be necessary. The petitioners claimed the manning tables could be converted into compensation proposals, and could turn the selection process into a competitive bidding process. The court found the submission of tables to be a reasonable requirement, held petitioner's claim to be mere speculation, and denied the injunction.

---

[31] 547 N.Y.S.2d 205 (1989).

§ 5.40 Conclusion

Contract is the primary means by which parties in the construction field regulate and control their relationships. When a contract is carefully drafted to accurately reflect the specific agreements, a party's likelihood of recovery is enhanced, should a dispute occur.

Although many of the contractual clauses that appear in construction contracts are common to all written contracts, there are a number that are peculiar to construction, such as changed site condition clauses, particular types of liquidated damages provisions, and bonding provisions. Careful thought, selection of appropriate contracts or contract clauses, negotiation, and discussion of specific contract clauses all promote certainty and reduce the possibility of disputes in construction contracts.

You should seriously consider consulting an attorney before executing a contract that contains unusual clauses, or clauses whose significance or meaning is not clear to you. The amount of money spent to consult an attorney will be trivial when balanced against the benefit of avoiding a potentially troublesome transaction.

CHAPTER 6

SUBCONTRACTING

§ 6.1 Purpose of Subcontracting

Subcontracting allows a prime contractor to have others perform some or all of the labor on a construction project. In theory, use of subcontractors (also called specialty contractors) increases the efficiency of construction projects. The assumption is that repeated performance of particular construction specialties by contractors will result in those contractors' being able to provide superior performance in those specialties at a lower cost.

Minimal licensing standards, and the low capital investment required for subcontractors, encourage proliferation of small contractors as independent businesses. Often, such businesses are handed down from one generation to the next. The main disadvantage in use of subcontractors on large jobs is the number of legal problems created by the subcontracting process.

Subcontracting usually reduces the necessity for the owner, architect, or other design professionals to deal directly with those who actually perform much of the labor on a project. Where there is a construction manager (CM) or a prime contractor, that entity has a contractual relationship with the subcontractors, directs their work, and is responsible generally to the owner for coordinating the overall progress of construction.

Construction projects produce a tremendous amount of documentation. Each contractor or subcontractor working on a project generates his own daily schedules, work diaries, change orders, and requests for information. Typically, all the subcontractors are working on the project at virtually the same time. Providing access to particular locations on the job site or establishing priorities for performing work can become a substantial problem for the prime contractor, owner, architect, construction manager, or other person responsible for coordinating the progress of the project.

Although many owners select a construction method that isolates them from interaction with the subcontractors, not all owners wish to be so isolated. An owner may direct the architect (rather than the prime contractor) to negotiate with and retain subcontractors. Under these circumstances, it may be unclear as to whether the owner or the prime contractor bears responsibility for any problems that arise with the subcontractors.

Under the traditional construction system, the role fulfilled by each party to a project is well-defined. If owners step over the bounds of the role usually assigned to them, problems may ensue. For example, in *E. C. Ernst, Inc. v. General Motors Corp.*,[1] the court held that, if warranted by the facts of a particular case, notice regarding certain matters described in the contract could be given to the prime contractor rather than to the owner, despite the contract's express requirement that such notice be given to the owner.

§ 6.2 Communication with Subcontractors

Even where there are intermediaries between the owner and the contractors providing labor for a project, the owner should take steps to ensure that lines of communication remain open to him from the subcontractors. Conferences among all the participants on a project should be held, even before construction

---

[1] 482 F.2d 1047 (5th Cir. 1973).

begins. The chain of command among the parties should be outlined. Those individuals who will be authorized to execute and approve change orders should be identified, in order to reduce misunderstandings when problems arise on a project. Subcontractors should be actively involved in preliminary meetings. This early involvement should contribute to a cooperative spirit for the project, which may prevent disputes or allow them to be resolved more readily.

During the construction process, project meetings should be held to monitor progress and to provide a forum for discussion of problems on a regular basis, as those problems occur. Minutes should be taken at the project meetings, and there should be a mechanism for approval of the minutes by those attending the meetings. A final set of approved minutes should be made available to all participants soon after each meeting.

Even more helpful for effective project management is preparation by the owner or prime contractor of a written policy manual, similar to an employee policy manual, regarding the appropriate procedures for the project. This policy manual should describe the change order process and outline the authority of various individuals on the project, such as superintendents, lead men, and design professionals. Subcontractors should have input into the contents of the written manual.

The more clearly the contract documents specify the respective roles and responsibilities of the owner, architect, engineer, CM (if one is used), or prime contractor, the less likely it is that performance will be delayed or disrupted because of communication problems.

The purpose of taking all these steps is to encourage communication among the subcontractors and with the owner, and to reduce the risk that misunderstandings or miscommunications will escalate into claims or, worse, into litigation.

§ 6.3 Performance of the Subcontract

As construction progresses, many changes in the original plans will have to be made. Change order requests usually originate with subcontractors or with the prime/general contractor. The change order requests require response from a representative of the owner, typically the architect. Clarifications of design drawings or design details have to be provided to subcontractors by the architect or engineer.

If a problem surfaces regarding a particular design feature of a project, the architect or engineer may consult with representatives of a subcontractor and work together with them to arrive at a solution. Cooperation between the subcontractors and the design professionals, or other representatives of the owner, is important if such problems occur.

If building the project as originally planned has become too expensive, the owner will institute cost-cutting measures. The owner and his representatives may meet with the subcontractors to solicit ideas for cost cutting, or may communicate to the subcontractors the method or methods selected to achieve that goal. The owner should closely monitor how the cost-cutting measures have been implemented, to ensure that they have been properly incorporated into the work.

In the event that an error is discovered in the drawings, the specifications, or the construction itself, all parties involved in the process will need to coordinate their activities to arrive at a viable solution for the problem. All of these procedures require interaction among the owner, his representatives, and the subcontractors.

§ 6.4 Assignment of Subcontracts

In order to control the cost of construction, some experienced owners execute subcontracts themselves with the subcontractors, and then assign the subcontracts to the prime contractor. In *Crown Construction Co. v. Opelika Mfg. Corp.*,[2] an owner was held not obligated to assign the subcontracts, although the prime contractor had the obligation of accepting the assignment.

AIA Document A201 (general conditions), at paragraph 5.4.1, allows for conditional assignment of subcontracts by the prime contractor to the owner (see **Appendix B**). The right to enforce performance under the subcontract may not be assigned if the subcontractor must perform under different conditions than if the right to enforce performance had remained with the prime contractor. In other words, the new prime contractor to whom the subcontract is assigned cannot require additional work from the subcontractor.

An assignment of a contract or contractual right does not require the tender of additional consideration. **Consideration**, the offering of money or something of value, is usually an important element in a valid contract. However, a prime contractor can assign the right to enforce the subcontractors' performance without paying additional or specifically identified consideration for that assignment.

If AIA Document A401 (subcontract) is used, assignment of subcontracts is expressly permitted at paragraph 7.3.1 (see **Appendix G**). If AIA Document A401 is used, the question of the assignability of subcontracts is specifically answered; however, not all projects use A401, not even those that use AIA Documents A101 and A201. If only the latter two documents are used, the issue of assignability should be addressed by an addendum, because AIA Document A201 does not dispose of the issue. In an individually drafted agreement, the parties should negotiate the issue of whether they wish to allow assignment of the subcontracts, and then draft an appropriate clause.

[2] 343 F. Supp. 1266 (N.D. Ga. 1972), *modified on appeal on another point,* 480 F.2d 149 (1973).

The Association of General Contractors (AGC) discourages involvement by the owner with subcontractors. However, the AIA, in its Documents A101 and A201, provides protection for subcontractors, by trying to ensure that the same benefits given to the prime contractor under the prime contract are granted to subcontractors. The customary involvements of an owner with subcontractors are for approval of the subcontractors and review of the subcontract terms, particularly those regarding timing and amount of compensation for subcontractors.

§ 6.5 Subcontract Clauses

Although the subcontractors do not normally have any contract relationship with the owner, the activity between the owner and the prime contractor may have a dramatic effect on the subcontractors. Failure of the owner to pay the prime often results in the prime's failing to pay the subcontractors.

§ 6.6 —Payment Clauses

Many subcontractor disputes focus on the issue of failure by the prime contractor to pay subcontractors. Prime contractors often argue that they are not required to pay subcontractors if the prime contractors are not paid by the owner. There is usually agreement by all parties that, if nonpayment of the prime contractor by the owner results from defective work by a subcontractor, or from failure of a subcontractor to submit the appropriate documentation for his work, it is appropriate for the prime to withhold payment from the subcontractor. However, if the prime contractor is unpaid because of the prime's failure to process necessary paperwork, the prime must still pay the subcontractors for work duly performed by the subcontractors.

The prime contractor's obligation to pay the subcontractors may hinge on whether a payment clause in the subcontract creates a condition for payment, or whether the clause merely designates a convenient time when payment will be made to the subcontractors.

It is common for the prime contractor to argue that the clause establishes a condition for payment. He will assert that the prime contractor and the subcontractor are "co-venturers" in the project and should share jointly the risk of nonpayment by the owner.

The effect of enforcing payment clauses as conditions for payment is to aid prime contractors with cash flow problems and to allow the prime contractors to transfer their financial pressures to the subcontractors. Because many subcontractors are small businesses with limited resources, a delay in payment to subcontractors can cause them financial hardship. A subcontractor may be forced to extend lines of credit, may be unable to pay his suppliers, or may lose his bonding status. In the worst-case scenario, the subcontractor may be forced to default under the subcontract or to file for bankruptcy.

Most disputes regarding payment clauses arise because there is disagreement over whether the prime is required to pay the subcontractor when the prime has not been paid by the owner. It is critical, in deciding this issue, to establish whether the payment clauses create a condition for payment of the subcontractor, or whether they are simply a designation of time for payment. The courts in different states have interpreted payment clauses both ways. The specific language of the particular clause may determine this issue.

More easily resolved are situations in which the owner has no excuse for failing to pay the prime contractor, or there is a breach by a subcontractor (other than the subcontractor from whom the prime is withholding payment). In these situations, the prime has no excuse for failing to pay the subcontractor.

AIA Document A201 (general conditions) is vague on whether payment of the subcontractors is dependent on payment of the prime contractor by the owner. At paragraph 9.6.2, A201 provides for payment of each subcontractor upon the prime's receipt of payment from the owner. This language is probably not enough to provide the prime contractor with support for the argument that payment of the subcontractor is conditioned on payment of the prime by the owner. In fact, if AIA Document A401 (subcontract) is used for the project, payment conditions are specifically negated at that document's paragraphs 11.3 and 12.1 (see **Appendix G**).

The courts have attempted to compensate for a perceived lack of choice by the subcontractor in negotiating the terms of the subcontract. The subcontractor has no means by which to assess the financial stability of the owner. For example, in *Thomas J. Dyer Co. v. Bishop International Engineering Co.,*[3] the court stated that subcontractors contract with prime contractors mainly on the basis of the prime contractor's solvency, not the solvency of the owner.

§ 6.7 —Payment of Funds for Work Performed by Others Considered Trust Funds

Certain states attempt to avoid nonpayment of subcontractors by designating funds received for the work of others as trust funds. Michigan, Minnesota, New York, New Jersey, Oklahoma, and Texas are among the states that so designate. In both New York and New Jersey, diversion of the funds is a crime. In addition, New York provides for civil penalties.

[3] 303 F.2d 655 (6th Cir. 1962).

§ 6.8 —Scope of the Work

The subcontractor should be as clear as possible as to the scope of the work he is to perform. The extent of his obligation should be clearly stated in the subcontract. Subcontractors should avoid incorporation by reference to the prime contract of terms between the prime contractor and the owner on scope of work, unless the subcontractor has actually had an opportunity to review the master contract and fully understands the obligations he is assuming.

§ 6.9 —Progress Payments

The subcontract should specifically describe a schedule of payments. In the absence of a schedule, payment to the subcontractor would usually not be due until completion of all work by the subcontractor.

§ 6.10 —Retention

It is common for the owner to hold back a certain percentage of the amount due under the contract, until all work is satisfactorily completed. Frequently, when a contractor claims the work is complete, the owner or owner's representative will conduct an inspection and compile a list of items to be completed or repaired, called a "punchlist." Prime contractors also usually withhold retention from their payment to subcontractors.

Some states have responded to improperly withheld retention, or delayed payment of retention, by enacting statutes that require prompt payment of retention to contractors. These states include Alaska, Arizona, California, Florida, Hawaii, Illinois, Louisiana, Oregon, Pennsylvania, South Carolina, Utah, Washington, West Virginia, Wyoming, and the District of Columbia. However, most of these laws apply only to public construction projects. Parties concerned about retention should consult an attorney regarding the existence, and the substance, of these laws in any particular jurisdiction.

§ 6.11 Mechanic's Liens

If problems occur on a project and the owner fails to pay the prime contractor, who then fails to pay the subcontractors, each party may resort to legal remedies for relief. A common remedy for a subcontractor, who has no formal contract with the owner, is to file a mechanic's lien against the property where the construction is occurring.

Most states have laws that permit someone who provides supplies or performs labor on property to file a lien on the property for the amount of the reasonable value of the supplies or labor.[4] If the unpaid party is a subcontractor, who is not in contract with the owner, he generally must provide a form of preliminary notice to the owner and construction lender prior to filing a lien.[5] Once the lien right is exercised by a subcontractor, the owner often pays the lien claimants to remove the liens, or requires the prime contractor to secure a bond, which will have the effect of removing the lien.

In anticipation that lien situations may arise, many owners demand the prior posting of payment bonds by the relevant prime contractor or subcontractors, to prevent the filing of liens by unpaid subcontractors or suppliers. If the owner is in direct contract with a contractor, the only way for the owner to avoid liability to unpaid suppliers is to pay the suppliers directly, or to use joint checks made out to both the contractor and the supplier to pay for the supplies.

* * *

Case 6.1
(a) Prelitigation

Roofbest, a small roofing contractor, was having financial difficulties. It fell behind on some of its payments to its primary suppliers. Nevertheless, it still was aggressively selling roofing jobs to residential customers. When a large pending project fell through, Roofbest was unable to continue its weekly payments to Tilemaster, its largest supplier. Tilemaster sued twelve individual homeowners for the cost of the supplies provided, for which it had not been paid. The amount owed by each homeowner averaged $6,000.

The homeowners were outraged that they were being called on to pay for supplies used on their roofs—for which they had already paid once. One homeowner, Jerry, had even shifted to Roofbest on Tilemaster's recommendation; initially, he had been intending to use a different roofer. He had been looking at tile in Tilemaster's showroom, and he asked the salesperson whether the roofer he was planning on using was competent. The salesperson said he knew nothing about the roofer Jerry had mentioned, but that Roofbest was very good. The owner subsequently decided to use Roofbest.

Frank, another owner, when served with the required Preliminary Notice of Lien, which, under the mechanic's lien laws, would enable Tilemaster to sue him for the cost of the supplies, called Tilemaster to inquire about the significance of the notice. Tilemaster assured Frank that Roofbest had always paid its bills on time, and that the notice was merely a formality. Because of this assurance, Frank took no further action to protect himself from Tilemaster.

Analysis. Tilemaster made a mistake in allowing Roofbest to fall behind in its payments. As the amount owed for supplies mounted, instead of collecting from

---

[4] Cal. Civ. Code § 3110 (Deering 1991).

[5] Cal. Civ. Code §§ 3114–3116 (Deering 1991).

Roofbest, Tilemaster sent out the preliminary notices regarding Tilemaster's lien rights on the property, and planned to collect the money from the owners. The result was that Tilemaster was required to accept discounted settlements from the owners. Tilemaster ended up being responsible for attorney's fees to prosecute the lawsuits, and these fees could not be collected from the owners. Tilemaster also was required to wait approximately one year to collect its money.

In the current sluggish economy, more and more contractors or suppliers are going bankrupt, or at least are having severe financial difficulties. Extending large lines of credit to contractors or suppliers may be seen as a worthwhile financial risk. On the other hand, the practice may result in losses greater than those that would have been suffered by refusing to continue to extend credit to marginal construction entities.

For the owners, use of joint checks to pay the contractor and supplier would have avoided double payment for the materials. Had the owners consulted with an attorney in advance of signing the contracts with Roofbest, they also could have secured lien releases from the supplier after the supplier had been paid.

(b) Post-Filing of Litigation

A California statute requires a contractor to defend and indemnify homeowners from suits by subcontractors or suppliers.[6] In this case, Roofbest failed to honor its legal obligation to protect the owners. The owners were required to obtain their own attorneys and to pay Tilemaster for the supplies. Most took the precaution of entering judgments against the principals of Roofbest, who failed to appear for trial. The owners hoped that, in the future, the principals of Roofbest would have assets that could be used to satisfy the judgments obtained by the owners against Roofbest for failing to indemnify them from the claims of Tilemaster.

Analysis. In this situation, the owners did not take sufficient steps to protect their interests. If Frank had called an attorney, or even another contractor, regarding the effect of the preliminary notice, or if he had attempted to discover how financially stable Roofbest was, he might have been able to insist that Roofbest pay his bill with Tilemaster before Roofbest ran out of money.

There were insufficient facts in this case to support an owner defense of fraud to defeat the mechanic's lien action. However, under certain circumstances, a court will deny recovery to a claimant. In *Burton v. Sosinsky,*[7] a court barred recovery from an owner by contractor and subcontractor lien claimants who had overbilled the owner.

Too frequently, people refuse to consult an attorney until the situation has reached an impasse and the only alternative is to file or defend a lawsuit. Even

[6] Cal. Civ. Code § 3153 (Deering 1991).

[7] 203 Cal. App. 3d 562, 250 Cal. Rptr. 33 (1988).

a brief consultation with an attorney might have saved Frank the considerable time and money he was required to spend defending against and settling Tilemaster's suit.

§ 6.12 Subcontractor Claims against Third Parties

In some circumstances, a subcontractor will try to make a claim for damages against another party on the construction project—the owner, a design professional, or another subcontractor—with whom the subcontractor has no contract. One of the barriers to a damages claim by a subcontractor against an owner is the fact that there is generally no formal written contract between the subcontractors and the owner. It is not uncommon, however, for a subcontractor to claim the existence of an oral contract between the subcontractor and the owner.

As an example, a problem arises on a job because the original specifications or plans make no provision for unknown subsurface conditions, such as an underground stream under the proposed location of the foundation for a 30-story building. No foundation or concrete work has begun on the site. The architects and engineers prepare a new set of plans and drawings, relocating the building 40 feet south on the same lot. All the necessary government agency approvals have been received. However, the grade of the soil on the new location is 5 percent greater than on the original location. The steeper grade requires that the foundations be placed at a greater depth. The subcontractor performing the foundation work is asked to perform his work at the new location. When he points out that his cost of performance now will be substantially greater, the design professionals tell him to submit a change order, which will be approved.

The subcontractor performs the work as requested in the new location, but the change order is not approved. The subcontractor performed the work because he was instructed to do so by the design professionals representing the owner. His claim will be for breach of the oral agreement between himself and the owner's agents.

In addition to a claim for breach of contract, the subcontractor may make a claim based on a negligence theory. A claim may also be submitted on the theory that the subcontractor is a third-party beneficiary of the contract between the design professionals and the owner. Under the third-party beneficiary theory, the proper performance by the design professional is intended to benefit the subcontractor. If the design professionals fail to properly perform under their contracts with the owner, this breach causes foreseeable damages to the subcontractor, for which the subcontractor is entitled to recovery.

Some of the standard form contracts attempt to prevent contract claims by subcontractors against owners (see AIA Document A201, paragraph 1.1.2 (**Appendix B**), and NSPE 1910-8, paragraph 6.9). Some jurisdictions, such as California, allow a plaintiff to bring an action where a third person attempts to interfere with a contract or prospective contract between the plaintiff and another. The above clauses do not, however, prevent claims based on a tort theory (breach of a noncontractual duty) such as negligence, or claims based on more unusual torts such as interference with contract or with prospective business advantage. If fully enforced by a court, however, paragraph 1.1.2 of AIA Document A201 may prevent a contract claim, because the paragraph specifically states that A201 does not create a contract relationship between the owner and any subcontractor.

§ 6.13 Sub-Bidding Issues and Laws

It is critical for subcontractors to examine all relevant available documents before bidding on a project. Failure to do so will result in a subcontract bid that is not fully informed, exposing the subcontractor to potential financial loss. Unforeseen or unanticipated costs incurred on the project because of unexpected subsurface conditions also present a grave risk for subcontractors bidding on projects.

For projects on which most of the labor is going to be performed by subcontractors, the prime contractor will not know his cost until he receives sub-bids from the subcontractors. Although the contractor may be able to accurately estimate the cost of the subcontracting work in areas such as masonry and roofing, the cost of the mechanical trades (electrical, plumbing, and heating/ventilation/air conditioning) may be harder for the prime contractor to calculate. The level of reliance by the prime contractor on the mechanical subcontractors' bids is higher, and the effect of a mistaken mechanical sub-bid may be more serious.

§ 6.14 —Bidding Mistakes

Clerical errors in sub-bids is another area that may result in financial loss. A subcontractor may be required to make a claim for additional compensation or to initiate litigation, in order to seek relief from being bound by an owner or prime contractor to an unprofitable bid. To obtain relief from the courts for a clerical error or mistake, the subcontractor must show:

1. The mistake is so material to the bid that to enforce it would be unconscionable;
2. It pertains to or is a material aspect of the bid;
3. It occurred in spite of the exercise of ordinary care by the bidder;
4. The circumstances of the mistake are such that the nonmistaken party may be placed in the status quo (the party's status before the mistake);
5. The subcontractor gave timely notice of the error to the nonmistaken party.

The clear message to subcontractors bidding is that, unless these factors are demonstrated, a court asked to decide whether to release a subcontractor from a mistaken bid is not likely to allow relief. A recent Utah case allowed a subcontractor to be released from his mistaken bid because the disparity between the subcontractor's bid and the next lowest bid exceeded 290 percent. The court found that it was not reasonable for the prime contractor to rely on the subcontractor's bid.[8]

§ 6.15 —Listing of Subcontractors

Many public contract systems require prime contractors bidding for contracts to list major subcontractors to be used on the proposed projects for which the prime contractors bid. These laws often state that listed subcontractors may be substituted only for specific reasons, which are enumerated. These bidding laws serve to protect subcontractors from bid-shopping, (described more fully at § **4.13**). The current AIA Document A201 (general conditions), at paragraph 5.2.1, requires the contractor to provide the owner, through the architect, with the names of subcontractors and suppliers as soon as practicable after the award of the prime contract. In the case of negotiated contracts, the applicability of paragraph 5.2.1, and the issue of whether the prime contractor is required to furnish a list of subcontractors for owner approval, is unclear.

The Engineers Joint Contract Documents Committee (EJCDC) addresses this issue in its Document No. 1910-8. Paragraph 6.8.2 states that (1) if supplementary conditions require the identification of subcontractors and suppliers, their names are to be submitted to the owner for acceptance by the owner and engineer; and (2) the contractor may not employ, either initially or as a substitute, a subcontractor or supplier to whom the owner or engineer reasonably objects.

Paragraph 6.8.2 also takes into account the fact that supplementary conditions may require the prime contractor to list subcontractors and suppliers in advance of the prime's bid, and to obtain prior approval of the prime's list by the owner and the engineer. The owner's approval can be revoked after subsequent investigation by the owner or engineer. In that case, the prime contractor would be paid the amount of any increased subcontractor cost.

To avoid problems with subsequent owner approval, a prime contractor may avoid executing subcontracts with the proposed subcontractors until the owner or his representatives has had a reasonable time to object. Another alternative for the prime contractor is to insert in the subcontract a clause that makes the contract conditional on receipt of approval of the owner or his representatives.

---

[8] Tolboe Construction Co. v. Staker Paving & Construction Co., 682 P.2d 843 (Utah 1984).

§ 6.16 —Subcontractor Quotes

One way for a subcontractor to avoid the issue of being bound to its quote is to label it a "quotation," or a "request for [the prime contractor] to make offers."[9]

The subcontractor may decide to submit a sub-bid by which it intends to be bound, but the subcontractor may seek also to bind the prime contractor to use the subcontractor. This can be achieved by including in the sub-bid language stating that use by the prime contractor of the sub-bid constitutes an acceptance of the sub-bid, which binds the prime contractor. In *Sharp Brothers Contracting Co. v. Commercial Restoration, Inc.,*[10] the sub-bid stated that the sub-bid must be accepted within ten days. The court held that such a limitation on an offer was appropriate. If no time limitation is stated in the offer, a court will assume the offer lapses if it is not accepted within a reasonable time.

The sub-bidder's post-bid behavior may be critical. In *S. M. Wilson & Co. v. Prepakt Concrete Co.,*[11] the sub-bidder lost his protection because he continued to negotiate after the expiration date stated in the sub-bid language.

§ 6.17 Flow-through of Prime Contract Clauses

When creating or selecting a set of contracts to be used on a project, an attempt should be made to have all the contracts use the same or similar major contractual provisions. It is common for the subcontracts to include language that incorporates certain provisions of the prime or master contract. This incorporation binds subcontractors to the obligations of the prime contract, and may give the subcontractors the benefits of the prime contract. If language in the subcontract refers to the prime contract but fails to incorporate it by reference, the subcontract will be interpreted by referring to the prime contract. The prime contract will take precedence if the two contracts conflict.[12]

Where contracts are individually prepared for a project, the subcontracts can use the prime contract as a model and use the same provisions on key issues such as claims procedures, arbitration, attorney's fees, and indemnity.

If standard form contracts are used, the prime contract usually contains a flow-through, or "conduit," provision. In AIA Document A201 (general conditions), the flow-through provision is found at paragraph 5.3.1 (see **Appendix B**). Because of the confusing wording of this clause, it is difficult to determine

---

[9] Leo F. Piazza Paving Co. v. Bebek & Brkich, 141 Cal. App. 2d 226, 296 P.2d 368 (1956); Cannavino & Shea, Inc. v. Water Works Supply Corp., 361 Mass. 363, 280 N.E.2d 147 (1972).

[10] 334 S.W.2d 248 (Mo. App. 1960).

[11] 23 Ill. App. 3d 137, 318 N.E.2d 722 (1974).

[12] Oxford Development Corp. v. Rausauer Builders, Inc., 158 Ind. App. 622, 304 N.E.2d 211 (1973).

which, if any, benefits and obligations apply to the relationship between the subcontractor and the prime contractor. The language of the clause also states that the prime contractor may effectively eliminate from the subcontracts the rights the prime contractor holds against the owner.

The flow-through provision in NSPE No. 1910-8, paragraph 6.11, is similar, in that it provides for flow-through of obligations from the prime contract to the subcontract. It differs in that it does not attempt to provide the subcontractor with the benefits enjoyed by the prime contractor as against the owner.

Different courts and jurisdictions vary on whether the incorporation by reference of other agreements that have not been executed or explicitly agreed to by a party is binding. In *C. H. Leavell & Co. v. Glantz Contracting Corp.,*[13] it was contended that a series of interrelated contracts for a construction project could create a single contract, which was binding on all the parties who executed the individual contracts. The *Leavell* court held that the architects were not liable under these contracts to the general contractor. However, in *Varco-Pruden, Inc. v. Hampshire Construction Co.,*[14] a California court expressed its opinion that incorporation by reference of a subcontract, which was followed shortly thereafter by a prime contract, might provide sufficient basis for enforcement of the two contracts as if they were a single agreement, despite the fact that the contract–owner agreement did not exist at the time the subcontract was executed. (It is important to note that this opinion was not central to the court's holding in this matter, and so the statement would not bind other California courts.)

Directly and explicitly linking all the contracts might solve the problem of binding all parties to the same set of obligations, but it would destroy the structure intended to be created by separating the owner from the subcontractors, and having each party work primarily with one other party in the contractual chain. It would also deprive the prime contractor of his ability to contract with subcontractors on different terms than he has with the owner. Part of the incentive for the prime contractor to take the responsibility for coordinating the work of the subcontractors is to derive profit from controlling the subcontractors' prices and costs.

Having multiple prime contractors is one way to attempt to avoid some of the problems inherent in the prime contractor/subcontractor system. This approach is discussed in some detail in § **4.18**.

§ 6.18 Subcontractor Default

A prime contractor's failure to coordinate work and to ensure cooperation among subcontractors increases the risk that subcontractors will default. All of

---

[13] 322 F. Supp. 779 (E.D. La. 1971).

[14] 50 Cal. App. 3d 654, 123 Cal. Rptr. 606 (1975).

the parties involved should be aware when a subcontractor is faltering, and should do their best to keep the subcontractor on the job.

The alternatives of replacing the subcontractor, or looking to the performance bond surety to pay for or complete the faltering subcontractor's work, are poor substitutes for the originally promised performance. Delay and disruption are the almost inevitable results of a subcontractor's default.

In some instances, rather than working to prevent subcontractor default, a prime contractor may be the cause of such default. The prime contractor's attempts to enforce a payment clause, which states that the subcontractor will not be paid until the prime contractor is paid by the owner, may aid a prime contractor with his cash flow problems, but it is likely to put extreme financial pressure on the subcontractor.

§ 6.19 Subcontractor Bankruptcy

Business bankruptcy filings through November 1991, nationwide, were up 148 percent over the previous year. In northern California, construction companies closing down or filing bankruptcy increased approximately 23 percent between 1988 and 1989.

Prime contractors face substantial problems if a subcontractor on a critical part of a project files bankruptcy. Once bankruptcy is filed, the bankruptcy system imposes an automatic stay on any legal actions against the bankrupt. Substantial legal fees will be incurred by the prime contractor, who is required to work with the bankruptcy courts. In addition, the trustee for the debtor, or the bankruptcy court, has the option of assigning the subcontract to another entity with whom the prime may be completely unfamiliar. Bankruptcy court procedures may take precedence over clauses in the subcontract that specify how the parties will deal with each other. Entitlement to retainages and progress payments will be an issue because of the bankruptcy filing. The subcontractor may even have the ability to recover payments made shortly before the bankruptcy filing to lower-tier subcontractors or suppliers. Those entities will most likely then look to the prime and the prime's payment bond for payment.

If litigation is underway, the subsequent bankruptcy filing may reduce or eliminate the ability of the parties to the litigation to obtain documents or other discovery materials from the bankrupt subcontractor.

For all of the above reasons, it is important to realize that the bankruptcy of a subcontractor can be costly and can delay completion of projects.

§ 6.20 Owner Removal of Subcontractors

With a specifically tailored contract (as opposed to a form contract), the parties may wish to address the issue of whether the owner can reject subcontractors in advance or terminate subcontractors after performance has begun. If the owner can terminate subcontractors, the circumstances for such action by the owner should be set out in as much detail as possible. For those projects where AIA Document A201 is used, paragraph 5.2.4 seems to prevent the owner from rejecting subcontractors if more than a reasonable period of time has elapsed since the list of subcontractors was submitted to the architect/owner.

If a prime contractor who has submitted a list of subcontractors to the owner then wishes to substitute an unlisted subcontractor for a listed subcontractor, the owner must not withhold reasonable consent. For example, in *Meva Corp. v. United States,*[15] the owner withheld permission to substitute, and the prime contractor retained the poorly performing subcontractor. After faulty completion by the subcontractor, the prime contractor was able to recover damages from the owner because of the unreasonably withheld consent.

§ 6.21 Conclusion

The use of subcontractors may provide higher quality construction at a lower cost to owners. However, the complexity governing the legal relationships among owners, contractors, subcontractors, and design professionals guarantees that disputes will occur on many projects.

If the contract documents plainly outline the necessary procedures to follow regarding change orders, requests for information by the subcontractors, issuance of architectural bulletins by design professionals, or other necessary procedures, many project disputes may avoid an escalation to litigation.

Holding frequent meetings on the job site, to review the progress of the work and the implementation of necessary changes, will help keep the channels of communication open among all the parties involved.

In subcontracting, the fragmentation of handling individual tasks and the increased use of specialty trades create special problems. Each party is concerned solely with pursuing his own profit. However, the failure or nonperformance of one party on a project inevitably causes harm to all. When a cooperative atmosphere is set and maintained on the construction site, and when claims and disputes are handled in a fair and predictable manner, the project is likely to be completed professionally, with a minimum of delay.

---

[15] 511 F.2d 548 (Ct. Cl. 1975).

CHAPTER 7

TORT CLAIMS

§ 7.1 Introduction

A tort is a civil wrong committed by one person against another, based on an obligation not arising from a contract. Generally, the duty breached by the commission of a tort is established by statutes or court-developed case law, rather than by prior agreement of the parties.

The statutes covering civil torts are independent from criminal statutes. Significant distinctions between the two are:

1. The criminal system is concerned with punishing behavior viewed as morally wrong, while the civil system is primarily concerned with compensating those injured or damaged by the actions of others;
2. The criminal system deals with fines and incarceration, while the civil system usually only makes money awards or grants injunctions.

There are intentional and nonintentional torts. A person alleging that another has committed an intentional tort must prove the act was committed deliberately. Examples of intentional torts include libel (written false statements), slander (oral false statements), assault (putting a person in apprehension of physical harm), and fraud (express or implied false statements made to induce reliance). Examples of nonintentional torts include negligence (acts falling below the normal standard of care), trespass (unpermitted entry onto the land of another), and battery (unpermitted and unwanted physical contact with another).

§ 7.2 Elements of a Tort

There are four basic elements to a tort: (1) duty, (2) breach, (3) proximate cause, and (4) damage. In order to win a tort case, the plaintiff must prove all of these elements by a preponderance of the evidence (that it is more likely than not that the defendant committed the acts with which he is charged).

§ 7.3 —Duty

The plaintiff in a tort case must establish that the defendant had a duty to the plaintiff. In most jurisdictions, either under court-developed common law or as a part of their statutes, the law holds that each person has an obligation to act nonnegligently toward others. A higher standard of care may apply to those in contract with each other, or where parties have a fiduciary relationship, or where some other relationship exists that causes a person to reasonably and foreseeably place more than the customary amount of reliance on the opinion or behavior of another.

The standard of care or duty that is applied to behavior of a tort defendant is objective, not subjective. The question is not whether the defendant tried, in good faith, to be careful, but whether his conduct was up to the standard of a reasonable person's conduct under the circumstances.

Between individuals who have no other relationship, there may be no duty. For example, a passerby has no duty to rescue a person in peril. However, once a passerby has undertaken to rescue a person in peril, the rescuer has the obligation not to place the imperiled person at greater risk.

Suppose a member of the public falls into a 20-foot-deep ditch at a construction site. A telephone repairman who is walking by hears the woman yell when she falls into the ditch, and sees her in the ditch. The repairman tells the victim that she need not worry, because he will get help. However, the repairman stops at a bar, forgets about the woman in the ditch, and fails to take any steps to get help. If the woman who fell into the ditch can prove that her situation was made worse, or that she sustained additional damage as a direct result of the failure of the repairman to follow through on his promise to get help, the woman may recover against the repairman, even though the repairman had no original duty to rescue or care for the woman. By undertaking to help the woman, he took on an obligation for which he may be liable, if he then fails to live up to the standard of care of a reasonable rescuer under the same circumstances.

§ 7.4 —Breach

After the plaintiff in a tort action has provided evidence that establishes that the defendant owed the plaintiff a duty, the plaintiff must prove the defendant breached that duty. In a negligence case, this would mean proving that the defendant's acts fell below the standard of care of a reasonable person. In the above example of the woman who fell into the ditch, and the repairman who failed to help rescue her after promising to do so, the woman must prove that the defendant failed to summon help, thereby breaching his duty as a self-appointed rescuer.

§ 7.5 —Proximate/Legal Cause

The plaintiff must prove that the acts of the defendant actually caused the physical harm or injury to the plaintiff. This physical harm is easier to define than the **legal cause** for which a court will impose liability. Many factors may be "causes in fact," or physical causes of harm or damage. In the above example, many factors could have been causes of the woman's falling into the ditch: (a) She may have been wearing high-heeled shoes while walking in an unpaved and uneven area, (b) she may have neglected to wear her glasses, despite poor vision, (c) the trench may have been inadequately marked or protected from public access, or (d) the construction contractor may have failed to create any means to exit the trench. Further injury or aggravation of the original injury may have resulted from the failure of the repairman to get help for the woman.

In this example, the woman would probably sue the contractor performing the work, the city, for failing to require better marking or protection for the public, and the repairman. To prove her case against the repairman, the plaintiff in this case must provide proof that his failure to rescue her was a significant factor in causing her damage. Such proof might include: (1) believing that help was on the way, the woman did not continue to call for help; (2) the woman did not try to climb out of the ditch; (3) the woman did not take steps to rescue herself, because she thought she would be rescued by the repairman, or by someone he summoned. In other words, because of his promise to get help, she failed to take other steps to rescue herself. All of these facts, if proven, would strengthen her case against the repairman for negligence.

§ 7.6 —Damage

It is only logical that the plaintiff in a tort case, as in any other suit, must prove damage. Without any show of damage, a court or arbitrator has no basis on which to calculate the amount of an award. If a framing contractor uses 30 nails to attach glu-lams to the frame of a building, instead of the 50 called for by the plans, the owner of the building may have a hard time recovering against the framer if the testimony of experts is that use of 30 nails is adequate to build a secure structure.

§ 7.7 Tort Claim as One of Several Legal Theories

When a lawsuit is filed, the plaintiff may assert that the defendants owe the plaintiff damages based on a variety of legal theories. The assertions contained in the complaint are called **allegations**. In many actions for breach of contract, for instance, the plaintiff may also allege that the defendants committed fraud, or negligence, and that these acts damaged the plaintiff. Typically, the same or similar facts are alleged to provide the basis for each of the different legal theories of the lawsuit. However, each of the different legal theories will have different necessary elements that must be pleaded and proven in order for the plaintiff to recover on the basis of each legal theory.

For example, an owner sues a prime contractor because the structure erected by the prime contractor has defects. The owner may allege the prime contractor breached the construction contract by not providing a defect-free building. The owner may also allege the prime contractor was negligent because it did not build a structure that was up to the standard of care expected of a competent contractor. The owner may also claim he is entitled to recovery under a fraud theory, based on allegations that the prime contractor built a defective building, knew it was defective, and nevertheless tried to convince the owner to accept the defective building. Another basis for a fraud allegation under these facts might be that the prime contractor never intended to fulfill his obligations under the contract, induced the owner to enter into the contract, and led the owner to pay the prime contractor for the construction.

Most of the above legal theories of recovery are based on the same essential fact: the prime contractor built a defective building. The representation by the contractor was that he would provide a defect-free building. If the plaintiff asserts alternative legal theories, other facts may be alleged. For instance, to support a fraud allegation, the plaintiff may allege that the defendant never intended to build a proper building, which may have caused the owner to suffer additional damages. The owner relied on the express representation that the contractor would build an appropriate structure, and would not have entered into the contract with the contractor if the owner had known that the contractor did not intend to perform properly. Different compensatory damages may be appropriate for this type of fraud allegation, and punitive damages may also be allowed. (See § **7.25**.)

§ 7.8 Parties to a Tort Action or Claim

Where a dispute arises under a contract, the persons or entities who executed and agreed to the terms of the contract are the parties to the lawsuit based on the contract. Where violation of a tort duty is the only legal theory of recovery, there is no agreement or contract between the parties that defines the duties of the parties. The basis for the claimed violation of the alleged duty would usually be found by consulting statutes, ordinances, administrative codes, or court-developed common law.

§ 7.9 —Plaintiffs

In some jurisdictions, the level of duty or care owed by a landowner to a plaintiff has depended on the classification of the plaintiff. The basic classifications of plaintiffs are **trespassers, licensees**, and **invitees**. Trespassers are those without permission to be on the land. Licensees are those with permission to be on the land. Invitees are those individuals who enter land with the intent of conferring economic benefit on the owner. The lowest standard of care applies to a trespasser, the next highest standard of care to licensees, and the highest standard of care to invitees. Some states have reduced the importance of determining the landowner's duty of care based on the classification of the plaintiff, in part because of difficulty in determining what category applies in each case, and in part because of public policy concerns about compensating injured persons.

The trend in the law has been to apply a general standard of care for landowners to all those entering on their land.[1] Ohio, however, has refused to follow this trend.[2]

The plaintiff in a tort case in the construction context may be a member of the public who is injured by construction activity. In the earlier example, a trench that was near a construction site, but was located on public property, was inadequately protected because of a contractor's failure to place an adequate physical barrier over the trench. A member of the public stepped on the plywood placed over the trench, fell into the trench, and was injured. In an example (**Case 5.1**), a worker who was an employee of a subcontractor was injured on a construction site because he failed to use a safety belt when working at height. Another category of plaintiff might be an owner of property adjacent to a construction site, who is upset because construction vehicles for the project are parking on his land without permission. The irate owner may sue the prime contractor, or the owner, for trespass.

Prior to initiating a lawsuit under each of these circumstances, an attorney for the plaintiff would usually advise the plaintiff to send a letter making a demand that the defendant pay damages, or cease the offensive activity (trespass). However, in some circumstances, a defendant's first awareness of a problem may be when he is served with a summons stating that he is being sued, with a complaint for damages attached.

§ 7.10 —Defendants

The plaintiff may have some choice about whom he will sue. There may be a number of defendants who are responsible for the plaintiff's damage or injury. If the plaintiff has a choice, he may sue only those defendants who have the assets to pay a judgment. As a general rule, multiple defendants will be joint actors, participating in the same act or set of acts that causes the damage or injury. However, under some circumstances, multiple defendants may be named because they all took some action—although not necessarily the same action, or set of acts—that caused or contributed to the plaintiff's damage or injury.

For example, a ten-story college dormitory is completed with sliding glass doors on some of the floors, but the balconies that had originally been a part of the design were deleted. An occupant of the dormitory opens one of the sliding glass doors, expecting to find a balcony, falls, and is injured. The injured party will sue the contractor and the college for failing to provide a safe building.

---

[1] Rowland v. Christian, 69 Cal. 2d 108, 443 P.2d 561, 70 Cal. Rptr. 97 (1968); Cal. Civ. Code § 1714; Cooper v. Goodwin, 155 D. Ct. App. 449, 478 F.2d 653 (1973) (classifications abolished except as to trespassers); Mile High Fence Co. v. Radovich, 175 Colo. 537, 489 P.2d 308 (1971); Marioenzi v. Joseph Di Ponte, Inc., 333 A.2d 127 (R.I. 1975).

[2] Moore v. Denune & Pipic, Inc., 26 Ohio St. 2d 125, 269 N.E.2d 599 (1971).

Because the doctor who treated the plaintiff at the hospital did not advise the plaintiff of an operation the plaintiff needed, to avoid having a permanent limp, the plaintiff also sues the doctor and the hospital for professional negligence. Even though the doctor had nothing to do with the fall from the dormitory, his failure to advise the plaintiff of the need for the operation has contributed to a part of her damage, the permanent limp.

A worker injured on the job will be unable to sue his subcontractor employer in civil court because the workers' compensation system (a separate court system) is where he will be compensated, except under very limited circumstances. However, the worker may sue the prime contractor in the civil court system in tort, alleging that the prime contractor was negligent for failing to properly supervise the construction, because the prime contractor failed to ensure that the subcontractor (the worker's employer) followed safe labor practices.

§ 7.11 —Joint and Several Liability

Where multiple defendants are named in a lawsuit, most jurisdictions use a legal doctrine called "joint and several liability." This doctrine provides that each individual defendant is liable for the entire amount of the damage to the plaintiff, regardless of the percentage of fault of each defendant. The rationale for this rule is that recovery for the injured person is deemed more important than any potential unfairness this rule may impose on an individual defendant. Where there is joint and several liability, a defendant who pays the entire amount of the judgment may sue his fellow defendants for contribution, based on their respective percentages of fault.

For example, a subcontractor employee fails to install the required number of bolts to the railing for a staircase in a building. After completion and occupancy of the building, a person with mobility problems uses the railing, and the railing comes loose, causing the plaintiff to fall and to be injured. The subcontractor who installed the railing was a three-person family-owned operation, with no insurance and no assets from which the plaintiff could collect a judgment. The prime contractor on the job, and the owner of the building, are businesses with both assets and insurance available for paying a judgment for the plaintiff.

The plaintiff may decide not to name the subcontractor as a defendant in the suit. Alternatively, the plaintiff may name the subcontractor in the suit, but not aggressively pursue the case against the subcontractor, because of problems in collecting the judgment.

The rule of joint and several liability originally developed because of those situations where the plaintiff could not distinguish the respective fault of the individual defendants. For example, a subcontractor follows an unsafe practice in erecting the steel girders for a building, and the general contractor allows the unsafe practices to be used. These unsafe practices cause an injury. If a defendant could provide evidence to show percentage of fault, this could be used to offset some of that defendant's liability. The burden of providing that evidence is placed on the defendants, not the plaintiff. This is contrary to the general rule that the plaintiff has the burden of proving all the facts necessary to establish his case. The reason for this shift of the burden of proof from the plaintiff was that it was believed that information regarding respective fault would be more likely to be in the hands of the defendants. The defendants would also have the incentive to produce such evidence, if any existed.

When multiple defendants are held responsible for cleanup of toxic waste, each will try to claim that he is not responsible for as large a share as may have been imposed on him. In *O'Neil v. Picillo,*[3] among the defendants named in a suit to recover cleanup costs were owners, operators, transporters, and generators of the waste found at a toxic site. Some defendants settled with the Environmental Protection Agency (EPA) before trial. The nonsettling defendants claimed that liability could be apportioned separately because the cleanup of the property had occurred in separate phases. During the first two phases of cleanup, no containers were found belonging to two defendant companies.

The court refused to accept the argument, or to apportion liability separately. The court stated that the manner in which the cleanup proceeded, whether it was in 1, 2, or 100 stages, was practically and theoretically irrelevant to the question of whether the damage to the site could be divided into distinct harms.

Some states, such as California and Washington, have enacted legislation that alters the rule of joint and several liability under certain circumstances. In 1986, California passed an initiative that became Civil Code §§ 1430–1432. This legislation abolished joint and several liability for noneconomic damages (compensation for pain and suffering). Each defendant is liable for noneconomic damages in direct proportion to that defendant's fault, and is entitled to contribution from other defendants or responsible parties if he pays more than his proportionate share.[4]

Joint and several liability still applies in California for economic damages. This alteration of the prior California rule of joint and several liability for all tort damages was based on the argument by "deep pocket" defendants such as large companies, or government agencies, that they were being required to pay excessive amounts in tort damages where they bore only a minor amount of responsibility. The fact that they were solvent defendants from whom the entire judgment could be collected encouraged plaintiffs to pursue only these defendants. The State of Washington has enacted legislation with a similar effect, regarding tort liability.

---

[3] 682 F. Supp. 706 (D.R.I. 1988), *aff'd,* 883 F.2d 176 (1st Cir. 1989), *cert. denied,* 110 S. Ct. 1115 (1990).

[4] Mills v. MMM Carpets, Inc., ______ Cal. App. 3d ______, 1 Cal. Rptr. 2d 813 (1991).

Where multiple defendants are sued, indemnity agreements and other contract clauses and legal rules of interpretation will be critical to a court's determination of the respective liability of the various defendants.

* * *

Case 7.1
(a) Prelitigation

An antique hotel in a major metropolitan area was being renovated. The basic plan was to gut the inside of the original structure, which was over 100 years old, and to retain the brick facade of the building. The interior would then be rebuilt to modern specifications. The structure had three stories.

After the interior was gutted, Woodco, the framing contractor, erected plywood spreader structures to hold up the exterior walls. Conco, the construction manager, was supervising the construction and coordinating the progress of the project. There was no prime contractor. Unfortunately, Conco did not have a representative on the job every day, because Steve, the Conco representative, was supervising other projects in the same area.

The work of Woodco was completed when it erected the plywood spreader structures. However, the timing of the project was behind schedule, and the spreader structures continued to be needed to hold up the outside walls as other work was performed. Woodco agreed to leave the spreader structures in place and to allow Conco to have some other framer remove the spreader structures after the other work was complete. In order to avoid future liability, Woodco insisted that Conco and the contractors agree to absolve Woodco from any liability for any injury, loss, or damage resulting from the spreader structures. Conco and the contractors signed a letter agreement to this effect.

Conco had erected signs warning all workers not to walk on the spreader structures, and to use the appropriate elevators or other ladders to descend from the higher levels of the building. Howard, a steelworker, was working on the third level of the structure, installing structural steel that would support the brick exterior and the rest of the structure after the plywood spreader structures were dismantled. Like many of the steelworkers, rather than take the construction elevator down from the third floor, Howard would walk along the plywood spreader structure to a convenient location, swing down from the top level to a ladder at the second-floor level, and then climb down. No representatives of Conco, or anyone else on the job, ever took any steps to discipline or admonish workers who used this method to descend from higher floors of the building.

On one occasion, when Howard swung down from the third floor, he grabbed a beam that was 2 inches by 6 inches by 6 feet and was wedged vertically between the floor and the ceiling of the spreader structure at the third-floor level. As Howard grabbed the beam, it gave way, not having been secured to the spreader structure in any way other than having been wedged in place to provide some strength to the floor and ceiling of the spreader structure.

Howard fell to the ground, sustaining severe injuries to his arms, back, and legs. He underwent multiple operations to various parts of his body to attempt to cure the various mobility, strength, and range-of-motion problems caused by the accident.

Insurance companies were involved in negotiations to try to resolve Howard's claim, but no overall settlement was reached prior to filing suit.

Analysis. Renovation of old structures carries special risks. The high cost of bringing an old structure into compliance with modern building codes may mandate a decision to tear down the old building, despite its quaint charm. In this case, the danger posed by the erection of the spreader structures was created solely by the fact that the renovation project was retaining the old exterior walls, rather than building an entirely new structure.

Daily on-site supervision again appears as a critical issue. Attempting to spread the cost of construction management by not having an employee on-site on a daily basis dramatically increases the risk that accidents or scheduling or coordination problems will occur. The fact that workers were not kept from walking on the spreader structures was one of the direct causes of the accident. The responsibility rested squarely on the construction manager who failed to fulfill his responsibilities.

Although the construction manager erected the signs warning that workers were not supposed to walk on the spreader structures, and stating that this use was dangerous, the signs alone probably were not sufficient to absolve the construction manager. Signs alone, without disciplining of workers caught violating the rule, and without having a supervisor remind the workers not to violate the rule, were insufficient to prevent workers' violating the rule. Additional steps may have been called for as being prudent actions under the facts of this situation.

(b) Post-Filing of Litigation

After some investigation regarding Howard's medical condition, and after exchanges of documents regarding the respective responsibility of the parties for the mechanics of the accident, certain facts were agreed on. It appeared that Conco had inadequately supervised workers by its failure to insist that they not use the spreader structures for pedestrian walkways or as a method to descend from the upper floors of the building. There was also some negligence by Howard in using this method to descend from the building. The beam inserted in the spreader structure had no identifying marks, but there were hammer marks from a carpenter's hammer at the ends of the beam. The only individuals who used that type of hammer on the project were Woodco employees. Based on this evidence, there was a strong argument that Woodco was primarily responsible for the accident.

At a judicial settlement conference, Hotelco, the developer, and Conco, the construction manager, agreed to a total amount of $1 million to compensate Howard for his injuries. Woodco refused to accept that number as representing the value of Howard's damages, and also refused to contribute anything toward Howard's settlement.

Conco and Hotelco negotiated a "sliding-scale" settlement with Howard. Under the agreement, Howard was guaranteed $666,000. If Howard obtained a verdict that exceeded $1 million from Woodco, the excess would be turned over to Conco and Hotelco. Conco and Hotelco would appear at trial on Howard's side, providing him with expert testimony and factual evidence in Howard's case against Woodco. The case was prepared and set for trial a number of times; finally, Woodco agreed to pay $300,000. This was only slightly less than Woodco would have paid under the settlement proposed at the settlement conference. Meanwhile, Woodco had incurred legal fees in excess of $100,000 during the subsequent litigation.

Analysis. Sliding-scale settlement agreements are a relatively recent concept, but they have been used extensively in the past few years to try and encourage settlement. Where one defendant is recalcitrant and the remaining defendants can agree on liability and damages, it is a useful tool to force the other defendant to either settle or risk heavy liability at trial.

This case also demonstrates the value of settlement conferences with knowledgeable judges who can convince the parties to compromise. The judge in this particular case had extensive experience in construction cases, and was able to command respect from the parties and their attorneys, who consequently accepted his recommendations. Even though the sliding-scale settlement was necessary to pull in the final defendant, the numbers and the proportions were worked out by the parties in cooperation with the judge.

California Code of Civil Procedure § 877 provides for the release of a settling defendant from cross-complaints for nonexpress indemnity. The basic test is that the settlement amount must be "in the ballpark," given consideration of various relevant factors, including the possible liability of settling defendants and the fact that settlements are usually less than the amount of trial verdicts. These factors are fully articulated in *Tech-Bilt v. Woodward-Clyde & Associates.*[5] California Code of Civil Procedure §§ 877.5 and 877.6 provide

[5] 38 Cal. 3d 488, 698 P.2d 159, 213 Cal. Rptr. 256 (1985). Numerous cases have arisen disputing the release of settling defendants: Singer Co. v. Superior Court (Cool-Temp Air Conditioning Co.), 179 Cal. App. 3d 875, 225 Cal. Rptr. 159 (1986); City of Los Angeles v. Superior Court (Hutcherson), 176 Cal. App. 3d 856, 222 Cal. Rptr. 562 (1986).

procedures whereby a party can affirm the good faith of a sliding-scale settlement. Constitutional objections to sliding-scale agreements have not been successful.[6]

* * *

§ 7.12 —Cross-Claims or Cross-Complaints

When a construction defendant is sued in a tort action, he will usually respond to the lawsuit by suing other defendants or third parties as cross-defendants. One of the common bases for suing cross-defendants in a tort case is a claim for indemnity or contribution from the cross-defendants. A prime contractor who is sued will usually, in turn, file a cross-complaint against the subcontractors on the project, claiming that the subcontractors are the parties who actually caused the unsafe condition that resulted in the plaintiff's being injured. Alternatively, the prime contractor will assert in his cross-complaint that, even if the prime contractor was negligent in failing to discover the unsafe condition, only a small percentage of fault can be attributed to the prime contractor, and the majority of the fault rests with other parties, such as subcontractors and/or design professionals.

§ 7.13 Tort Theories

As described previously, different legal theories are used in civil lawsuits based on tort. In most cases, more than one legal theory will be asserted in a complaint.

§ 7.14 —Negligence

The basis of tort negligence theory is that a person is liable to another if he has not been as reasonably careful as would be expected of a prudent person under similar circumstances, and his failure to act carefully has caused damage. What is often called "malpractice," where a professional such as an architect or engineer is sued, is really professional negligence. Certain professionals, such as engineers, architects, attorneys, or doctors, will be held to the standard of reasonable care of a professional within the local community. For a professional who holds himself out as a specialist, the standard is correspondingly higher.

---

[6] Riverside Steel Const. Co. v. William H. Simpson Const. Co., 227 Cal. Rptr. 424, *review granted and ordered citable,* 709 P.2d 1309, *cause transferred,* 748 P.2d 307, *transferred to* 199 Cal. App. 3d 1051, 245 Cal. Rptr. 477 (1986); City of Los Angeles v. Superior Court (Hutcherson), 176 Cal. App. 3d 856, 222 Cal. Rptr. 562 (1986).

Next to breach of contract, negligence is probably the most commonly asserted legal theory in construction disputes.

* * *

Case 7.2
(a) Prelitigation

Tidy Roofing was performing repair work on the roof of low-income housing units in a large metropolitan city. Prior to beginning work, Joe, the more senior member of the roofing crew, advised the occupant of a unit that had a leaking roof that the work would be performed on the roof. Vanessa, the resident, asked whether she should leave the unit. Joe assured Vanessa that it was not necessary for her to leave. He asked her to show him where water was coming into the unit, so he would know which areas of the roof needed patching.

Joe and Steve, the second member of the roofing crew, had nearly finished patching the roof when Steve stepped on a soft spot on the roof's surface. Steve's left leg went through the roof and penetrated into the ceiling area of Vanessa's apartment. A small amount of debris fell into the apartment. Because it was beginning to rain, Joe and Steve patched the hole that had been made. They then went to the front door of the apartment and informed Vanessa of what had occurred. She acted surprised. Joe and Steve went into the room where the debris had fallen, and offered to clean it up. Vanessa refused to allow them to clean up the debris, and became abusive.

Vanessa filed suit against Tidy Roofing and the city (owner of the apartment building) for negligence. She alleged Tidy Roofing and the city had caused her personal injuries and property damage. Vanessa claimed that she had been in the room where the debris fell, and that, when some of the debris had hit her on the head, it had knocked her to the floor and she had injured her neck and back. In addition, she said she had received medical treatment for the injuries and had sustained damage to personal property from the debris that fell into the apartment.

Analysis. There is a question as to whether it is a safe practice for a roofer to work on a roof with occupants in the areas immediately below. The risk of injury to occupants seems to indicate that it would be prudent to have the occupants vacate the units during this work.

If the roofers had ensured that the apartment was vacant when they repaired the roof, it would have been impossible for Vanessa to make a claim for personal injury. The property damage might still have occurred, but this was a small element of Vanessa's claim. In all likelihood, a property damage claim alone would have been resolved by insurance adjusters rather than proceeding to trial.

The defense of the case would have been strengthened by good documentation by the roofers concerning the work performed, and notes of what happened, prepared immediately after the events. Documentary evidence should have been created and kept available. Tangible evidence, such as the debris that fell through and was cleaned up, or photographs of the debris, would have helped to provide Tidy Roofing with a better defense at trial.

(b) Post-Filing of Litigation

The case went to trial. Vanessa was a very poor witness on her own behalf. Most of the evidence indicated that she had not been in the room where the debris had fallen. She had thrown away the items of personal property claimed to have been damaged, rather than attempting to have them repaired. The medical treatment she received for the alleged injuries appeared excessive.

One issue at trial was whether Joe and Steve were negligent. The question of whether they should have known that the roof could give in at that particular spot was explored in their testimony. Joe, the more experienced roofer, testified that when he first went to work on a roof, he always walked very carefully around the roof, and knelt down, to observe the contour of the roof, which would show whether the roof was sagging. Sagging in the surface of a roof is an indication that the plywood structure under the roofing material is weak and could give way. Joe said he had performed his usual observation on this roof and had observed no sign that the roof structure was sagging. Joe also testified that he had seen some roofs that had collapsed under the weight of workers even where there was no visible sagging.

The trial judge awarded Vanessa minimal damages. If the judge believed Vanessa had fabricated the injury, she should have awarded no damages for personal injury. The judge could also have found that Tidy had not been negligent and therefore was not liable for any property damage suffered by Vanessa.

In all likelihood, the doctor who had treated Vanessa for her injuries had taken a lien on the lawsuit for payment for his services, and the attorney had taken the case on a contingency basis. After the doctor, attorney, and costs were paid, little or nothing was left for Vanessa.

Analysis. The credibility and the appearance of witnesses are important considerations in determining the strength of a case presented at arbitration or at trial. Arbitrators and judges are affected by the appearance and verbal facility, and by the sympathy evoked by witnesses. A good attorney will advise a client that the case is not as strong if the client cannot recall events well or does not appear credible. Unfortunately, the prejudices of arbitrators or judges are reflected when they are called on to determine the credibility of witnesses and to decide whether a plaintiff or defendant should receive or pay an award.

* * *

§ 7.15 —Fraud

Plaintiffs often allege fraud as one of their legal theories of recovery. To prevail, a plaintiff must prove in a fraud case that: (1) the defendant made an express or implied representation that was false and known to be false by the person making the statement, (2) the plaintiff relied on the false representation, (3) it was reasonable under the circumstances for the plaintiff to rely on the false representation of the defendant, (4) the plaintiff was damaged as a result of the plaintiff's reliance on the false representation.

For example, a subcontractor substantially underbids a project. He knows that he cannot perform the work for the bid amount, and does not expect to. Instead, he begins work. During construction, he demands compensation from the prime contractor which is beyond the bid amount, claiming that extra work, or changes in scope from the original work, entitle him to additional compensation. In an action by the prime contractor against the subcontractor, the prime will allege that: (1) the subcontractor made a false representation (that the subcontractor intended to perform the work for the bid amount), (2) the prime contractor relied on the artificially low bid of the subcontractor, which was reasonable for the prime to do since the subcontractor was in the best position to determine the price of his performance, and (3) the prime contractor will be damaged if he has to pay, for the subcontractor's performance, an amount that is higher than he bargained for. As with any other contract breach, the damaged party must take reasonable steps to mitigate the effect of damages he suffers.[7]

§ 7.16 —Strict Liability (Liability without Fault)

Strict liability, or liability without fault, is a legal doctrine that allows a plaintiff to recover against a defendant who has not acted intentionally or negligently. The rationale that justifies the imposition of strict liability on defendants in certain inherently dangerous situations is that, even where a person exercises the greatest standard of care, damage is likely to occur. In those circumstances, liability can be imposed without any showing that the defendant intended the

---

[7] For cases on the obligation of a party to mitigate damages, see: C. A. Davis, Inc. v. City of Miami, 400 So. 2d 536 (Fla. App. 1981) (defendant contractor argued against recovery of owner based on owner allegedly having spent more than was necessary to correct contractor's work; court stated contractor could only dispute completion cost based on waste, extravagance, or lack of good faith by owner); Great Lakes Gas Transmission Co. v. Grayco Constructors, Inc., 506 F.2d 498 (6th Cir. 1974), *cert. denied* (Feb. 24, 1975), 95 S. Ct. 1330 (contractor unsuccessfully contended that owner could pass its loss to customers through rate increases); S. J. Groves & Sons Co. v. Warner Co., 576 F.2d 524 (3d Cir. 1978) (breaching contractor unsuccessful in argument that prime should have mitigated damages by firing subcontractor).

harm or was careless. Strict liability may be imposed for damage caused by storage of hazardous materials, or on defendants who keep wild animals or domestic animals with vicious tendencies, such as pit bull dogs. For these cases, the plaintiff must prove the defendant's ownership or control of the dangerous instrumentality (the toxic waste or other hazardous element, or the dog), and the plaintiff must prove that his damage was caused by the dangerous instrumentality owned or controlled by the defendant.

Many jurisdictions also provide for imposition of strict liability where a defendant places into the stream of commerce a defective product that, when used as reasonably intended by the defendant, causes damage.[8] The *Restatement of Torts* was generated by a group of legal scholars, lawyers, and judges who determined in 1963 that one who sells any product in a defective condition that is unreasonably dangerous to the user or consumer, or to his property, is subject to liability for physical harm thereby caused to the ultimate user or consumer, or to his property, if (1) the seller is engaged in the business of selling such a product, and (2) the product is expected to and does reach the consumer without substantial change in the condition in which it is sold. This rule applies even when (1) the seller has exercised all possible care in the preparation and sale of his product, and (2) the user or consumer has not bought the product from, or entered into any contractual relation with, the seller. Most courts have adopted this rule in judicial opinions, making it binding law in most courts today.

The rule allows those injured by defective products to sue where there is no contractual relationship and no negligence by the manufacturer. One justification for imposing strict liability in these circumstances is the fact that the manufacturer of a product may be best able to control the design of that product, and it is foreseeable that an unsafe product will injure consumers. If the manufacturer knows that it will have to pay for injuries caused by its unsafe product, the profit incentive for making an inexpensive, unsafe product may be reduced. The manufacturer of the product can spread the cost of making the product safer by incorporating it into the cost of each unit sold, and passing a small portion of the total additional cost on to each consumer.

As a general rule, this theory has been applied to personal injuries sustained when an ordinary consumer uses a product. In California, however, in *Kriegler v. Eichler Homes, Inc.,*[9] this theory of strict liability was applied to a home builder for building a defective home. The residence in *Kriegler* had passed through a number of owners before being acquired by Kriegler. During the time Kriegler owned the house, the heating system failed. The appellate court allowed Kriegler to recover, based on a theory of strict liability.

The *Kriegler* court based its decision on the fact that a home buyer has no real ability to evaluate the details of construction. The buyer relies on the skill

---

[8] Greenman v. Yuba Power Products, Inc., 59 Cal. 2d 57, 377 P.2d 897, 27 Cal. Rptr. 697 (1963); Restatement (Second) of Torts § 402A.

[9] 269 Cal. App. 2d 224, 74 Cal. Rptr. 749 (1969).

of the developer, who implicitly represents that the house was erected properly and is habitable. Because improper construction foreseeably results in damage to owners, public interest dictates that the cost of the injury be borne by the developer, who created the danger and is in a better economic position to bear the loss than the individual homeowner.

The home builder was held liable not only to the first owner, but also to Kriegler, who bought the house six years later. There was no finding that the builder was negligent in its construction of the home. The court *did* say that the test for whether liability would be imposed would be one of reasonableness, not perfection.

Other cases have not applied strict liability to parties in the construction context.[10] Two additional cases have limited strict liability in the construction area. In *Allied Properties v. John A. Blume & Associates, Engineers,*[11] a court held that an engineer was not strictly liable to owners because he merely sells services and is not subject to the doctrine of strict liability. In *United States Financial v. Sullivan,*[12] a developer was held not to be strictly liable to a construction lender because the lender was not a consumer, but instead was a link in the marketing chain that mass-produces homes.

Traditionally, California courts and laws have been more liberal than other jurisdictions in sheltering consumers from risk and in imposing liability on large corporations (or, in effect, insurance companies), which are assumed to be better equipped to spread the cost of these losses.

Some other states also have imposed strict liability on those in construction. In *Ernest W. Hahn, Inc. v. Armco Steel Co.,*[13] a court imposed strict liability on a manufacturer of roof joists. In *Adobe Building Center, Inc. v. Reynolds,*[14] a manufacturer of mortar mix was found strictly liable.

For contractors working with hazardous materials such as asbestos, PCBs (polychlorinate biphenyls), or other toxic materials, federal laws such as CERCLA (Comprehensive Environmental Response, Compensation, and Liability Act of 1980),[15] SARA (Superfund Amendments and Reauthorization Act of 1986),[16] and RCRA (Resource Conservation and Recovery Act of 1976)[17] may

[10] Gagne v. Bertran, 43 Cal. 2d 481, 275 P.2d 15 (1954) (engineer not strictly liable to prospective owner for failure to properly test soil, although liability could exist under theories of negligence or deceit); Del Mar Beach Club Owners Association v. Imperial Contracting Co., 123 Cal. App. 3d 898, 176 Cal. Rptr. 886 (1981) (engineer and architect not strictly liable to owner because those who sell their services for the guidance of others are not liable, absent negligence or intentional misconduct).

[11] 25 Cal. App. 3d 848, 102 Cal. Rptr. 259 (1972).

[12] 37 Cal. App. 3d 5, 112 Cal. Rptr. 18 (1974).

[13] 601 P.2d 152 (Utah 1979).

[14] 403 So. 2d 1033 (Fla. Dist. Ct. App. 1981).

[15] 42 U.S.C.A. § 9607 et seq. (West 1991).

[16] 42 U.S.C.A. § 6924 et seq. (West 1991).

[17] 42 U.S.C.A. § 6928 et seq. (West 1991).

also impose strict liability. In addition, environmental laws, such as those listed above, provide civil or criminal penalties for actions that may not have been illegal at the time the actions were taken. Although these measures may seem unfair, justification for harsh penalties on those involved with toxic waste is found in the seriousness of the health risk posed by these substances, and the high cost of their cleanup.

* * *

Case 7.3
(a) Prelitigation

In this matter, Bill, a seller of wooden stakes, was sued by Roger, a landscape architect. Bill had provided lodge pole pine stakes, at a cost of approximately $400, for use by Roger in a retaining wall that was part of the landscaping at a residence. Roger had specified soak-treated pine wooden stakes, and had described the application he intended: a wooden retaining wall. Bill stated his belief that the stakes suggested by Roger were suitable for this application. Several years after the stakes were installed, the retaining wall failed because the stakes had rotted. The problem was discovered at the time the property was being inspected prior to purchase.

Roger told the owners of the home that he would fix the problem. He called Bill. Bill inspected the wall, and the rotted stakes. Bill said he would pay to "take care of it." Roger spent $11,000 to redo the wall, and then billed Bill for that amount. Bill refused to pay that amount, having assumed that he would provide only new wood, or pay for the cost of new wood. He did not think he would be billed for the entire cost of redoing the retaining wall. In addition, the cost of redoing the wall was higher because fencing had been installed after the original installation around the area where the wall was located. Some of the fencing had to be removed and reinstalled, to allow access to the wall area for some of the heavy equipment required to replace the retaining wall.

A critical factor from Bill's perspective was that he was dealing with Roger, who had been a landscape architect for over 20 years at the time he bought the stakes from Bill. He had specified soak-treated pine wooden stakes—pine stakes that had been dipped in a preservative consisting of copper napthanate and diesel fuel. The wall would have failed eventually, no matter what type of wooden stakes were used, because the existence of earth-to-wood contact results in rotting due either to wet rot or infestation by termites or other insects. However, pressure-treated wooden stakes would have lasted longer. Roger thought he was getting pressure-treated stakes, and he claimed Bill had sold him a defective product.

Analysis. In a product liability claim, negligence of the seller, the manufacturer, or the consumer is not directly relevant. The relevant questions are: whether the product was being used for its intended use, and whether it met the reasonable expectations of the consumer of the product. The lodge pole pine stakes are suitable for some types of uses. Even in this case, they lasted a certain number of years.

The question of the expertise provided by Bill in recommending the stakes was significant. For example, if Roger had called Bill and simply ordered the same stakes, without describing how he intended to use the stakes, Roger would not have won the case.

By putting himself in the role of "expert" seller, Bill allowed his expertise in recommending the stakes for use in the retaining wall to become an issue. The fact that Roger should have known, better than Bill, whether this product would be suitable for this use was not as relevant as the fact that Roger apparently relied on Bill's expertise in deciding which stakes to purchase for this particular application. Roger stated at trial that he relied solely on Bill's recommendation. He did not call other suppliers or consult any other sources to determine whether the soak-treated lodge pole pine stakes were suitable for use in a retaining wall.

Bill erred in offering to take care of the problem without establishing a dollar limit, or without stating more precisely what he intended to pay for, to help fix the wall. If he intended to provide new stakes, or to pay only for the purchase of new stakes, he should have stated those facts specifically. Roger and Bill had known each other for many years, which may have been part of the reason that they did not handle the issue in a more formal way. Had Roger known that Bill did not intend to pay to repair the entire wall, he might have been more careful about how much he promised to do to satisfy the homeowner.

(b) Post-Filing of Litigation

There was some pretrial testimony and exchange of relevant documents. One of the important pieces of evidence that came to light was that the drawings done by Roger for the project contained the letters "DFPT" next to the drawing of the stakes. Everyone, including Roger, agreed that this stood for "Douglas Fir Pressure Treated." According to this detail of the drawing, Roger had not even followed his own specifications. He claimed he thought the product he purchased from Bill was pressure-treated, and that the difference between Douglas fir and pine was not significant. He also stated that he thought soak treatment and pressure treatment were the same process.

One offer of settlement, in the amount of $3,000, was made by Bill to Roger. That offer was rejected, and the matter proceeded to trial. Both sides put on expert witnesses. Roger's expert testified that the stakes provided were inadequate for the intended purpose. Bill's expert testified that he thought the stakes were appropriate for that use. The judge admitted evidence of Bill's settlement offer, and found for Roger. The judge concluded that the standard of care for a seller should not differ simply because the buyer was, in essence, an expert in the field, and was more knowledgeable than the seller, regarding application of the product. Although the judge's conclusion was not logical, it would not be disturbed by an appellate court unless the judge had abused his discretion or committed a gross error of law. Rather than spend the money to appeal, Bill paid the judgment.

Analysis. The trial result in this case is a strong argument for the wisdom of settling cases before trial. A critical issue in the trial was the judge's admitting evidence regarding Bill's offer to settle. Most judges would not have allowed evidence on this issue to be admitted. As a general rule, offers to settle are not admissible, because of a public policy of encouraging settlement. If every defendant knew that an offer to settle would be admitted at trial, no offers would ever be made.

In this case, the judge said that the statement by Bill that he would take care of the wall was not an offer to settle, but a new contract, which Bill failed to perform by not paying the money to repair the wall. Statements, actions, or failures to act by parties to contracts can be used by courts to interpret whether contracts exist, and the terms of those contracts.[18]

Bill's attorney tried to argue that, if the offer to pay for the cost of the wall was a new contract, the contract failed because an essential element of the contract was not agreed on: the price. The judge refused to accept this argument, and the evidence was admitted, severely damaging Bill's defense.

* * *

In the above case, breach of contract, negligence, and strict liability were all alleged as legal theories of recovery. In more complex cases, different legal theories may be necessary because there are multiple factual causes of damage, such as faulty equipment, negligent supervision, and failure to enforce safety regulations.

* * *

Case 7.4
(a) Prelitigation

At a site where a large office building was being erected, some of the laborers were required to use scissor-lifts to install rivets in the steel girders, at approximately third-floor height. A worker performing this task was supposed to move the scissor-lift into place with the lift fully compressed. The worker would raise

---

[18] For example, in Giem v. Searles, 470 S.W.2d 327 (Ky. 1971), the making of two progress payments, despite a lack of a showing that the work complied with the contract terms, demonstrated to a court's satisfaction that the contractor was entitled to payment despite the noncompliance. The court noted that the cost estimate originally submitted was 10 percent of the actual prices. During the project, the actual prices were billed on an open account and those prices were paid by the general contractor. In another case, a contractor's failure to object to direction for additional work by an owner provided a basis for a court to find that the contractor agreed with the owner's interpretation of the required work. Bulley v. Andrews, Inc. v. Symons Corp., 25 Ill. App. 3d 696, 323 N.E.2d 806 (1975).

his cage to the appropriate height, lock the transmission from the cage, and perform the necessary riveting. The worker would then cause the cage to descend. He repeated the sequence in each location where he was installing rivets.

One of the laborers, Harold, was impatient with the length of time it was taking him, using the approved procedure, to perform the riveting. The project was behind schedule, and he wanted to speed up the process. Instead of lowering the cage after he had riveted in one location, he decided to manipulate the scissor-lift forward to the next location, 10 feet away, with the cage in the raised position. After he had done this on three occasions, he moved to the fourth location and began riveting. At that point, the lift suddenly jerked forward. He was thrown from the lift and severely injured. Harold could not recall whether he had touched the controls before the lift jerked forward.

The lift came to rest some distance away, after crashing into one of the exterior girders of the steel skeletal structure of the building. The manufacturer of the lift, the prime contractor, and the workers' compensation carrier for Howard's employer were unable to agree on a settlement.

Analysis. The primary issue in this case is the pressure put on workers to get work finished on delayed projects within deadlines that may be unrealistic. When the pressure to complete work results in unsafe construction practices, contractors risk heavy liability for worker injuries.

(b) Post-Filing of Litigation

The claim against the prime contractor was that the prime had failed to properly supervise Harold's work, and that Harold should not have been allowed to perform the riveting without lowering the cage. The claim against the lift owner, which had rented the lift to the prime contractor for the project, was that the lift had been improperly maintained, allowing an electrical short that caused the lift to move forward by itself. The claims against the lift manufacturer were that the buttons that operated the lift were defective, and that the controls were defectively designed or manufactured, allowing some sort of electrical short, which caused the sudden forward movement that injured Harold.

After numerous depositions, the parties were unable to agree on a settlement for Harold. The manufacturer and the lift owner claimed that Harold had used the lift improperly by moving it forward without retracting the cage mechanism, and that he had not set the brake properly before beginning his riveting. Another allegation was that Harold had actually touched the button that propelled the lift forward. In testing similar scissor-lifts, Harold's experts were unable to "recreate" an electrical short that would reproduce the forward movement that had occurred in the accident without any of the buttons being activated.

Although the parties were ultimately able to agree on an amount that represented compensation for Harold's injuries, they could not agree on their proportional contributions to a settlement.

The attorneys hired a retired judge to conduct a private settlement conference. The judge asked that counsel for each party, and a representative of each party, also be present. The retired judge asked each party to present arguments regarding each party's position on liability and damages in this case. The judge then asked to speak to each party separately, with each party's counsel and insurance adjuster present. The judge discussed some of the merits of each party's case, in these private sessions, and offered his opinion on the likelihood of success for each party's argument. He also offered his opinion of the best and the worst that Harold could recover if the case went to trial. After approximately 7 hours of this mediation process, the parties agreed to an overall settlement.

Analysis. Product defect claims are difficult to prove at trial; the technical complexities are not easily explained to a jury. Multiple causes may have contributed to a construction accident. In this case, the fact that the experts for the plaintiff were unable to recreate the electrical malfunction, which they believed had caused the lift to jump forward, posed a problem for the plaintiff's case against the manufacturer and the owner of the lift.

The design and shape of the handles, and the location of the controls, were important factors as to whether it would be easy for a person to mistakenly cause the lift to move forward when raised. Safety might require a lock-out device, so that the lift could not be moved forward when the cage was raised.

Some safety engineers might believe that the cage should have contained a dead man's pedal similar to those on locomotives, so that the lift could not have been moved forward at the same time as another operation was being performed, such as the riveting.

Product defect cases often call for expert testimony by engineers or by others who have special knowledge. The basic test applied is whether that testimony will help the trier of fact to understand the case and to reach a verdict. In other construction areas, the facts of each case may be central in determining whether expert testimony is necessary. In a claim against an architect for professional negligence, expert testimony will be required unless the conduct challenged can be evaluated using ordinary human experience and knowledge.[19]

The rule applied in many states regarding liability for unsafe or defective products is articulated in Section 402A of the *Restatement of Torts,* as described above.[20]

---

[19] Nelson v. Commonwealth, 235 Va. 228, 368 S.E.2d 239 (1988). However, when an architect sued for fees in Hess v. Zoological Society of Buffalo, 134 A.2d 824, 521 N.Y.S.2d 903 (1987), no expert testimony was allowed because the court found (1) that the language of the contract was unambiguous, and (2) that the issue in the case was interpretation of the contract, which was a matter of law for the court to determine.

[20] Cases applying these standards include: Vandermark v. Ford Motor Co., 61 Cal. 2d 256, 391 P.2d 168, 37 Cal. Rptr. 896 (1964); Bexiga v. Havir Mfg. Corp., 60 N.J. 402, 290 A.2d 281 (1972).

In some cases, purchasers of services, rather than products, have attempted to hold the sellers of services liable under a strict liability standard. Typically, these cases have not been successful.[21]

The retired judge who settled this case had previously served in the geographical area. The attorneys representing the parties in this action respected his opinion, and this respect was a strong contributing factor to the settlement.

* * *

§ 7.17 Vicarious Liability

Vicarious liability is the concept that one person may be held liable for the acts of another, although the person being held liable has not personally committed any negligent or intentional acts giving rise to liability, other than employing a person who commits a tort. The most common situation where this principle is applied is where an employer is held vicariously liable for the acts of his employee. This is also called **respondeat superior**. The primary justification for imposition of vicarious liability on the employer is that the employee was acting as an agent for the employer at the time the act giving rise to liability was committed.

The rationale for imposition of liability under this theory is that the employer was deriving a benefit from the acts of the employee and should also bear the risk of harm caused by the employee. Imposition of this rule also satisfies a public policy concern that an employer be careful about whom he chooses to work for him, providing an incentive for the employer to develop and implement procedures for ensuring that his employees are careful.

The traditional rule was that a person was not liable for the acts of independent contractors working on his behalf. This rule has been largely eroded. In many situations, the more control the principal exerts over the independent contractor, the greater the likelihood that vicarious liability will be imposed on the principal (or employer) on whose behalf the independent contractor was working.

In addition to vicarious tort liability, an employer may be directly liable where an employee commits acts that cause damage. This type of liability is usually called **negligent employment** or **negligent entrustment**. For example, a cement contractor hires a driver who has had five vehicle accidents within the past year. If the driver is carrying cement to a construction site, and causes an accident, the employer may be directly liable. He knew the driver was unsafe

---

[21] *See* Schipper v. Levitt & Sons, Inc., 44 N.J. 70, 207 A.2d 314 (1965); *but see also* Kriegler v. Eichler Homes, Inc., 269 Cal. App. 2d 224, 74 Cal. Rptr. 749 (1969); Wright v. Creative Corp., 30 Colo. App. 575, 498 P.2d 1179 (1972), *reh'g denied* (June 6, 1972). An article that explores this subject in detail is *Comment, Strict Liability in the Building Industry,* 33 Emory L.J. 175 (1984).

and he should not have had the driver working in that capacity; or, he should have removed the driver once he learned of the driver's prior accidents. If the employer is vicariously liable, a finding of direct liability on the part of the employer may not be important. However, if the allegations of vicarious liability cannot be proven, it may be important to establish direct liability under a theory that the employer himself was negligent for employing the individual as a driver at the time of the accident.

§ 7.18 Tort Defenses

For each tort theory asserted by a plaintiff, a variety of defenses may be asserted by each defendant. The most common defense is that of contributory or comparative negligence, that is, the defendant was one of the causes of the plaintiff's damage, in whole or in part.

§ 7.19 —Comparative or Contributory Negligence

In the not too distant past, under the theory of **contributory negligence**, a plaintiff was prevented from receiving any recovery if the plaintiff himself were negligent. Even if a defendant's acts were 99 percent of the cause of the plaintiff's harm, and the plaintiff's acts were only 1 percent of the cause of the harm, no recovery would be allowed because it would be said that the plaintiff was guilty of contributory negligence and therefore did not deserve to recover. Some jurisdictions attempted to modify this rule by establishing that, if the plaintiff was less than 50 percent responsible for the injury, he could recover for the remainder. Because of the hardship this rule worked on injured plaintiffs, most jurisdictions now use a rule called **comparative negligence**. Under the rule of comparative negligence, even if the plaintiff is 99 percent negligent, he can collect the dollar amount represented by the remaining 1 percent from the defendant or defendants who are responsible.

§ 7.20 —Assumption of the Risk

In cases brought under a strict liability theory, contributory negligence (usually now called comparative negligence) is not usually available as a defense, because negligence of either the plaintiff or the defendant is not relevant. However, a defendant in a strict liability case can use as a defense the fact that the plaintiff **assumed the risk** of the harm that occurred. The defendant's argument is that the plaintiff proceeded to act in the face of a known, recognized, and obvious danger, and therefore the plaintiff does not deserve to be compensated.

If a laborer is told not to go into a basement to perform work because there are dangerous levels of radiation, but the laborer ignores the risk and proceeds into the basement, he would be said to have assumed the risk of being injured by being exposed to the radiation in the basement.

A determination of whether the defense of assumption of the risk will be successful in any given case will depend on the specific facts of each case. A critical issue may be how clearly and definitely the plaintiff actually understood the extent of the risk posed. Under some circumstances, a person may be injured, but other factors are the cause of his injury. In the example above, if the laborer is exposed to harmful radiation by proceeding into the basement and he is injured, he can be said to have assumed the risk of that occurrence. He has not, however, assumed the risk that the basement will contain other dangers, such as a pit bull dog that might bite him. The owner of the dog may still be strictly liable for the injury caused by the dog's biting the laborer, if the plaintiff went into the basement and was not warned of the risk of being bitten by the dog.

§ 7.21 —Immunity from Tort Liability

In a variety of situations, parties may be immune from liability for torts. Employers are immune from suit by employees for tort damages (except under very limited circumstances) because the workers' compensation system is the sole remedy for an employee injured while working or acting for his employer.

Traditionally, government entities could not be sued, because of the legal doctrine of **sovereign immunity**. The basic rule of sovereign immunity is that the government cannot be sued, except in those situations where it consents to be held liable. The passage in 1946 of the Federal Tort Claims Act gave persons the right to sue the federal government, with two exceptions: (1) no suit could be brought for liability based on the commission by government officials of certain intentional torts, and (2) the federal government could not be sued in tort for acts that involved performance of certain discretionary government duties.

In *Dalehite v. United States,*[22] tort liability was not allowed against the government for conduct based on a policy or planning decision. *Dalehite* also stated the rule that the government could not be held liable unless it has been shown to be negligent. This means that no strict liability, or liability without fault, may be imposed on the government.

Because of criticism of the rule of sovereign immunity, more than one-third of the states have abolished it. Many of these states have also passed legislation similar to the Federal Tort Claims Act, giving citizens the right to sue the state government, except under certain circumstances.

---

[22] 346 U.S. 15, 73 S. Ct. 956, 97 L. Ed. 1427 (1953).

§ 7.22 —Sovereign Immunity and the Construction Process

Local government agencies are heavily involved in modern-day construction. They check drawings and specifications, and they issue building permits. The government agencies would prefer immunity from tort liability for any mistakes they may make as a part of these procedures. However, a recent case in Florida refused to allow immunity to a government agency for negligent checking of drawings and specifications. In *Manors of Inverrary XII Condominium Association v. Atreco-Florida,*[23] review and approval of plans and specifications prior to issuing a building permit and conducting on-site inspections prior to issuing a certificate of occupancy were "operational activities." A municipality was not protected by doctrine of sovereign immunity from liability for negligence of a building inspector in approving plans, specifications, and construction that did not comply with building codes.

A supreme court decision in California in 1961 abolished unlimited sovereign immunity in that state.[24] However, design liability presents particular problems in construction. Evaluations of cost-effectiveness play a large part in design decisions, and knowledge regarding the safety of construction design is continually evolving. Certain statutes were enacted in California in 1963, to balance the relevant interests regarding design liability of the state.[25]

If a person is injured by a product defectively designed by the federal government, or by a contractor hired by the federal government, the federal government is immune from suit. The contractor or manufacturer who designed the product for the government also is likely to claim as defenses: (1) the contractor followed the government specifications (if the contractor followed apparently adequate government specifications,[26] and (2) the **government contract defense**, which gives the government contractor the same immunity based on sovereignty as is allowed to the government.

Government contractors performing shipbuilding during World War II claimed these defenses in many of the cases brought by shipyard workers for the asbestos exposure that had occurred during that time. The shipyards maintained that they had followed the specifications set out by the Navy in building the ships, which included using an overwhelming number of asbestos-containing products. Evidence strongly suggests that the shipyards knew of the health hazards associated with asbestos during this time. The success of this defense may depend on whether the shipbuilder followed the specifications exactly, and

---

[23] 438 So. 2d 490 (Fla. App. 1983).

[24] Muskopf v. Corning Hosp. District, 55 Cal. 2d 211, 359 P.2d 457, 11 Cal. Rptr. 89 (1961).

[25] *See* West's Anno. Cal. Govt. Code § 830.6 (1992).

[26] Boyle v. United States Technologies Corp., 487 U.S. 500, 108 S. Ct. 2510, 101 L. Ed. 2d 442 (1988), *reh'g denied,* 109 S. Ct. 1182 (1989).

whether the government can be shown to have been as aware of the dangers of asbestos exposure as were the shipbuilders.

* * *

Case 7.5
(a) Prelitigation

An amusement park was constructed adjacent to a state highway. The property was located in a 100-year flood zone which means that, historically, every 100 years, the strongest storm likely to occur in a 100-year period would cause flooding to the zone. Large-scale flooding occurred one winter, because of the severity of heavy rains. The state highway police diverted water from the state highway, which resulted in flooding to the amusement park. The park also suffered flooding because of lack of maintenance of nearby storm drains and culverts, which were controlled by local and state agencies.

The amusement park presented an administrative claim (a prerequisite to filing a civil suit) to the local and state government agencies responsible for maintenance of the culverts and storm drains, and against the state highway police for diverting the water onto the property containing the park. The claim stated that the park should be compensated for the damages suffered to the property because of the negligence of the named agencies. The administrative claims were denied, and the park filed a suit against the government agencies.

Analysis. Public entities, in addition to sovereign immunity, have a broad ability to take actions, free of liability to the public, based on the constitutional grant of **police power**. This enables public agencies to take actions to protect the public health and welfare using a variety of means, even if those means result in damage to some component of the public. For example, during the 1906 earthquake and fire in San Francisco, firemen dynamited houses that were standing in the path of the fire, in an attempt to prevent the spread of the fire to other homes. Some homeowners sued the city for the destruction of their homes. Those suits were unsuccessful because courts determined that San Francisco was properly exercising its police power to take steps to protect the general health and welfare of the inhabitants of the city.

Under certain circumstances, if less restrictive means of achieving the same protection of the public are available, public entities may be liable if they elect to take more drastic steps. In the 1906 San Francisco fire, if hosing down the homes had been possible and would have achieved the same end, without destroying the homes, that option might have provided a basis to support recovery by the owners whose homes were deliberately destroyed by the city.

However, the more exigent the circumstances, the more likely it is that the public agencies will be allowed wide latitude in determining what are the appropriate measures to take.

(b) Post-Filing of Litigation

After suit was filed in the amusement park case, the government agencies asserted sovereign immunity as a defense, and alleged comparative responsibility on the part of the owner of the property: the owner had chosen to locate his amusement park in a flood zone. The owner had not provided any barriers to protect his property from flooding from the highway, because the owner wished motorists to be able to view the amusement park as they passed. If the owner of the property had erected walls, they would have protected the land from flooding, but the miniature golf course and other attractions of the park would not have been visible from the highway. In fact, the exits from the highway were placed after the park came into view, to allow those attracted by the view of the park to have easy access from the highway.

At trial, the court denied the claim of the owner of the property. The court based its decision on the sovereign immunity defense of the government, the police power claim of the government agencies to act as they deemed fit in the face of public danger (the flooding of the highway), the owner's comparative negligence in locating his park in the flood zone, and the owner's failure to take greater precautions to protect his property from flooding.

Analysis. There was substantial evidence in this case for finding comparative negligence by the owner. The owner chose to locate in a flood zone. He also made a conscious determination not to protect the property from flooding from the highway, in order to increase the visibility of the amusement park to passing motorists. The sovereign immunity and police power defenses were also persuasive, because the public agencies have to balance their obligation to the public at large (the motorists on the highway) with their responsibility to an individual member of the public, such as the park owner.

The issue of the duty owed by public agencies was addressed in a recent Louisiana construction case, *Sunlake Apartment Residents v. Tonti Development Corp.*[27] An architect being sued for malpractice, based on the existence of building code violations, was held entitled to bring a third-party action against the state. An apartment building had burned down. The architect was sued by the residents for his design, which violated fire codes. The architect sued the state for approving his plans, which failed to meet code requirements. The state argued that it had a duty to the general public, and that this duty did not extend to the architect, who was an individual. The *Sunlake* court found that the public duty defense did not apply where a public official had a specific public duty. The case was remanded for a determination of the facts, to decide whether the state had a specific duty to ensure the safety of buildings for which it issues permits.

[27] 522 So. 2d 1298 (La. App. 1988).

The state also contended it was immune from liability under the theory that review of the architectural plans was a discretionary act. Legislation passed subsequent to approval of the plans appeared to provide immunity for the defendants. Although the legislative amendment was retroactive, the court ruled that it only granted immunity to the state for policy-making or discretionary acts. The court found the state had no discretion as to whether to review the plans, and therefore no immunity existed.[28]

* * *

§ 7.23 Tort Remedies

The most common remedy for torts, as for breach of contract, is money damages. Under special circumstances, equitable relief, such as injunctions, may be available, as outlined in § **5.39**.

§ 7.24 —Compensatory Damages

Money damages for those injured because of the tort of another are based on the amount of money that will fully compensate the plaintiff—that is, put the plaintiff in the situation he was in before the loss. For a personal injury, the total amount of compensatory damages would be an amount calculated by adding together the medical bills, loss of income, degree of temporary or permanent disability, and an amount to compensate the plaintiff for the pain, suffering, and emotional distress endured because of the injury.

For a tort that causes purely economic losses, the money damages awarded are an amount that would compensate the plaintiff for those economic losses, plus any sufficiently foreseeable and provable lost profits.

§ 7.25 —Punitive Damages

Civil law is primarily concerned with compensating those parties who are damaged, not with punishing wrongdoing. However, if the acts of the defendant are

---

[28] Regarding liability for negligent inspections, *see also Note, Local Government Liability in Virginia for Negligent Inspection of Buildings, Structures and Equipment,* 18 U. Rich. L.R. 809 (1984). Regarding negligent issuance of building permits, *see Comment, Municipal Tort Liability for Erroneous Issuance of Building Permits,* 58 Wash. L. Rev. 537 (1983); Holman v. State, 53 Cal. App. 3d 317, 124 Cal. Rptr. 773 (1975); Morris v. County of Marin, 18 Cal. 3d 901, 559 P.2d 606, 136 Cal. Rptr. 251 (1977) (injured worker allowed to sue county, which had issued building permit to worker's employer which, in violation of state law, had failed to file workers' compensation certificate).

sufficiently repugnant, and deliberate or intentional behavior by the defendant is proven and has harmed a plaintiff, **punitive damages** might be awarded. These are damages not intended to merely compensate the plaintiff for his damage suffered because of the acts of the defendant. Punitive damages represent an amount of damages sufficient to punish the defendant, or to deter such future conduct of that defendant or of others who are similarly situated.

If an employee of a subcontractor is injured because the superintendent on the job (a representative of the prime contractor) did not enforce safety regulations, the worker can receive compensatory damages. However, if the prime contractor has a history of failing to enforce safety regulations, and numerous other workers have been injured as a direct result of the prime contractor's deliberate or negligent failure to enforce safety regulations, punitive damages might be awarded. The case for an award of punitive damages might be even stronger if various other workers had complained about the condition, and the superintendent for the prime contractor willfully and knowingly ignored the risk posed by the failure to follow the safety regulations.

Determining the amount of punitive damages presents problems. One way of determining the appropriate amount of punitive damages is to examine the outrageousness or "willfulness" of the defendant's behavior. Another approach is to examine the defendant's net worth and calculate what dollar amount will be sufficient to punish the defendant, deterring him from behaving in a like manner in the future. In theory, the amount of compensatory damages should not in any way determine or limit the punitive damage award. In practice, if a punitive damage award exceeds the amount awarded for compensatory damages by too great an amount, a judge is likely to reduce as excessive the amount of the punitive damages award.

There are those who criticize the jury system, claiming that juries are swayed by passion or prejudice, which causes them to award excessive punitive damage awards. Such critics ignore the fact that most large punitive damage awards are later reduced in amount. Either the judge refuses to allow the jury's verdict to stand, or the defendants appeal the large awards, and a later, far smaller, settlement is negotiated between the plaintiff and defendant.

§ 7.26 Conclusion

Many disputes or lawsuits that arise in the construction context are based on noncontractual duties. Tort claims of negligence, fraud, trespass, or strict liability may be made even where parties have a contract that describes their duties and obligations. These tort claims provide alternative, noncontractual, legal theories under which the plaintiff may recover.

In some circumstances, claims must be made based on tort theories because no contract describing the relationship of the parties exists. Such claims can arise between different subcontractors on a project, or between an owner and a subcontractor, or for an owner who buys a defective house from someone other than the original developer who built it.

For third parties to the project, such as adjacent landowners, or members of the public who are damaged by actions of those in the construction process, tort claims may be the only legal theory of recovery.

Where liability is joint and several, one defendant who has assets may end up paying an entire claim, even though the defendant is responsible for only a small percentage of the damage to the plaintiff.

In most jurisdictions, a plaintiff is not precluded from bringing a suit because he himself was negligent. In a claim based on a defective product, however, the defendant can protect himself if he can successfully prove the plaintiff assumed the risk of the harm that damaged the plaintiff.

Common criticisms of the tort system are that the system is too expensive and makes American businesses uncompetitive. It is important to realize, however, that, without many of these legal theories, the majority of valid construction claims would go uncompensated.

CHAPTER 8

INSURANCE

§ 8.1 Introduction

Insurance is an important and integral part of allocation and avoidance of construction risks. Many construction contracts require that prime contractors or subcontractors obtain insurance policies to cover risks during a construction project. These include policies of comprehensive or commercial general liability insurance, builder's risk insurance, all-risk insurance, and professional liability insurance.

General liability insurance is intended to cover the risk that a company may incur liability for some act that is incidental to its primary business. In other words, a framing contractor who has general liability insurance will be covered if one of its workers accidentally drops a beam onto property neighboring the job site and damages a structure.

Builder's risk insurance is intended to protect a contractor or owner from some of the risks that may cause losses during the construction of a project. If the structure is destroyed by a fire or other natural disaster during construction, this insurance is intended to protect the policyholder from having to absorb that loss.

Professional liability insurance is intended to protect clients of professionals, or third-party beneficiaries of those professional services, against damages resulting from nonintentional acts of the professionals that cause damage.

When a project is being built, if poor architectural design causes increased cost to the owner and the contractor, both the owner and the contractor may file claims against the architect under the architect's professional liability policy. The owner hired the architect, and the architect therefore owes the owner a contractual duty of care. The contractor is both a third-party beneficiary of the architect's contractual acts for the owner, and a claimant against the architect under a theory of negligence.

Secondary types of insurance include excess and umbrella insurance policies. These types of insurance are intended to protect a party from the risk that claims against other insurance will exceed the policy limits of the primary insurance. Given the uncertainty of economic conditions, an additional consideration is that secondary insurance will cover claims if the primary company goes bankrupt or is out of business at the time a claim is made. Under normal circumstances, these secondary types of insurance do not usually pay for claims until all primary insurance limits are exhausted.

Applicable statutes may require construction entities to obtain liability insurance, unemployment insurance, and workers' compensation insurance. The obligation to obtain these types of insurance may also arise from contract clauses specifying that the prime contractor or subcontractor is to comply with all applicable codes, ordinances, and statutes.

States do not have laws requiring that professionals carry professional liability insurance. However, clients of design professionals may insist on such insurance. Many design professionals carry professional liability insurance to protect their assets and to protect those who may be damaged by their work.

§ 8.2 Conditions Precedent for Effectiveness of Insurance

The insurance premium, or a portion thereof, must usually be paid before an insurance policy will become effective. Delivery of the policy terms to the insured (the person or persons covered by the insurance) may also be required to activate the policy of insurance. If a contractor arranges for insurance through a broker, but the broker fails to secure the insurance, no coverage exists, although the contractor will have a right to sue the broker for professional negligence.[1]

The insurance policy may not be effective if the applicant fails to provide the insurance company with information requested by the insurance carrier. The policy may be initially effective, but may be rescinded by the insurance carrier if it discovers that an applicant made false representations on the application, concerning a material fact on which the insurance carrier relied in issuing the policy.[2] In the event of a loss where a material misrepresentation was made, a carrier may rescind the policy and refund the premium, or it may exclude the claimed loss from coverage. If the policy is rescinded for misrepresentation, the insured may find it very difficult in the future to secure other liability insurance.

Construction contract terms may also specify which insurance policies are primary, which are secondary, and under what circumstances claims may be made under different insurance policies. For example, if the prime contractor on a project has general liability insurance, and the prime requires all the subcontractors to obtain insurance, the subcontract may specify that the subcontractors' insurance policies are primary. The prime contractor's insurance policy will only be used after all available subcontractor insurance has been exhausted.

Determining what insurance covers a given loss, or whether any insurance covers a loss, depends on the terms or conditions that trigger coverage.

§ 8.3 "Occurrence" Policies

Occurrence policies cover acts and omissions that occur during the period of the policy. If a contractor obtains a comprehensive general liability (CGL) policy for the period January 1, 1991, to January 1, 1992, any accident or property damage that occurs during that 12-month period will be covered, if all other relevant conditions are met.

Occurrence policies are almost unavailable, particularly for professional liability policies. Insurers maintain it is too difficult for them to calculate an appropriate premium based on the risk, for example, that an act performed in 1991 will result in a claim in 1997. **Claims made** policies are therefore the most common type of insurance available for professional negligence.

§ 8.4 "Claims Made" Policies

Claims based on acts performed by those in the construction field may not be made until many years later. With a claims made policy, coverage exists for all

---

[1] Couch on Insurance 2d §§ 25:34–25:46 (2d Ed. 1983 & Supp. 1987).

[2] Alperin v. National Home Life Assurance Co., 32 Ill. App. 3d 261, 336 N.E.2d 365 (1965).

claims made during the policy period. For example, a design professional has been in business since 1977. For the first five years of his practice, he was covered by Home Insurance. From 1982 to 1992 he was covered by Transamerica Insurance. Both of these policies were claims made policies. If a claim was made in 1984 for an act committed in 1978, Transamerica, not Home, would have been the applicable insurance carrier. If both policies had been occurrence policies, Home would have been the carrier to cover the loss.

Most courts have sustained the validity of claims made policies, although, in *Jones v. Continental Casualty Co.,*[3] a New Jersey court found the claims made clause in a policy void because retroactive coverage for prior acts resulting in present claims only existed if the insured had a policy for the previous time period with the same company that was insuring the risk at the time the claim was filed.

Some hybrid policies require (1) that the claim be made during the policy period and (2) that the act for which the claim is made must have occurred in the policy period.

§ 8.5 Requirement of an Insurable Interest

To recover for a loss under an insurance policy covering damage to property, a person must have an insurable interest in the damaged or destroyed property. When a contractor is engaged on a construction project, the contractor has an insurable interest in the property on which the project is located, as long as the contractor is working on the project and has not turned control of the premises over to the owner. The contractor's insurable interest in the property may be terminated, depending on the facts of a case, by the owner's acquisition of insurance for the property.[4] Abandonment of the contract by the contractor, by mutual consent or by reason of one of the parties' exercising a right under the contract, may terminate the contractor's insurable interest.[5] However, a defaulting contractor retains an insurable interest and continues to have a risk of loss if damage occurs at the property subsequent to his default.

§ 8.6 Types of Insurance

A wide variety of types of insurance policies is available. Many are tailored to specific needs of those involved in construction.

---

[3] 123 N.J. Super. 353, 303 A.2d 91 (1973).

[4] American & Foreign Insurance Co. v. Allied Plumbing and Heating Co., 36 Mich. App. 561, 194 N.W.2d 158 (1971) (termination not found).

[5] U.S.F.& G. v. Rob Homes, Inc., 323 So. 2d 105 (Miss. 1975).

§ 8.7 —Comprehensive General Liability

Comprehensive general liability (CGL) insurance covers claims, brought against contractors, that are related to performance under the construction contract. The CGL policy does not provide a warranty of the quality of the work of the contractor. It covers claims for bodily injuries and consequential property damage, and is not intended to cover ordinary business losses. Common exclusions in a CGL policy are for work products, for property in the control of the insured, and for work performed by or on behalf of the insured.

The work product exclusion, usually found in a general liability or builder's risk policy, means that the product produced by the work of the contractor is not covered. In other words, there is no insurance coverage to pay an owner who receives a structure with defective framing. The owner will have to sue if he wishes to have the contractor pay for the differential between acceptable framing and the damaged framing, or for the cost of repairing or redoing the framing, unless that work is covered by other insurance or by a performance bond.

Insurance exclusions for property controlled by the insured, or for work performed by the insured, are common exclusions because the insurance is not intended to warrant the quality of the contractor's work or to cover foreseeable risks that are with the control of the insured. General liability insurance is intended primarily to cover unforeseeable risks such as accidents in which people are injured or natural disasters in which property is destroyed.

Most cases have held that the negligent work of contractors or subcontractors is not covered by CGL policies.[6]

Some cases, however, have allowed coverage for negligent work. In *Colard v. American Family Mutual Insurance Co.,*[7] coverage for negligent work was allowed under a CGL policy despite the presence in the insurance policy of a work product exclusion. In *Emasco Insurance Co. v. L&M Development, Inc.,*[8] coverage for negligent work was allowed because of an implied warranty exception to the work product exclusion.

The length of time during which coverage is effective under a CGL policy is often an issue. The standard language in most CGL policies provides that coverage under the policy is terminated after final payment is made for the construction. Broad-form property damage endorsements are available to cover the property after construction is over. These endorsements may also be obtained to cover the quality of the work of subcontractors. Court interpretation of policy language may differ greatly from state to state. For instance, broader coverage

---

[6] Reliance Insurance Co. v. Mogavero, 640 F. Supp. 84 (D. Md. 1986); Wechsler v. Beall & Associates Roofing Corp., 442 So. 2d 1104 (Fla. App. 1987); Vernon Williams & Son Construction Inc. v. Continental Insurance Co., 591 S.W.2d 760 (Tenn. 1979).

[7] 709 P.2d 11 (Colo. App. 1985).

[8] 372 N.W.2d 908 (N.D. 1985).

for risks was not granted under this type of an endorsement in *Knutson Construction Co. v. St. Paul Fire and Marine Insurance Co.*[9]

* * *

Case 8.1
(a) Prelitigation

A prime contractor agreed to build a guest house on land also occupied by the luxury residence of a wealthy owner. When the owner presented building plans to the prime contractor, the prime contractor noted that the plans had not been approved by local planning and building agencies. The owner told the prime contractor to proceed and said that he, the owner, would obtain the necessary approvals. In fact, the owner never obtained these approvals.

In addition, the contractor knew that the guest house violated applicable zoning. Zoning requirements mandated that the structure not contain living space. The contractor knew it would be used for residential purposes, because he installed plumbing and electricity consistent with use as a guest house. During the construction phase of the guest house, after the roof had been installed, it was discovered that the height of the second story of the guest house was approximately 6 inches shorter than required by applicable building codes. It cost the prime contractor approximately $60,000 to raise the ceiling of the second story.

As an inducement to the contractor to continue to perform the work, the owner told the prime contractor that the owner would have the prime contractor build a home on an adjacent lot after the guest house project was completed. Counting on the promised future project to absorb this present additional roofing cost, the contractor did not ask for additional payment for the roof-raising work. He also did not ask for payment for many other items for which he believed he was entitled to be paid beyond the contract amount.

After he excluded the above items, when all the work had been performed on the guest house, the prime contractor still had change orders and extras outstanding from the owner in the amount of $40,000. One subcontractor hired by the prime remained unpaid, in the amount of $15,000, and filed a mechanic's lien action against the property.

Analysis. This case points squarely to the responsibility of a contractor to ensure that the plans and specifications he uses are in compliance with local ordinances and regulations. The contractor knew that the owner was building a guest house, which violated zoning. He built a structure that contained plumbing and electricity, which indicated that the structure could be used as a guest house. It is difficult for any contractor to defend himself in court when he has failed to comply with basic building codes. Maintaining that the contractor

---

[9] 396 N.W.2d 229 (Minn. 1986).

failed to comply because of direction by the owner is unlikely to persuade a court that the contractor should not pay the owner for the cost of bringing the structure into compliance with building codes.

(b) Post-Filing of Litigation

The subcontractor's mechanic's lien claim was filed in municipal court, because of the amount involved ($15,000). The response of the owner to the subcontractor's claim was to retain a large law firm and remove the case to superior court. The owner cross-complained against the prime contractor for $300,000 in damages suffered by the owner, because of the structure's lack of code compliance. Numerous defects were claimed as to the work performed by the prime contractor.

The prime contractor had some insurance when the project began, but then had a lapse of insurance. Subsequently, he obtained an insurance policy with a liability limit of $300,000. Both insurance companies refused to defend the prime contractor, based on the policy's exclusions for defective work. Each also claimed, as an additional fact defeating coverage, that some of the work was performed outside the effective dates of the insurance.

Eventually, the prime contractor settled the matter by paying off the subcontractor's mechanic's lien claim and making a small settlement with the owner.

Analysis. It is risky for subcontractors or prime contractors to try to financially recover from an unduly low bid by means of change orders or requests for extras. These demands for additional compensation beyond the contract amount are resisted vigorously by owners, who feel they have agreed to pay a certain price, and who do not want to be pressured into paying additional money to complete the project.

Contractors seeking additional money for extras should maintain extensive documentation, to avoid being forced to accept losses for the extra work performed as part of the cost of doing business. Occasionally, litigating disputes that involve large amounts of money, where the contractor has strong evidence of entitlement to these claimed amounts, may also be necessary to maintain a position of strength in negotiating future projects and payment. If the other parties know that a particular party is serious about recovering money to which that party is entitled, litigating one matter may enable that party to settle five others for a fair amount.

* * *

§ 8.8 —Builder's Risk Insurance

Builder's risk insurance covers risks beyond those of a conventional CGL policy. It covers ordinary risks to the structure and to those who come into contact with the structure. It also covers the risk of damage to the contractor's tools and equipment. Other items that may be covered by builder's risk insurance include: loss of structures during the course of construction, loss of landscaping, recovery for the cost of cleanup of debris in the event of destruction of the structure, loss of business income, cost of preparation of new blueprints, and cost of a temporary storage site.

§ 8.9 —Professional Negligence Policies

Insurance policies that cover the professional negligence of design professionals are distinguishable from other insurance policies in construction. A significant distinction between these policies and others used in construction is that they *are* intended to cover the beneficiaries of the policies for negligence of the professional. CGL policies for contractors do not cover the owner for negligence of the contractors.

When a design professional is sued for negligence, all available insurance policies should be consulted to determine whether any policies cover the acts alleged in the pleadings. In *Comstock Insurance Co. v. Thomas A. Hanson & Associates, Inc.,*[10] the architect represented to the owners that certain work could be performed at a certain site, for less than a specified amount. These representations were untrue. The architect was found liable for breach of contract and negligence. The insurer tried, however, to claim that no coverage existed for the losses.

The policy language stated that it covered all sums that the insured (the architect) became legally obligated to pay as damages, as a result of claims made against the architect in the performance of, or failure to perform, professional services. Excluded from coverage were express warranties, guarantees, estimates of probable construction, or overages of cost estimates. The court held that coverage existed for the acts of the architect because negligence was one of the causes of the loss.

It should also be noted that other negligent acts of architects, or other design professionals, may be covered under a policy for professional negligence, despite this type of exclusionary clause. Before concluding that coverage does not exist for any particular loss, careful examination of the policy language, coverage, and exclusions should be conducted. Claimants should consult an attorney familiar with the law of the particular state to ensure that possible coverage is not overlooked.

---

[10] 77 Md. App. 431, 550 A.2d 731 (1988).

§ 8.10 Format of a Typical Insurance Policy

The first page of an insurance policy is usually called the **declarations sheet**. This page will be either fully or partially typed out, in contrast to much of the rest of the policy, which is usually a preprinted standard form. The first page will describe the location of the property, identify the insured, define the policy period, list the policy limits for the different types of coverage, state the amount of the premium to be paid and the applicable deductible amount, and identify the endorsement pages that apply to the policy. Endorsement pages are the method insurers use to add or delete certain types of coverage, or to exclude unusual risks.

Most policies contain a definitions section, which lists the meaning intended by the insurer for words or phrases that will be used in the policy. In some situations, the insurer may intend that a common English word will have a different meaning than is usually assigned to it. Definitions are listed for words or phrases such as "occurrence," "loss," "insured," "structures," "property," "property damage," and "accident."

Generally, policies will state affirmatively those risks that are covered, and will specifically identify the risks that are excluded. The affirmative representations are likely to be contained in a section that identifies the property to be covered and the type of losses that will be covered. The excluded risks, or causes of loss that are excluded, may be listed separately. Certain items of personal property or real property may also be excluded from coverage. Additional endorsements, or extra insurance coverage, may be available from an insurance company. For example, most current policies exclude from covered risks any claims for the cost of cleanup of toxic waste or pollutants. Some insurance companies may provide coverage for this type of risk upon payment of an additional premium.

Insurance policies are likely to contain an arbitration provision. The arbitration provision commonly uses language such as the following:

> In case we fail to agree with you as to the meaning or effect of any provision of this policy, said disagreement shall be resolved by binding arbitration in accordance with the statutory rules and procedures of the state in which the property is located or in accordance with the Commercial Arbitration Rules of the American Arbitration Association. Should an arbitration under the terms of this clause result in a determination in your favor, we will reimburse you for expenses actually incurred by you with respect to the arbitration, including reasonable attorney's fees, in a sum to be determined by the arbitrator(s). The provisions of this clause are mandatory and may be enforced either by you or by us.

The issue of whether such a binding arbitration provision in an insurance policy will be enforced varies from state to state. Some states have a strong policy of encouraging parties to dispose of their disputes without going to court. Enforceability of the clause may also depend on whether the insured actually saw the policy language before the insurance contract was effective, and on the relevant judicial attitude toward insurance company contracts.

The general rule of interpretation of any contract is that ambiguities will be interpreted strictly against the drafter. This rule of contract interpretation is commonly applied to insurance contracts, which are usually prepared by insurance companies without input or alteration possible by consumers. A more critical level of scrutiny may be applied to those clauses in a contract that deprive a party of its right to take a dispute to court, because the party is being deprived of a constitutional right to due process and judicial relief.

§ 8.11 Policy Exclusions

Most insurance policies used in construction exclude contractually assumed risks. In other words, if a contractor agrees in the construction contract to ensure that all hazardous waste is handled properly, but fails to do so, a general liability policy will not cover this type of claim.

Policies covering design professionals for negligence may exclude coverage for work not customarily performed by the design professional as a part of his normal duties. For example, if an architect provides legal advice to an owner, his professional liability insurance would not cover any damage resulting from those services. Architects should refer an owner to an attorney when the owner asks for advice that is legal in nature.

Where tasks performed by a design professional or contractor are not normally a part of the tasks usually done by that person, such as legal advice by an architect, careful attention should be paid to policy language. If coverage is desired for work not covered under the standard language, it may be necessary to obtain special endorsements, or to obtain specially drafted policies to cover the proposed activities.

§ 8.12 Insured's Duties under Policy

If a party is notified of a claim or potential claim resulting from work on a construction project, the party should inform his insurance broker of the potential liability. The broker will then either notify the insurance carrier or have the party speak directly to a claims representative of the company. In some cases, the claims representative will take information from the insured, but will take no further action to investigate the claim until the party is served with a lawsuit. This delay is based on the fact that most policies provide the insured with a defense when legal actions are filed, but do not cover mere claims. The filing of the action against the insured triggers the obligation of the insurance company to take formal steps to protect the rights of the insured. It is not uncommon, however, for a carrier to conduct some investigation of a known claim prior to a lawsuit being filed against the insured.

Insurance policy language usually states that notice of a claim or potential claim is to be given to the insurer "as soon as practicable." The purpose of this notice is to allow the insurer the opportunity to evaluate the claim and to assemble the evidence necessary to defend a lawsuit. Even if the insured believes that the claim is unlikely to result in liability, failure to notify the carrier of a claim may cause a court to relieve the insurance carrier of the obligation to cover the loss under the policy.[11]

The portion of the policy that describes the duties of the insured following a loss usually also provides that the insured is to provide documents and information pertaining to the claim, and that the insured is to cooperate in any reasonable manner with the requests of the insurance company. The insured may have the obligation to file a written report of the facts and circumstances surrounding the claim, or to file a detailed, sworn statement of loss, within a specified period of time following the loss.

A claims adjuster may ask the insured to give a tape-recorded statement to a claims representative, or to submit to questions and provide answers to the insurance company in front of a court reporter, in a manner similar to a deposition. An insured should consult with an attorney before providing any detailed information to the insurance company. This is particularly true if the insured is asked to provide orally any information that will be tape-recorded, or transcribed by a court reporter. If an insured is not properly advised of the effect of providing such information, he could find himself defeating his own insurance coverage, while remaining personally liable for losses.

It is also common for policies to include a provision that the insured is obligated to assign, or **subrogate** (pass on) its right to make a claim against third parties to the insurer that has paid for an insured's loss. For example, if a prime contractor recovers under his CGL policy for property damage resulting from the negligence of a subcontractor, the CGL carrier for the prime will probably take steps to seek reimbursement from the negligent contractor. The prime contractor will have waived any right to recover from the negligent subcontractor by accepting the coverage under his own policy, except to the extent that the loss was not covered by the prime's insurance. The prime will continue to have a right to sue for his deductible under his policy, or for any other sums that were lost because of the culpable acts of the third party and were not covered by his insurance.

§ 8.13 Limitation of Action under Policy

Another standard clause in insurance policies is a statute of limitations. For claims based on a contract, the period allowed is usually shorter than for the legislative statutes of limitations. A typical statute of limitation for a written

---

[11] Empire City Subway Co. v. Greater New York Mutual Insurance Co., 35 N.Y.2d 8, 315 N.E.2d 755, 358 N.Y.S.2d 691 (1974).

contract is four years. Insurance policies usually shorten that period with this type of language:

> No suit, arbitration, or other action on this policy for recovery of any claim shall be sustainable in any court or other forum unless all the requirements of this policy shall have been complied with and unless commenced within twelve (12) months after the inception of the loss.

Courts have commonly upheld the shorter statutes of limitations contained in insurance policies. Disputes may arise regarding the definition of "inception" of the loss, because the insured may not have notice of the claim until years after the policy has been in effect. It is not uncommon, when an insured notifies an insurance carrier of a claim, for the carrier to list among its reasons for denying coverage or defense the fact that more than 12 months have elapsed since "inception" of the claim.

Alternatively, insurance companies may provisionally agree to defend and provide coverage for a claim, under what is called a **reservation of rights**. This means that defense and coverage of the claim are provided but they are subject to a later denial of coverage and defense. Under these circumstances, if the insurance carrier later determines that there was no obligation to defend or to cover the loss, the insurer may request reimbursement from the insured for the amounts the insurer has paid for defense or settlement.

§ 8.14 Insurer's Duty to Investigate

The insurance company has a duty to investigate claims pertaining to those it insures. The insurance company also has an express or implied duty to expeditiously negotiate a reasonable settlement within the policy limits. Failure to investigate claims adequately, or to pay meritorious claims promptly, has resulted in insurers being held liable to their insureds and third parties for breach of "the covenant of good faith and fair dealing."[12] These types of lawsuits are also called **bad faith** suits.

In 1979, in the case of *Royal Globe Insurance Co. v. Superior Court,*[13] the California Supreme Court found an insurance company liable for punitive damages for bad faith, based on violations of Insurance Code regulations that set out the duties of insurance companies. Following that decision, many cases were filed in California, as well as in other states, both by policyholders and by third parties (plaintiffs in the underlying tort action) for breach of the covenant of good faith and fair dealing. Numerous multimillion-dollar verdicts were awarded to plaintiffs. The basis for third-party bad-faith suits was that third

---

[12] Gruenberg v. Aetna Insurance Co., 9 Cal. 3d 566, 510 P.2d 1032, 108 Cal. Rptr. 480 (1973).

[13] 23 Cal. 3d 880, 592 P.2d 329, 153 Cal. Rptr. 842 (1979).

parties were deemed to have rights as beneficiaries to the insurance contract between the insured and the insurer.

Between 1979 and 1988, many third-party plaintiffs used the right set out in *Royal Globe* to sue insurers for bad faith, based on violations of the duty set out in the Insurance Code. The usual claim was that the insurer had failed to settle a claim for a reasonable amount when liability was clear, thereby causing damage to the third party.

However, in 1988, a more conservative California Supreme Court decided the case of *Moradi-Shalal v. Fireman's Fund Insurance Cos.*[14] In that case, the court specifically overruled *Royal Globe's* finding that third parties had the right to bring actions for bad faith against insurers, based on violations of the Insurance Code. The court held that the insurance regulations were enacted to allow the insurance commissioner to regulate the conduct of insurance companies, but were not intended to provide private plaintiffs with a tort cause of action.

The *Moradi-Shalal* court stated that a claim by third parties for bad faith against insurance companies continues to exist, based on the common-law theory of bad faith. However, the message of *Moradi-Shalal* in California, and to other parts of the country as well, has been that insurance companies risk little by delaying settlement. Third-party bad-faith suits against insurers have largely disappeared in California.

The majority of states do not allow third parties damaged by activities of insurers to sue insurers directly for bad faith.[15]

Because insurance policy forms are complex and are drafted by insurance companies, the clauses or exclusions set out in the policies will usually be strictly interpreted against the insurance companies. Most courts will find that coverage exists where the policy language regarding coverage or exclusions is ambiguous.

§ 8.15 Duty to Defend

Comprehensive general liability (CGL) policies and professional negligence policies, as well as most other insurance policies, specify that the insurer is obligated to provide the insured with a defense when a claim is made that is

---

[14] 46 Cal. 3d 287, 758 P.2d 58, 250 Cal. Rptr. 116 (1988).

[15] States allowing, other than California, include: *Pennsylvania,* Shearer v. Reed, 428 A.2d 635 (Pa. Super. 1981); *Montana,* Lough v. Insurance Co. of North America, 789 P.2d 576 (Mont. 1990). States refusing to allow third parties to sue insurers for bad faith include: *Louisiana,* Security Insurance Co. v. Deshotels, 458 So. 2d 186 (La. App. 1984); *Illinois,* McCarter v. State Farm Mutual Insurance Co., 130 Ill. App. 3d 97, 85 Ill. Dec. 416, 473 N.E.2d 1015 (1985) (rule referred to in dicta; this case allowed action against insurer on fraud cause of action); *Florida,* Cardenas v. Miami-Dade Yellow Cab Co., 538 So. 2d 491 (Fla. App. 1989); *Texas,* Hart v. Aetna Casualty & Surety Co., 756 S.W.2d 27 (Tex. App. 1988).

covered under the insurance policy. Because lawsuits may allege liability under a variety of different legal theories, or causes of action, if the insurer has the obligation to defend an insured under one cause of action in a lawsuit, the insurer must defend on all causes of action asserted. The duty to defend is also broader than the duty to indemnify or cover losses.[16] Some states, however, allow an insurance carrier to provide a partial defense.[17]

The duty to defend a suit or claim is triggered only by those actions that arise when coverage is in effect, and for those losses that are covered under the policy language or are not specifically excluded in an all-risk policy. In the event that the carrier refuses to defend a matter, the insured should secure counsel and defend the action. It is unlikely that the insured would be allowed to sue the carrier for the damages if the insured let the matter go by default.

The actions that can be taken by an insured where the insurance company refuses to defend or pay for defense costs, and such defense is appropriate, are the following: (1) a court may be asked to enter successive orders that the insurance company pay for defense costs, (2) a court may be asked to grant an injunction ordering the insurer to pay defense costs as they are incurred, (3) a court may be asked to order specific performance (an order by a court that a party perform a specific act other than the payment of money).

Although an insured can wait until after the claim or lawsuit has been resolved to sue the insurer for defense costs, this delayed demand for reimbursement has been determined to be an inadequate remedy for the insured.[18]

A learned treatise on insurance law has stated that the duty to defend becomes "useless and meaningless," if the insured is unable to obtain immediate enforcement of the contractual obligation to defend, and if the insured is required to wait for reimbursement.[19]

§ 8.16 Other Insurer Duties

As a general rule, an insurer has the obligation to advise insureds of the existence of coverage under different or alternative provisions of the insurance policy. Because of the superior level of sophistication of insurance claims adjusters, as compared to the typical consumer of insurance, some states describe the relationship of insurer to insured as a "special" relationship that imposes a fiduciary relationship on those representing the insurance company.

---

[16] Gray v. Zurich Insurance Co., 65 Cal. 2d 263, 54 Cal. Rptr. 104, 419 P.2d 168 (1966); Couch on Insurance § 51:47 (Rev. Ed. 1982 & Supp. 1987).

[17] *See* Couch on Insurance, § 51:51 (Rev. Ed. 1982 & Supp. 1987).

[18] Gon v. First State Insurance Co., 871 F.2d 863 (9th Cir. 1989); *see also* Gray v. Zurich Insurance Co., 65 Cal. 2d 263, 54 Cal. Rptr. 104, 419 P.2d 168 (1966); Jarchow v. Transamerica Title Insurance Co., 48 Cal. App. 3d 917, 122 Cal. Rptr. 470 (1975).

[19] Appleman, 7C Insurance Law and Practice § 4684 at 83 (Berdal Ed. 1979).

The significance of this special relationship is that it imposes the highest possible standard of care on the insurance representatives.

§ 8.17 Coverage Issues

Factual disputes are common between insurance companies and their insureds regarding coverage. In *Domingue v. H&S Construction Co.*,[20] a member of the public was injured when she slipped and fell. She alleged the cause of her fall was a guidestring left by a paving contractor who was hired by a city. The plaintiff contended that the guidestring constituted a dangerous condition. The insurance company for the contractor refused to defend the case for the contractor. The court ruled the insurance company was required to pay for the cost of defense, plus interest and penalties, as well as to pay for the attorney's fees incurred in the defense, because the contractor was found to be negligent, and negligent acts were covered under the policy.

Twenty-five years ago, most comprehensive general liability (CGL) insurance policies covered loss or damage to an insured for pollution or toxic waste. Beginning in the late 1970s, and continuing into the early 1980s, insurance companies became more aware of the magnitude of the risk posed by providing coverage of toxic waste discovery and disposal. In response, the insurance companies drafted policy language for the CGL insurance policies that attempted to exclude coverage for slow leaks of toxic materials, but still covered insureds for a sudden or accidental discharge of toxic materials.

Today, insurance policies typically exclude coverage for any type of toxic waste damage. In recent years, insurance companies have fought with insureds about whether an insurance policy that calls for coverage for certain types of toxic waste problems is intended to cover the circumstance where an insured is required to perform cleanup of toxic waste. The insurance companies have argued that the insurance policy language is intended to cover "damages." They maintain that the term damages covers only those situations where the insured has to pay another party money for the damage caused by toxic waste. The insurers argue that a governmental order to clean up toxic waste is not a claim for damages, and therefore no coverage exists under the policies for this risk.

Some courts have interpreted the word "damages" narrowly, holding that it only applies to money awards in contested lawsuits for money damages. Other courts have expanded the definition of "damages" in insurance policies to include cleanup costs. A few recent cases suggest factors that will be considered in the future.

In *Aerojet General v. Superior Court*,[21] the United States Environmental Protection Agency (EPA) brought an action against Aerojet under the provisions of

[20] 546 So. 2d 913 (La. Ct. App. 1989).

[21] 211 Cal. App. 3d 216, 257 Cal. Rptr. 621 (1989), (Aug. 10, 1989).

the Comprehensive Environmental Response, Compensation, and Liability Act (CERCLA)[22] and other federal statutes. The EPA alleged that it was entitled to recover from Aerojet costs it had incurred in eliminating present and future environmental and health hazards created by Aerojet practices. The federal district court agreed with the plaintiff EPA and ordered Aerojet to pay for those costs.

Aerojet then sued its own insurers for reimbursement for the ordered payment of the cleanup costs. Aerojet based its claims against the insurers on the damages clauses in the comprehensive general liability (CGL) policies. The policies provided for payment of sums that the insured became legally obligated to pay as damages because of injury to, or loss, destruction, or loss of use of property. The policy did not define damages. The trial court held the insured's CGL policies did not cover environmental cleanup and restoration costs imposed by the federal and state governments. This decision became final when the California Appellate Court refused to rehear the case.

However, the result in the above case was flatly contradicted and reversed by the California Supreme Court in the case of *AIU Insurance v. Superior Court.*[23]

In *AIU,* the insured had been ordered by state and federal agencies to clean up toxic waste. The insured asked for reimbursement for these costs under the damages provisions of several insurance policies. The policies stated that the insurer was to pay "on behalf of the Insured all sums which the insured shall become legally obligated to pay as damages because of injury to or loss, destruction or loss of use of property." Some of the policies added to "damages" advertising liability, "ultimate net loss," and expenses for litigation, settlement, adjustment, and investigation of claims.

On November 15, 1990, the California Supreme Court unanimously rejected the trial court's holding that damages did not include restitutionary relief or the cost of compliance with mandatory injunctions sought by government agencies under CERCLA. The California Supreme Court ruled that the cost of government-ordered cleanup constituted covered damages under the comprehensive general liability (CGL) policies. The California Attorney General's Office, which had filed a brief in the case as a "friend of the court," welcomed the ruling, stating that this decision will help state and federal governments enter into settlement agreements providing for private money to be used for cleanups rather than public funds.[24]

This 1990 decision followed the existing California Supreme Court bias toward finding coverage for those insured. The issue of recovery for cleanup costs

---

[22] 42 U.S.C.A. §§ 9606(a), 9607 (West 1991).

[23] 213 Cal. App. 3d 1219, 262 Cal. Rptr. 182 (1989), *rev'd,* 51 Cal. 3d 807, 274 Cal. Rptr. 820 (1990).

[24] *Los Angeles Daily Journal,* Nov. 16, 1990.

for negligent pollution is resolved in California, when these types of policies and the same language are used.

However, insurance companies in California may be able to assert a defense to coverage or liability under the "damages" clauses in CGL policies, if the insurer can show that the pollution was willful, not accidental or negligent. That apparently provided the rationale for a California Superior Court jury to conclude that there was no insurance coverage in a subsequent case involving Aerojet and toxic waste.[25]

In the latter case, there was no dispute as to the basic facts. Aerojet's test site for rocket engines and fuels had been contaminated over a period of decades with a variety of toxic substances. These substances had infiltrated an estimated trillion gallons of groundwater. Estimates for cleanup costs ranged up to $500 million.

Aerojet alleged the insurers were liable, because the pollution was neither expected nor intended by Aerojet. The insurers argued that there should be no insurance coverage for the cleanup costs, because the pollution was the *foreseeable* consequence of improper hazardous waste disposal techniques, rather than a result of a sudden accident. The jury found against Aerojet, and in favor of the insurance companies.

Other courts, using a strict interpretation of the word "damages," have interpreted the insurance policies as excluding the cost of government-ordered cleanups. In *Maryland v. Armco, Inc.*,[26] the policy language in issue defined damages as sums where the insured has been "legally obligated to pay damages because of injury or destruction to property." The *Armco* court defined "damages" in a legal, technical sense, and concluded that the cost of government-ordered cleanup was not covered by the policy. The court described the cleanup measures as "prophylactic," and unconnected to harm to specific third parties.

In *Mraz v. Canadian Universal Insurance Co., Ltd.*,[27] the policy in issue defined occurrence as "an accident, including injurious exposure to conditions, which results during the policy period, in bodily injury or property damage, neither expected nor intended from the standpoint of the insured." The *Mraz* court found no duty to defend the insured from a suit for cleanup costs brought by the government. There was no "occurrence" as defined by the policy, the court said, because discovery of the damage occurred outside the policy period. The court also held that recovery for response costs alleged in the government complaint were not "property damage" as defined by the policy language.

---

[25] Aerojet General Corp. v. Transport Indemnity Insurance Co., No. 262425, Calif. Daily J., Jan. 14, 1992.

[26] 822 F.2d 1348 (4th Cir. 1987), *cert. denied,* 484 U.S. 1008, 108 S. Ct. 703, 102 L. Ed. 2d 43 (1988).

[27] 804 F.2d 1325 (4th Cir. 1986).

§ 8.18 Use of Contract Clauses Requiring Naming as Additional Insured

Contract clauses may also require that owners be named as additional insureds under the prime contractor's general liability insurance policies, and that the prime contractor be named as an additional named insured under subcontractors' CGL policies. If the insurance carrier believes the addition of a named insured increases the risk for the carrier, the insurer may charge an increased premium.

§ 8.19 Failure to Obtain Insurance as Required as Contract Breach

If an owner fails to insist on the contractor's providing proof of insurance prior to allowing the contractor to begin performance, the owner may be precluded from using the contractor's failure to obtain insurance as the basis for a breach of contract claim. In *Whalen v. K-Mart Corp.,*[28] an owner allowed a subcontractor to work on a project without enforcing a condition of the contract that required the subcontractor to provide proof of insurance. The subcontractor performed his work and was paid by the owner. When the owner sued the contractor, one of the owner's claims was for breach of contract, based on the subcontractor's failure to provide proof of insurance before beginning work on the project. The court denied the claim for contract breach. The owner was presumed to know about the lack of insurance, because reasonable diligence would have brought the matter to the owner's attention. The owner's failure to insist on the proof of insurance effectively waived the contract requirement to provide proof of insurance.

If an insurance broker is to obtain insurance from a given carrier for an insured and fails to do so, the insured may have a claim against the broker for professional negligence. If the broker handled insurance only for that company, or if the broker has authority to bind the company to a policy without approval on an application, the insurance company may be required to provide the insurance. If the broker is an independent broker who handles other insurance companies as well, or who does not have the authority to bind a company to a policy without further approval, the insured may have a claim only against the broker.

§ 8.20 Conclusion

Insurance is essential to any modern construction project. Insurance for various parties in the construction process may be required by law, or by provisions of

---

[28] 166 Ill. App. 339, 519 N.E.2d 991 (1988), *cert. denied,* 121 Ill. 2d 587 (1989).

the contract, or both. When contractual provisions require that insurance be obtained, and that proof of insurance be provided, they should be strictly enforced by owners and prime contractors in order to avoid a later court finding that these conditions have been waived.

If parties to the construction process plan to allocate responsibilities in a manner that varies from the traditional roles, insurance policies should be scrutinized carefully to avoid the risk that insurance may not cover losses. Specially drafted binders or endorsements may be necessary to ensure that coverage exists for foreseeable risks. Although the cost of obtaining special insurance may be high, that cost should be balanced against the danger of liability posed by foreseeable uncovered losses, and the likelihood that such losses will occur.

Contract clauses commonly specify that certain policies or types of insurance shall be primary, and that others will be secondary, in the event of specified losses. It is important to review language in insurance policies that describes which policies will be primary in the event of losses.

Insurers are obligated to investigate claims thoroughly, and to defend and indemnify insureds if coverage is found to exist. Insurers who fail to meet these contractual obligations may be subject to tort or contract liability and to the payment of punitive damages, which may far exceed the amount involved in the claim.

In the event of a loss, all insurance policies covering aspects of the construction process should be reviewed to determine whether coverage exists.

CHAPTER 9

BONDS IN CONSTRUCTION

§ 9.1 Role of Bonds in Construction

Bonds protect the public from risks posed by the work of subcontractors, material suppliers, and laborers on public projects. By using bonds, private owners and other parties to private construction can proceed with projects with a degree of protection and security.

The Miller Act[1] requires government prime contractors to obtain performance bonds to protect the government, and payment bonds to protect suppliers of labor and materials to public projects. Under the Act, the contracting officer establishes the amount that is believed to be adequate. As in the private sector, this figure is generally the same as the contract amount.

Some states have enacted laws that permit use of construction bonds; others have enacted statutes that require the use of bonds. These rights may be a substitute for mechanic's lien rights, or may exist in addition to mechanic's lien rights. Court opinions in states that require bonds have occasionally held the statutes to be invalid. Consultation with an attorney in the state where a project is located, regarding the applicable statutes governing bonds or the legality of a contract requirement that a bond be obtained, is essential if this issue arises.

[1] 40 U.S.C.A. § 270a *et seq.* (West 1991).

Persons who supply labor or materials for a construction project are entitled to recover under the bond for the owner's benefit if the bond contains conditions for their benefit and protection. This right arises because those supplying labor or materials are third-party beneficiaries under the bond.[2]

Subcontractors and suppliers can recover if the bond contains an express statement of their right to recover. Conversely, if the bond language expressly states that it is not for the benefit of third parties such as subcontractors or suppliers, the intent of the parties to the bonding agreement is likely to control, and no recovery will be allowed.

§ 9.2 Role of Sureties in Construction

Sureties should carefully monitor the progress of each project for which they hold bonds, to avoid losses. They also must be aware of and knowledgeable about the laws of each relevant jurisdiction, to avoid forfeiting rights or risking liability. Various provisions of the bankruptcy laws will be relevant if the sureties are required to pursue defaulting contractors who may file for protection under bankruptcy laws to avoid performance or to avoid indemnifying the sureties for construction defaults.

§ 9.3 Performance Bonds

Most construction contractors are required to provide performance bonds to ensure completion of the contractor's work on a construction project. The performance bond provides the owner with a promise from the surety that the surety will either complete the work of the principal contractor under the bond, or will pay the owner for the cost of completion of that work.

The Miller Act[3] requires performance bonds for any public contract that exceeds $25,000 in amount, for the construction, alteration, or repair of any public building or public work of the United States. Most other public entities also require performance bonds. For private work, owners will demand a performance bond from contractors. The AIA form of performance bond, contained in AIA Document A312 (1984) is widely used by private owners. (See **Appendix E**.)

AIA Document A312 (Performance Bond) provides that, if the contractor performs the construction contract, the surety and the contractor shall have no

---

[2] *See* D&L Building Inc. v. Maltby Tank & Barge, 747 P.2d 517 (Wyo. 1987) (legislative intent was that benefit of contractor's bond was not limited to those in privity with the principal); for similar holding, *see* Lozier Brothers, Inc. v. Fremont County School District, 747 P.2d 515 (Wyo. 1987).

[3] 40 U.S.C.A. §§ 270a–270d (West 1991).

obligation under the bond, except to participate in conferences in the event that the owner is considering declaring a contractor default (paragraphs 2 and 3.1). The owner may not declare termination of the contractor for default under paragraph 3.2 until 20 days have elapsed since the owner gave notice of intent to declare a default under paragraph 3.1. Once termination is declared, the owner must agree to pay the balance of the contract price to the surety, or to a contractor selected by the owner to complete the work, in accordance with the terms of the contract the owner had with the original defaulting contractor (paragraph 3.3).

After the above steps, the surety is to arrange (1) for the contractor, with the consent of the owner, to perform and complete the construction contract, or (2) for the surety to undertake to complete the contract itself, or through its agents, or (3) for the surety to obtain bids or negotiate proposals to complete the work (paragraphs 4 through 4.4).

The surety may also waive its right to perform and complete, by investigating and determining the amount it owes the owner and tendering payment to the owner (paragraph 4.4.1). Alternatively, the surety may deny liability for the principal contractor under the bond, and notify the owner of its reasons for doing so (paragraph 4.4.2).

Fifteen days after the surety receives from the owner the notice of the contractor's termination for default, the surety itself will be deemed in default, and the owner may enforce any remedy available to him (paragraph 5).

Any action in equity or law under the bond may be brought in a court in the location where the work is being performed. Action must be brought within two years after the contractor's default, or within two years after the contractor ceased work on the project, or within two years after the surety refused or failed to perform its obligations under the bond. If these provisions are held to be void, the applicable limitation of actions will be that which is applicable to sureties as a defense in the relevant jurisdiction (paragraph 9).

Other sources for performance bond forms are Standard Indemnity Agreements, such as those referred to in Construction Control Bond Indemnity Agreement Provisions.[4] Standard provisions in those agreements are similar to those contained in the AIA Performance Bond. Many of these provisions are statements of the surety obligations and rights contained in various relevant state codes.

In some circumstances, a court may refuse to enforce a contractual requirement that a performance bond be posted. In *Terra Group Inc. v. Sandefur Management, Inc.*,[5] a contractor was the lowest qualified bidder for work on a project, and was accepted verbally by the developer. The contractor's surety subsequently refused to provide a performance bond because the surety believed the developer did not have the financial strength to pay the contractor the

---

[4] MacNamara and Mungall (Defense Research Institute, Inc., 1975).

[5] 527 So. 2d 849 (Fla. Dist. Ct. App. 1988).

amounts called for under the contract. The developer then refused to sign the formal contract with the contractor. The court determined that a valid contract existed from the time the developer verbally accepted the bid, regardless of the surety's refusal to provide a performance bond.

§ 9.4 Payment Bonds

Payment bonds perform a function similar to that of performance bonds, except that they are intended to secure the payment of suppliers or subcontractors in the construction project, rather than to secure the performance of work.

The AIA form for payment bond is also contained in AIA Document A312 (1984). This document binds the contractor and the surety to the owner, to pay for labor, materials, and equipment that are furnished for use in performance of the construction contract (paragraph 1). The obligation of the contractor is fulfilled if the contractor pays claimants directly or indirectly, or if the contractor indemnifies and holds the owner harmless from claims, demands, liens, or suits for labor, materials, or equipment furnished for use in the construction project (paragraphs 2.1 and 2.2).

The surety has no obligations under the bond until the claimants who are employed directly, or who have a direct contract with the contractor, (1) give notice to the surety, which notice sets out the amount of the claim, and (2) send a copy of the claim or notice to the owner that they are making a claim under the bond (paragraphs 4 and 4.1).

The surety's obligation under the bond cannot exceed the amount of the bond, and the amount of the bond is to be credited for any payments made in good faith by the surety to relevant parties under the construction contract (paragraphs 7 and 9).

Suit or legal action may not be brought under the bond if more than one year has elapsed since (1) the date on which the claimant gave notice required under the bond, or (2) the last date on which labor or service was performed by anyone, or (3) the last date on which materials or equipment were supplied under the construction contract (paragraph 11). In the event that the statute of limitations set out in the bond is void, the applicable limitation for actions shall be the minimum period of limitation available to sureties as a defense in the jurisdiction where suit would be brought (paragraph 11). However, statutes of limitations in bond agreements are typically upheld.[6]

---

[6] *See* Ribeira & Lourenco Concrete Construction Co. v. Jackson Health Care Associates, 231 N.J. Super. 16, 554 A.2d 1350, *cert. granted,* 563 A.2d 826 (1989), *aff'd,* 571 A.2d 1311 (1990).

§ 9.5 Licensing Bonds

Most states require that contractors be licensed to perform construction. In order to obtain a contractor's license, the contractor will be required to post a bond for a small amount of money, typically $5,000. California required most contractors to post a bond in the amount of $2,500 until 1980, when it was raised to $5,000. Swimming pool contractors must post a bond in the amount of $10,000.[7]

In the event of a dispute with a contractor, an owner may be able to make a claim against the contractor with the licensing authority. This claim could result in the contractor's losing his license and forfeiting his bond. If a lawsuit is filed against a contractor, one of the legal theories asserted in the complaint is likely to be a claim against the contractor's bond. Unfortunately, recovery of the full bond amount will not provide satisfactory compensation for any except the smallest construction disputes. In addition, if a contractor has a number of claims filed against him, even that small amount may be unavailable to an owner who is not first in line to claim against the licensing bond.

Recovery under a licensing bond may be denied, depending on the type of claim. In *All Bay Mill & Lumber Co. v. Surety Co.*,[8] a supplier who was not paid by a contractor was denied recovery under the bond because the supplier failed to establish that it was damaged due to a willful and deliberate violation of state licensing law by the contractor.

§ 9.6 Surety Defenses

After learning of the default of a contractor under a payment or performance bond, the surety must investigate the claim, and will explore all possible defenses to the claim. Verification that the claim is appropriate under the terms of the bond is an obvious first step. Establishing that the claim arises from the project for which the bond was obtained is also important.

The surety may have defenses to a claim under the bond. The parties have the obligation to keep the surety aware of facts and circumstances that arise during construction and may forseeably cause problems. For example, in *National Union Indemnity Co. v. G.E. Bass & Co.*,[9] an argument was successfully made that the owner/obligee under the bond made payments to the principal under the bond, which materially altered the risk secured by the bond. These facts were held to successfully relieve the surety of its obligation under the bond. Material

---

[7] Cal. Bus. & Prof. Code § 7071.6 (Deering 1992).

[8] 208 Cal. App. 3d 11, 255 Cal. Rptr. 790 (1989).

[9] 369 F.2d 75 (5th Cir. 1966).

alterations to the scope of the work or method of payment may also allow a surety to avoid its obligation.

Where an engineer obtained a performance bond for a contract with an owner, and then entered into a separate subcontract with a subcontractor, a court held that the performance bond only covered the first contract with the owner, and not the second with the subcontractor.[10] The court stated that the additional risk that arose as a result of execution of the subcontract was not covered, because the engineer had not informed the bondholder of the subcontract and the engineer had not obtained the bondholder's consent to the subcontract.

Depending on the circumstances, common contract defenses such as fraud, duress, illegality of the contract, or impossibility of performance under the construction contract may all be viable arguments that the surety owes no obligation under the bond.

§ 9.7 Bankruptcy Problems for Sureties Relating to Defaulting Contractors

Bankruptcy laws provide that any action or proceeding against a bankruptcy debtor is prohibited, unless the creditor receives a determination from the bankruptcy court that grants relief from the automatic stay.[11] Problems that may be faced by sureties under a construction contract are complicated and expensive to resolve, because of the complexity of bankruptcy laws and the high cost of specialist legal fees.

As a general proposition, once an entity or person files for and receives protection under the Bankruptcy Code, executory (not yet performed) contracts of the debtor are subject to rejection, assumption, or assignment, depending on which of these courses of action the trustee in bankruptcy believes would be to the benefit of creditors.[12] The trustee is likely to attempt to complete those contracts that appear to be profitable. If the trustee estimates incorrectly, the surety's exposure under the bond may be heightened, because it may have to pay amounts beyond the bond to complete the contract.

§ 9.8 —Performance Bonds and Bankruptcy

In the case of bankruptcy by a contractor under a performance bond, the surety may attempt to enter the bankruptcy proceedings to establish (1) that the trustee cannot provide adequate assurance that it can perform particular contracts,

---

[10] Employer's Insurance v. Construction Management Engineers, 297 S.C. 354, 377 S.E.2d 119 (Ct. App. 1989).

[11] 11 U.S.C.A. § 362 (West 1991), Rule 7062; Fed. R. Civ. P. 62.

[12] 11 U.S.C.A. § 365(b)(1) or 365(f)(1) (West 1991).

(2) that the court should reject the trustee's attempt to assume contracts, because such assumption would burden the bankruptcy court, and (3) that it is in the best interest of all concerned that the surety be allowed to take over the work of the debtor. The surety is likely to argue that, if the trustee is allowed to assume the burden of the contractor's performance, mistakenly believing the surety can complete the contract profitably, the trustee will dissipate the funds of the bankrupt contractor and cause delays that will increase the cost to the surety of honoring its obligation under the bond. In bankruptcy proceedings, the trustee may try to secure contract balances and retainages that are, arguably, due the debtor, but that have not yet been received by the debtor at the time of the bankruptcy filing. This course of action by the trustee may also prejudice the rights of the surety. *Pearlman v. Reliance Insurance Co.*[13] held that the surety has a common-law right to contract balances upon completion of the work and its payment of subcontractors and material suppliers. Once bankruptcy has been filed, all legal actions by or against the debtor are stayed (halted). In *Pearlman,* because the court held the amounts due the bankrupt contractor under the contract were not a part of the debtor's estate (assets subject to control by the bankruptcy court), the surety could pursue its rights to secure the contract balances without violating the automatic stay of the bankruptcy court.

However, in *In re Glover Construction Co.,*[14] the bankruptcy court found that progress payments owed to the contractor by the owner on an uncompleted contract were property of the estate under Section 541 of the Bankruptcy Code. Section 541(a)(1) broadly defines the estate property as all legal or equitable interests as of the commencement of the case. The court's determination on this issue will depend largely on the facts of each case.

Another issue may arise regarding the right of the surety to advance payments that have been made to the bankrupt contractor. The line determining which advance payments are property of the bankruptcy estate, and which advance payments may be obtained by the surety, may rest on whether the construction contract was substantially completed by the defaulting contractor. In *Balboa Insurance Co. v. United States,*[15] the court determined that, if the project was unfinished at the time notice of default was received, the surety may be able to obtain the advanced funds.

Although the surety may wish to wrest control of the construction performance from the trustee and the defaulting contractor, the surety may need substantial cooperation from the defaulting contractor. The surety may be unable to assume control of the project immediately, and it may be impossible to complete the work without help from the defaulting contractor. It may be helpful for the surety to negotiate an agreement by which the contractor will provide the surety

---

[13] 371 U.S. 132, 83 S. Ct. 232, 9 L. Ed. 2d 190 (1962).

[14] 30 Bank. 873 (W.D. Ky. 1983).

[15] 775 F.2d 1158 (D.C. Cir. 1985).

with consulting or administrative services during the period of completion. The profits derived by the contractor from the agreement will become a part of the bankruptcy estate, when performed after filing of bankruptcy. However, even after negotiating and executing such an agreement, the contractor may reject the agreement, and the bankruptcy trustee may be successful in rejecting a consulting agreement as executory, with the result that the agreement would not be enforceable.

§ 9.9 —Payment Bonds and Bankruptcy

Payment bonds under bankruptcy law are treated in a similar manner to performance bonds under bankruptcy law. Payment bond sureties take over the rights of a secured lender of the principal under the bond.[16] The payment bond surety also has priority over the trustee in bankruptcy of the debtor's estate.[17]

Because the surety steps into the shoes of the defaulting principal under the payment bond, the surety may obtain rights to contract funds. Some laws allow the surety to obtain contract funds that were owed to persons supplying labor and materials in connection with public works.[18] The time for a surety to claim rights under these statutes must be carefully examined for each state. For a surety to be able to demand contract funds, many statutes call for presentation of these claims in substantially shorter time frames than the time frames allowed for presentation of a bond payment claim. The above circumstance is by no means an exhaustive description of the problems encountered when a contractor defaults under a performance bond and subsequently files for bankruptcy. The complexity of the issues that can arise demonstrates the necessity for a surety, working with a defaulting contractor who has filed for bankruptcy, to seek the advice of an attorney familiar with both the field of construction and relevant bankruptcy laws, as well as with the contract laws of the applicable jurisdiction.

Consideration of these problems also underlines the fact that judicious selection of contractors, and familiarity of all concerned with the terms of the construction contract and the terms of the performance bond, may avoid expensive entanglement with bankruptcy proceedings.

---

[16] Framingham Trust Co. v. Gould-National Batteries, 307 F. Supp. 1008, 427 F.2d 856 (1st Cir. 1970).

[17] 31 U.S.C.A. § 3727(c) (West 1991).

[18] *See* Ind. Code. Ann. § 5-16-5-1 (Burns 1987); Mass. Gen. Laws Ann. ch. 30, § 39F (West 1979).

§ 9.10 Conclusion

To ensure that those with a potential claim under a bond do not forfeit their rights, claimants should obtain copies of all bonds that apply to their projects. A failure to follow all of the enumerated procedures contained in the bonds may result in an inability to recover under the bond. In particular, the bond may contain language that provides for a short statute of limitations.

The rights and obligations of sureties may be overlooked when reviewing contract documents, and may not even have been considered until a substantial problem arises with performance under the construction contract. Prior knowledge, by all parties to the construction contract, of the benefits and obligations concerning bonds and the sureties is critical. In particular, as called for in AIA Document A312 and under most bond forms, the surety should be advised as early as possible about the existence of a potential problem that may trigger the surety's obligations under the bond agreement. Involvement of the surety in the construction process may also provide the other parties to the agreement with the participation of a relatively independent party whose lack of bias may aid in helping solve problems that arise. The surety's primary concern will be to have the job fully performed without resort to the bonds. The surety will then make a profit from the bond premium and avoid the risk of unsuccessfully pursuing the bond principal for indemnity. This narrow focus may allow some disputes to be resolved without resort to costlier dispute resolution measures, such as arbitration or litigation.

Although licensing bonds are routinely required by law in many states, the amount of money available under the bond is of minimal significance for all but the smallest projects. If a lawsuit is filed, the plaintiff will ask for recovery of the amount of the licensing bond, but the dollar amount available under the bond is so small that its inclusion may be only a matter of form.

The more closely a surety is involved in the entire construction process, from contract negotiation to completion of performance, the less likely it is that a construction default will occur that will require the surety to step in and take over performance or payments under the bond agreement.

CHAPTER 10

GOVERNMENT ENTITIES AND CONSTRUCTION PROJECTS

§ 10.1 Introduction

The effect of government regulation and control may be seen in nearly every aspect of construction. Unfortunately, when those involved in the construction process attempt to comply with government regulations at the local, state, and federal levels, they may require legal or specialized advice to successfully perform and complete even the most basic construction projects.

§ 10.2 Government as Regulator

The government regulates the construction process in order to protect the safety and welfare of construction workers and professionals. Regulation is also directed toward the protection of members of the public who may be affected either during the construction process or after it is completed.

For example, if adequate fencing is not placed around a construction project, neighborhood children may trespass onto the area at times when work is not being performed. Children may be injured because of the hazardous conditions posed by the job site, of which they are unaware or from which they are unprotected. Regulations to require fencing of construction projects, and to secure the construction site during nonoperating hours, help to protect the public from injury.

Protecting members of the public by regulating construction methods and procedures is equally important. Ensuring that concrete placed on a construction site is cured properly is essential, if the concrete is to have the necessary strength and integrity. If concrete in a subfloor of an office building is defective, an accident could occur. The floor could collapse five years after the building is constructed, causing loss of life to occupants or passersby, and damage to the building and adjacent property.

§ 10.3 Zoning and Land Use

Zoning of land by government agencies is intended to protect the public by providing an organized system for regulating residential, commercial, and industrial uses within a community. Governmental zoning power is based on the inherent police power described in federal and state constitutions. The police power allows the government to take steps it deems reasonably necessary to protect the health and welfare of its citizens. There are some restraints on this power, but, generally, government is allowed to institute a broad range of rules and regulations.

Zoning determines the types of uses that will be allowed in each area of the community. Most of the early zoning statutes in the United States were passed in the first decades of this century. These statutes dealt with the uses to which property could be put. The U.S. Supreme Court case of *Village of Euclid v. Ambler Realty Co.*[1] approved the use of a zoning plan that assigned contemplated uses to areas of a city. The level of density of building (structures per acre) is also addressed by zoning regulations.

Until the advent of the civil rights movement in the late 1950s and early 1960s, an integral part of transfers of land involved the exclusion of certain ethnic and religious groups from ownership or occupancy of property. Jews, Blacks, Hispanics, and others, were some of the excluded groups. This type of

[1] 272 U.S. 365, 47 S. Ct. 114, 71 L. Ed. 303 (1926).

exclusion was sometimes achieved by the use of restrictive covenants regarding property, created by early homeowner associations or by developers when tracts of land were developed. Each subsequent owner took the land subject to the restrictive covenants contained in articles and bylaws of the homeowner associations.

It is common for a subdivision of a city, or a group of homes in an area, to have, as a part of the documents publicly recorded in the county recording system, covenants that dictate the conditions under which each owner in that area or subdivision takes title, and by which each owner agrees to abide. If the covenants are not obeyed by one owner, the other owners in the area may seek relief from a court to enforce the covenants. Each owner has taken the property subject to those covenants or restrictions, and that owner is in breach of a contractual agreement with all the other owners if he fails to abide by the covenants or restrictions.

In the 1960s, restrictive covenants that discriminated against particular ethnic groups were attacked legally. Racial or ethnic discrimination by individual or subdevelopment covenants was held to be unconstitutional under both state and federal constitutions. Not only will the courts no longer enforce discriminatory covenants regarding the ownership or occupancy of land, but, today, owners using discriminatory practices in selling or renting property can be sued for such practices.

§ 10.4 —Variances

Under certain circumstances, an owner of land may be able to effect a change in the zoning or cause a change in the permitted use of land. This type of permission is called a **variance**. To obtain a variance, the landowner must prove that the use originally permitted is not economically feasible, and that to limit the owner to that use of the land will cause economic hardship to the owner. Generally, however, public policies regarding permitted uses for land will not be disturbed. For instance, commercial or industrial buildings, or commercial or industrial activities, are not normally allowed in purely residential areas. The noise level emitted by many types of heavy industry might interfere with the sleep of residents in such an area. If commercial or manufacturing uses permit emission of noxious fumes in a residential area, few persons would wish to live in that area. The quality of life for residents who remain would deteriorate, and prices for homes in that area would decline.

Occasionally, an area originally zoned solely for one type of use will have shifted in nature, and the purpose for the zoning protection will no longer exist. For example, an area that was completely residential 50 years ago is now occupied solely by industries, with the exception of several residences. Under these circumstances, there is no reason to refuse to allow other types of compatible industries to enter the formerly residential area. Conversely, in an area where uses were formerly entirely industrial, residential use may have grown, because of the area's proximity to places of work for persons who work in downtown areas of the city.

In the 1950s and 1960s, many people moved out of the inner city areas; they sought refuge from urban problems by moving to suburban areas. Recently, there has been a shift; people have moved back into downtown areas as commuting problems have become acute. The appeal of a residence that is within walking distance of the workplace increases as the magnitude of problems associated with commuting grows.

Industries that emit noxious fumes, or may pollute air or water, usually will not be allowed near residential areas because of the health risks posed to residents by these industries. Even in purely industrial areas, a significant factor to be considered is the potential for damaging the soil or groundwater, which can cause health problems for the workers or nearby residential areas. Environmental concerns have increased the public level of awareness regarding the need for regulation of the use and handling of hazardous or toxic materials.

When zoning was first developed, not all property was being used in a manner consistent with proposed zoning plans. The concept developed of allowing a **nonconforming use** to continue after a zoning plan was enacted. The theory behind allowing nonconforming uses to continue was based on an assumption that such uses would die out with the passage of time, resulting in areas that would eventually contain only the uses described and intended by the zoning plans that had been developed. Arbitrary prohibition of preexisting, nonconforming uses raised issues of fairness. Allowing such uses to continue for a period of time seemed to be a more equitable way to achieve the ultimate goal of homogeneous community areas. However, in practice, these preexisting, nonconforming uses did not disappear.

Amortization ordinances were an alternative method developed to rid communities of ongoing nonconforming uses. Under this system, properties that did not conform were actually ordered to end such uses after the lapse of a specified period of time. This method also had limited success.

Most communities contain areas where diverse uses that conflict with the zoning plans established for those areas still survive.

§ 10.5 Regulation by Federal Agencies

In 1970, the federal government enacted the federal Occupational Safety and Health Act (OSHA). This law sets standards for workplace safety. Violation of these standards can result in imposition of fines on employers, and/or the placement of a stop work order on a project until full compliance is effected. Where there is an injury in the workplace, and there is noncompliance with OSHA regulations, liability may be found. The violations themselves can be used as evidence of negligence.

§ 10.6 Regulation by State Agencies

Corresponding state Occupational Health and Safety Act (OSHA) agencies require that contractors within a state comply with state-specified safety requirements to protect the health and welfare of workers, and to protect members of the public who may be in areas where construction is occurring. State OSHA agencies require, for example, that workers use hard hats on construction projects, and that they wear steel-toed shoes. Steelworkers, or other laborers working at height, must use safety belts to avoid a fall or injury if they lose their balance. Because of the danger of an injury or accident occurring if a scissor-lift is moved when it is in the extended position, a worker using a scissor-lift in a number of different locations must lower the scissor-lift every time he moves from one location to another, even though the work could be performed more rapidly without lowering the lift.

§ 10.7 Regulation by Local Agencies

Owners must obtain building permits from local agencies before construction can begin. In most states, the owner or developer must file building plans, engineering drawings, and detail drawings with various local agencies, before permits will be granted. Frequently, agencies will demand alterations in plans for safety reasons. In California, for instance, where there is a high risk of earthquakes, new construction is required to have structural integrity that can withstand earthquakes of a specified magnitude. Areas of the country that have a high risk of other natural disasters, such as fire, flooding, hurricanes, or tornadoes, have similar building standards intended to protect the public from injury or damage.

Before building permits are issued, local agencies may require that developers provide environmental impact reports (EIRs) to analyze the effect of the proposed construction on the animals, plants, and other natural resources in the vicinity of the proposed structure or business. If the proposed location of a particular building or business will increase the burden on local roads, highways, or public transportation, local agencies may impose, as a condition for granting a building permit, a requirement that a developer contribute money toward the upkeep of those systems. Another way to cover the cost of greater use of the public systems is to require that new homeowners pay special assessments when they settle in a particular area. Yet another method is to require that the developer make payments to the community to reimburse it for the increased use of local resources.

§ 10.8 Government as Regulator of Hazardous Materials and Toxic Wastes

In the past twenty years, both the public and the government have become aware of the need for regulation and cleanup of hazardous or toxic waste, and for regulation of industries that create such waste. In 1969, the National Environmental Policy Act (NEPA) was enacted by Congress.[2] Under NEPA, the citizens and the government are described as "trustees" of the environment for future Americans. As trustees, they are charged with the task of preserving nature and wildlife in America.

NEPA requires that federal government agencies use organized planning to protect natural resources. Congress has also enacted other legislation to protect the environment, including legislation to protect the air.[3] The generation of solid waste must be controlled,[4] and sites containing dangerous hazardous waste are to be cleaned up.[5]

Local water districts, air pollution control districts, and the Environmental Protection Agency (EPA) have regulatory powers to control the handling and disposition of toxic or potentially toxic materials, and have the authority to impose heavy civil and criminal penalties for violations.

§ 10.9 Hazardous Waste Regulation and Environmental Protection at the Federal Level

The Environmental Protection Agency (EPA) is the primary federal agency charged with the task of cleaning up toxic waste and protecting the environment. Since NEPA was passed, and since the EPA was formed, large amounts of public and private funds have been channeled into attempts to clean up existing toxic waste, and to prevent the health risks to people, animals, and natural resources posed by future disposition of toxic waste. In 1976, Congress amended the Solid Waste Disposal Act. This amendment became known as the Resource Conservation and Recovery Act (RCRA).[6]

These laws empower the administrator of the EPA to enforce RCRA provisions, to issue an order assessing civil penalties for past or current violations, and to require compliance with RCRA by temporary or permanent injunction. Hundreds of sites have been designated as toxic waste locations that require cleanup and have been given priority for the allocation of public funds for

---

[2] 42 U.S.C.A. § 4331(c) (West 1991).

[3] 42 U.S.C.A. § 7401 (West 1991).

[4] 42 U.S.C.A. § 6901 (West 1991).

[5] 42 U.S.C.A. § 9601 (West 1991).

[6] 42 U.S.C.A. §§ 6901–6907, 6911–6916, 6921–6931, 6941–6949, 6951–6954, 6961–6964, 6971–6979, 6981–6987 (West 1991).

cleanup. Legal proceedings may be instituted by either the EPA or private plaintiffs against polluters of sites containing toxic wastes, to recover costs expended or to demand that the polluters undertake cleanup of the sites according to the standards demanded by the government.

Those in the construction industry who make a business of toxic waste cleanup may view the actions of the government in requiring private parties to clean up toxic waste as a financial opportunity. However, many contractors may not be willing to undertake the financial and health risks associated with cleaning up hazardous waste. Some enterprising contractors or engineers may contract to remove toxic materials, such as solvents or heavy metals, from soil or groundwater. Other contractors may enter the field of abatement of asbestos, or undertake the elimination of other types of pollution or contamination from commercial, industrial, or residential structures. However, most laws provide for strict liability against the contractors who undertake such cleanup efforts. If those contractors' handling of the waste causes damage or injury, liability may be imposed, whether or not the contractor handling the waste acted negligently.

Government regulation regarding hazardous or toxic waste has increased the cost of nearly every step of the construction process. Additional controls on construction procedures are necessary to avoid incurring liability for creation or improper handling of hazardous waste when participating in construction. The financial risk posed by potentially running afoul of federal, state, or local agencies charged with the responsibility of regulating and enforcing rules regarding hazardous waste must be included in the costs of construction passed on to owners.

§ 10.10 Minority Set-Asides and Preferences

As a matter of public policy, all levels of government have used certain preferences in determining which contractors will be awarded government contracts. Local, state, and federal governments have attempted to correct past injustices to minority male contractors and to female contractors in the award of construction contracts. Current rules about minority contract set-asides, in which certain specified percentages of public contracts must be awarded to minority contractors, have been a source of many lawsuits.

In the recent U.S. Supreme Court case on this issue, *Richmond v. J.A. Croson,*[7] the city of Richmond, California, required that at least 30 percent of the dollar amount of city construction contracts be awarded to minority business enterprises. This standard was adopted after a public hearing at which no evidence was presented regarding past discrimination, although the plan declared that it was intended to be remedial. In *Croson*, the U.S. Supreme Court determined that a generalized finding of past discrimination within the entire construction

---

[7] 109 S. Ct. 706, 102 L. Ed. 2d 854 (1989).

trade was insufficient to demonstrate a compelling government interest justifying the 30 percent allotment for minority contractors. Factors that could justify such a plan include: (1) gross statistical disparities between the number of prime contracts awarded to minority firms, and the minority population of the city, and (2) where special qualifications are required for award of a contract, the number of qualified minorities available. The court concluded that any such preferential plan for award of public contracts must be narrowly tailored to achieve the goal of breaking down patterns of minority exclusion.

In *Carl Bolander & Sons Co. v. City of Minneapolis,*[8] the City of Minneapolis bid invitations required that all bidders meet a goal of subcontracting 10 percent of the total bid price to minority businesses and 5 percent to women-owned businesses. The specifications for the bidding stated that a bidder's failure to meet the minority or female business utilization goals, or to show meaningful good faith to meet the goal, would be grounds for finding the bid nonresponsive. The bid form contained a representation that the bidding firm would use its best efforts to enter into subcontracts with women and minority business enterprises (W/MBEs); space was provided for listing of the W/MBEs. No W/MBEs were listed by McCrossan, the lowest bidder.

Upon the opening of the bids, McCrossan was declared the low bidder, with a bid of $2,141,622.80. Bolander was the second lowest bidder, at $2,168,347.50. Other bidders were only slightly higher. On its form regarding equal opportunity in contracting, McCrossan listed "MBE Inc." as a W/MBE it would utilize, and promised to use and document its best efforts to comply. Shortly after the bids were opened, Bolander discovered that no women-owned businesses were listed on McCrossan's bid. Bolander submitted a written protest to the City. Bolander was informed that the City had awarded the contract to McCrossan.

Bolander asked for an injunction to prevent performance of the contract or for a declaratory judgment to invalidate the contract, on the ground that McCrossan's bid was nonresponsive because it failed to name a women-owned business enterprise.

The appellate court held that the contract must be set aside. It based its decision on its finding that the purpose of competitive bidding is to deprive or limit the discretion of contract-making officials. Although McCrossan's deviation did not affect the amount of the bid, the bid was unresponsive and could prejudice the government's broad policy objective of minority participation.

In *Bridgeport Restoration v. Petrucci Construction Co.,*[9] the plaintiff was a roofing and siding contractor, qualified as a disadvantaged business enterprise (DBE) (also known as a minority business enterprise). The transit district invited bids for a contract to build a garage, which was to be funded in part by the federal and state governments. In accordance with the requirements of

---

[8] 438 N.W.2d 735 (Minn. App. 1989).

[9] 211 Conn. 230, 557 A.2d 1263 (1989).

government funding, the bid notice required at least 10 percent DBE participation. The defendant was awarded the contract, having attached to the contract, as well as to the bid, a "Schedule of DBE and WBE [women's business enterprise] Participation" that listed the plaintiff as a DBE company that would perform the roofing work, sheet metal work, and miscellaneous other work for $400,000, which was 4.9 percent of the total contract.

The president of the minority business and an agent of the defendant agreed on a bid price of $400,000 in telephone negotiations, before the defendant submitted its bid on the prime contract to the transit district. The defendant's project manager also told the president of the plaintiff company that if the defendant were awarded the contract, the plaintiff would get the subcontract. After the transit district awarded the contract to Petrucci, Petrucci asked for adjustments in the subcontract, which Bridgeport refused to accept. Petrucci then awarded the subcontract to a different qualified bidder.

The trial court found for the minority business plaintiff, and awarded it $60,000 for the loss of the contract and $7,500 for attorney's fees. The trial court found that: (1) the parties had formed a conditional contract as alleged in the complaint, (2) plaintiff incurred lost profits of $60,000, and (3) there was sufficient evidence to award legal fees. The trial court's decision was upheld on appeal.

In *Milwaukee County Pavers Association v. Fiedler,*[10] plaintiffs brought an action (pursuant to 42 U.S. Code Section 1983), challenging the constitutionality of a Wisconsin statute providing for reservation of $4 million of highway and construction contracts for disadvantaged businesses (MBEs/DBEs). Plaintiffs sought a permanent injunction against enforcement of the statute, a temporary injunction enjoining two specific bid lettings scheduled for the near future and execution of bids under the statute, and a declaration that the statute was unconstitutional under the standards for reviewing affirmative action programs articulated in *Richmond v. J.A. Croson.*[11]

Plaintiffs were highway contractors doing business in Wisconsin. They were qualified to bid on most of the state highway construction projects, but they were precluded from bidding on projects under the challenged statute because they were not "disadvantaged businesses." Approximately 180 firms qualified in Wisconsin, according to the guidelines established to comply with federal regulations.

The court held that the statute was constitutionally suspect, although the State of Wisconsin was to be lauded for its efforts to ensure that businesses owned by disadvantaged people would enjoy equal business opportunities in the construction industry. After discussing *Croson* thoroughly, the court found that plaintiffs met the applicable tests, and granted plaintiffs the requested injunctive relief to prevent enforcement of the statute.

---

[10] 707 F. Supp. 1016 (W.D. Wis. 1989).

[11] 109 S. Ct. 706, 102 L. Ed. 2d 854 (1989).

Government contracts frequently contain requirements that bidders be minority-owned or qualify as a small business. Litigation may ensue, regarding the validity of a bidder's status. In *Service Engineering Co. v. Southwest Marine, Inc.*,[12] plaintiffs, Service and Triple A Machine Shop, and defendant, Southwest, competed for contracts to repair Navy vessels. The set-aside contracts they sought were designated for "small businesses," pursuant to the Small Business Act of 1958.[13] For such contracts, a business was "small" if it had fewer than 1,000 employees.

Prior to 1976, the statute allowed a business to calculate its size based on the number of persons employed during the pay period ending nearest the last day of each of the past four quarters. Businesses would understate their true size by firing employees just before the last pay period of the quarter, then rehiring them after the end of the quarter.

To correct this abuse, the SBA changed to a system that used the average of the number of employees during each pay period of the preceding year.

In October 1981, Southwest consulted with counsel regarding the appropriate method for calculating business size. After reviewing both methods, counsel advised Southwest that it should use the quarterly method, and that Southwest should state on the application form that the quarterly method of calculation was used to calculate its business size. On the application, Southwest stated it was using Schedule B, Note 1 (the quarterly method), but Southwest did not state explicitly in its application that the quarterly method was being used. The SBA regional administrator issued a size determination that upheld the status of Southwest as a small business on the basis of this application.

For roughly the next two years, Southwest periodically fired and hired employees to stay below an average of 1,000 employees, using the quarterly method. In September 1983, Triple A challenged Southwest's status as a small business. The SBA required Southwest to complete the application form again, and to attach any additional evidence that supported its position as a small business. Southwest resubmitted the application form, stating that it believed the application form resolved Triple A's allegations. In this renewed application, Southwest did not disclose that it used the quarterly method of calculation, or that it was firing and rehiring employees to maintain its "small" status.

There was an additional request by the SBA that Southwest certify its total number of employees for certification as a "small business." Southwest again failed to disclose its use and manipulation of the quarterly method. In May 1983, Service, the plaintiff, and defendant Southwest both submitted bids for a large contract for repairs and modernization of Navy vessels, certifying they were small businesses. The Navy awarded the contract to Southwest. On December 6, 1983, the SBA reversed its earlier finding, and determined Southwest

---

[12] 719 F. Supp. 1500 (1989).

[13] 15 U.S.C.A. § 631 *et seq.* (West 1991).

was not a "small business." The SBA decertified Southwest, preventing it from obtaining further set-aside contracts.

Service then sued Southwest, asking that the award of the Navy contract to Southwest be set aside. As a matter of law, the court found that Southwest committed mail fraud, based on its manipulations of the quarterly method for counting its employees, which it had not disclosed to the SBA. Southwest had been awarded the large contract because of the government's reliance on the fraudulent representation that Southwest was, in fact, a small business, as defined by the SBA. The court found that the undisputed evidence established intentional fraud. Only after being awarded the large Navy contract, and upon further demand of the SBA, did defendants disclose to the SBA that the declaration was completed by using the quarterly method to determine its size. The facts also demonstrated fraudulent intent, a necessary element under California law.[14] Southwest's Navy contract was set aside.

§ 10.11 Conclusion

Some government control is necessary to protect workers from careless or reckless employers, to protect members of the public from the effect of unsafe construction practices, and to protect the eventual tenants of buildings from the hazards posed by unsafe construction. However, the tendency of government to expand its requirements, and to use public contracts to implement social policy, has resulted in public construction in the United States being extremely costly.

Political debate is unending about the extent to which government should control activity, as compared to the alternative of leaving control in the hands of consumers and market forces. The past 15 years or so have seen growing skepticism toward the belief that government must protect us from ourselves.

---

[14] Cal. Civ. Code §§ 1572, 1710 (West 1991).

CHAPTER 11

AVOIDING OR TERMINATING CONSTRUCTION LITIGATION

§ 11.1 Introduction

Disputes, litigation, and expensive delays in construction may be avoided or minimized with thorough preproject planning, appropriate contract documentation, and judicious selection of the other parties to a construction project.

§ 11.2 Relevant Prior Considerations for Contractors and Design Professionals

Many considerations will determine whether a contractor or design professional will decide to participate in a particular project. These considerations may include: (1) the priority of the project within the owner's organization, (2) the likelihood that the project will obtain the necessary approvals to proceed to completion, (3) the contractor or design professional the owner is planning to use on the project, and whether the contractor or design professional has had any prior experience with the other entities, (4) the contracts being used between the owner and the other parties, (5) the contractor's or design professional's experience with the particular type of construction being used for the project, (6) the financial strength of the owner relative to its ability to pay for completion of the project, particularly if the project should encounter delays and increased costs, (7) the nature of the fee arrangement to which the owner will agree, and (8) the bonding required and/or obtained by the owner, design professional, and contractor.

Where an organization is handling many projects at one time, the relative priority assigned to a particular project will be important in determining whether it will proceed smoothly to completion. If it is a small project being built by an organization that typically handles large projects, the contractor or design professional may have a difficult time in obtaining from the owner the prompt responses that are essential to orderly progress of the work. If the completion of other projects is delayed, a small project may be temporarily shelved or neglected. If the owner has extreme problems with other projects, a small project may even be abandoned, resulting in large losses to the others involved in that project.

Approvals for construction are required for almost any work. The owner must secure permits and approvals from local governing authorities, from organizations such as the EPA and OSHA, and from other state and federal entities. The attitude of the necessary authorities toward the owner personally, and toward the type of use to which the property will be put, may determine whether approval will be received for construction. An owner may wish to develop property in an area where growth is disfavored. The contractor or design professional should ascertain whether it will be possible to obtain the necessary permits from the various relevant government agencies.

The contractor or design professional should have experience or background knowledge about the others who are going to perform work on the project. This experience or knowledge will be helpful in keeping the work progressing on schedule. If the contractor or design professional has worked in the past with the other parties to be used on the project, the other parties may already trust his integrity and competence. Similarly, those working on the construction project who have established a reputation with the design professional or contractor, or within the industry, will be relied on and trusted more heavily by him.

If the contractor or design professional is familiar with the type of contracts to be used by the owner on the subject project, the risk may be reduced that problems will occur on the project because of confusion about the rights and duties of the parties. If the procedure used for contractors to present claims to the design professional for extra compensation is well-established, it may not be necessary to go to litigation or arbitration to solve most disputes.

Different types of construction may have widely varying customs. In commercial construction, for example, the pattern for each of the different trades to perform their work may be less tightly formatted than the pattern used in residential construction. In particular, concentrated residential work, such as planned unit developments or condominiums, may require each trade to come onto the project in a precise sequence to perform its work. If the owner, design professional, or contractor has only built commercial projects previously, these customs may be unfamiliar.

Financial capacity of the owner is a concern for the contractor and design professional, not only as to the issue of payment to the contractor and design professional for their work, but also in regard to the owner's ability to see the project through to completion. Few projects proceed precisely as scheduled, and few fail to encounter "unforeseen" difficulties. Problems nearly always occur on projects: rises in the cost of materials, rises in the cost of labor, subcontractors who abandon the job, new requirements imposed by government agencies, unexpected subsurface conditions. The possibilities are almost endless. It could even be said that the only certain prediction about a construction project is that it will encounter some "unforeseen" difficulties. If the owner's financial status is precarious, the design professional risks investing substantial uncompensated time and energy in a project.

Closely related to concern about the owner's financial viability is the manner in which the contractor or design professional will be paid. The more that payment relies on completion of the project, the greater the risk that the contractor and design professional face if the project has problems. The more concern the contractor and design professional have about the financial strength of the owner, and the greater their uncertainty as to completion of the project, the more the contractor and design professional should insist on a compensation arrangement that provides for progress payment at frequent intervals.

On some projects, the contractor or design professional also may be faced with the choice of whether to walk away from a troubled project and give up a chance to collect fees, or to see the project through to completion and risk the investment of further time and energy in hopes of collecting the amounts due for services.

The bonding obtained or required by the owner provides the owner with security that the project will be completed, and that payment will be made to subcontractors and suppliers for services or goods provided to the project. This subject is explored in depth in **Chapter 8**. Briefly, the common types of bonds used in construction are licensing bonds, payment bonds, performance bonds, and bid bonds.

§ 11.3 Negotiating Solutions: Resolving Disputes without Outside Aid

It is important to retain a pragmatic perspective in attempting to resolve conflicts that occur on construction projects. A workable negotiated solution is almost always preferable to incurring the transaction costs of litigation, arbitration, and mediation. The attitude of the parties toward a negotiated solution will be more positive than their response to an imposed solution. It is far more likely that the negotiated solution will be implemented smoothly.

Most books and literature on business negotiation and dispute resolution stress the value of assigning priorities to the various issues and positions that arise when there is a dispute. The first step is to determine which of the positions taken may be modified, and which positions are nonnegotiable. By realistically evaluating the priority of the concerns, parties can clarify in their own minds how they wish to approach the dispute.

Another useful step in negotiation is to begin a session by establishing a personal relationship with the opposing party. If a personal relationship previously existed, an attempt can be made to resurrect it. This could involve beginning a negotiating session with discussion about family, hobbies, or other subjects unrelated to the dispute. In this way, the parties may be able to establish some rapport.

The next step is to try to outline the limits of the dispute, and to discuss those areas related to the conflict about which there is little or no dispute. This could involve reviewing the facts that led to the dispute, without assigning blame for problems, or discussing the responsibility or involvement of other parties who are not present at the negotiating session. Without allocating responsibility for payment, discussions may begin regarding dollar amounts associated with the delay, defects, or other problems.

Once the framework is established regarding facts, and there is some agreement regarding relevant issues, the following advice will usually aid resolution: approach the negotiation by describing your perception about who is responsible for the problem. At this point, it is helpful for you to acknowledge some responsibility for the problem. Even if you feel you have no responsibility, acknowledging a small amount of responsibility may be a helpful tactic. By accepting some of the responsibility, and thereby accepting some of the cost, you may be able to defuse some of the defensive anger or hostility that has arisen. This may be an important step toward resolving the situation.

The truth is that most construction disputes *do* involve some responsibility on the part of each party involved. For one party to deny any responsibility for the problem usually is unreasonable, and it does not further the process of negotiating a solution. For a party to take an unreasonable position in addressing a particular problem, or for a party to attempt to solve the problem without input from the other party is likely to result in litigation.

* * *

Case 11.1
(a) Prelitigation

A finish concrete subcontractor was hired to perform finish concrete work on concrete that encircled a swimming pool at a residence. The contract amount was nominal. After the work was completed, the prime contractor was advised by the owner that the owner was dissatisfied with the color of the surface of the concrete. The prime contractor tore out all the concrete and installed new concrete.

The prime contractor back-charged the original finish concrete subcontractor an amount that exceeded the contract amount by approximately 18 times. In the back-charge itemization, the prime contractor listed an amount for the re-do of the finish concrete work that was twice as much as the original contract for that work.

The subcontractor was never provided with an opportunity to try to patch or otherwise repair the concrete, nor to perform the finish work when the concrete had been replaced. Instead, the prime contractor charged the subcontractor for having another person perform that work. The items on the back-charge included: (1) removal of concrete, (2) cost of disposing of concrete, (3) cost of new concrete, (4) cost of installing new concrete, and (5) cost of new finish concrete work.

The prime made numerous demands to the subcontractor for payment for the back-charges. When the subcontractor consulted with his insurance company, the company advised him that it would not cover the claim, because it was for defective work, which was not covered under the insurance policy.

Analysis. Inspection and monitoring of the concrete work, if done early in the progress of that work, would have alleviated this situation. Even if the prime contractor was not taking the time or effort to inspect, the finish concrete contractor could have saved himself time and money by doing a small portion of the work and verifying that the prime and the owner were satisfied.

A clause in the subcontract establishing liquidated damages, including a representation that the finish subcontractor would not be liable for any amount in excess of the contract amount, might also have avoided the dispute's growth into a lawsuit.

(b) Post-Filing of Litigation

After receiving no response to his demands, the prime contractor sued the subcontractor for the back-charges. During pretrial discovery, the fact was established that the subcontractor was never given an opportunity to cure the problem with the concrete work, and never received timely notice of the problem. Other issues were: the excessive amount being charged by the prime contractor for the cost of redoing the concrete, and the question of whether repair or patching of the offensive sections of the work would not have been less costly alternatives that would have reduced the damages of the prime.

Eventually, the case went to judicial arbitration. The arbitrator was presented with the evidence discussed above. The defendant finish concrete subcontractor testified that he could have cut out the portions of the concrete objected to by the owner and patched the concrete to match. Taking into account the 30 years' experience of the finish concrete subcontractor, and the fact that the prime contractor appeared to have acted hastily and carelessly, the arbitrator awarded the prime contractor only the cost of the original contract.

Analysis. At arbitration, a crucial issue in this case was whether the prime contractor had acted appropriately under the circumstances. He failed to work with the subcontractor to resolve the problem. It appears he did not exercise good judgment about how to solve the problem, because there were adequate, less costly alternatives to complete removal and replacement of the concrete.

Even if one party to a contract fails to meet its obligations under the contract, the damaged party has a responsibility to **mitigate**, or reduce, the amount of damages he or she may suffer. This usually involves taking reasonable steps to minimize and contain the effect of the problem.[1] Because the prime contractor failed to act reasonably under the circumstances, he was unable to recover the cost of the removal and replacement of the concrete.

* * *

Case 11.2
(a) Prelitigation

Buildright, a developer, built a residential project in an area overlooking a bay. The project consisted of 36 luxury homes. The streets and lots were staked out, foundations were laid, and the homes were built. Within a few years after the homes were sold and occupied, two major categories of problems developed.

One problem was major water intrusion around the window frames of most of the homes. The area was a windy location, and leaks started appearing after a number of heavy storms occurred. Wet rot was found in the textured drywall located around the window frames. Various solutions were attempted, to locate the source of the problem, and none was satisfactory.

The second problem was at a home owned by Jack: major earth movement was causing the structure to separate. The surveyor had made a one-degree error, which resulted in all the lots being staked improperly. Jack's lot was at the

[1] For cases on the obligation of a party to mitigate damages, *see* C. A. Davis, Inc. v. City of Miami, 400 So. 2d 536 (Fla. App. 1981) (defendant contractor argued against recovery of owner based on owner's allegedly having spent more than was necessary to correct contractor's work; court stated contractor could only dispute completion cost based on waste, extravagance, or lack of good faith by owner); Great Lakes Gas Transmission Co. v. Grayco Constructors, Inc., 506 F.2d 498 (6th Cir. 1974), *cert. denied* (Feb. 24, 1975), 95 S. Ct. 1330 (contractor unsuccessfully contended that owner could pass its loss to customers through rate increases); S. J. Groves & Sons Co. v. Warner Co., 576 F.2d 524 (3d Cir. 1978) (breaching contractor unsuccessful in argument that prime should have mitigated damages by firing subcontractor).

end of the project, and the one-degree error had resulted in his lot's being approximately 40 feet away from the originally designated location.

The land in the area around and near Jack's home had been subjected to both cutting and filling, to achieve the necessary grade for some of the homes. The market value of the home before the problem developed was approximately $400,000. In essence, Jack's home was now worthless because the structural damage made it unsafe for occupancy.

Negotiations between the developer and the homeowners' association did not result in resolution of either the water intrusion problem occurring at many of the homes, or the soil settlement problem affecting Jack's home.

Analysis. Errors had occurred at the early stages of construction of this residential project. Although, at the time, it seemed a cost-effective alternative not to restake the property after the surveying error was discovered, hindsight suggests that ignoring it was not the right action to take. The effect on the market value of the other homes, and the damage to the developer's reputation, were high costs to pay for that one error.

Water intrusion is a common problem in planned unit subdivisions. One possible reason for the recurrence of this problem in large-scale residential developments is the speed with which the homes are erected. Another contributing factor may be that these projects frequently suffer financial reverses, requiring the use of successive, different prime contractors on the project. Overall sloppiness in coordinating work, or in inspecting the quality of supplies or workmanship, may be additional factors that accompany increasing problems and diminished morale on the project.

(b) Post-Filing of Litigation

With regard to the water intrusion problem, a number of factors were involved: (1) the type of windows selected was not intended for use in high-wind-and-storm areas, (2) the flashing around the windows had been installed improperly because the prime contractor had failed to ensure that all the locations had the flashing properly installed, and (3) some of the windows had been installed with the weepholes plugged at the bottom instead of at the top. This allowed water to enter the windowframes at the top but did not provide an outlet for water at the bottom, which proper installation would have allowed.

The windows were replaced, the flashing was redone as necessary, and the weepholes of the windows were corrected. The cost of redoing the flashing and repairing the weepholes was borne equally by the prime contractor and the framing contractor; the homeowners' association paid for the cost of new windows where these were necessary.

Resolving the issue of the soil settlement on Jack's property was more difficult. One possible cause of the settling was that a part of the land obtained for the project had previously been used as a quarry. The soil used to fill the quarry was settling because it had not been adequately compacted at the time the quarry was filled. Another factor was the improper staking of the properties because of the one-degree surveying error.

When the project supervisor was asked about this problem, he admitted that the developer had become aware of the surveying error after all the lots were staked. He stated that restaking the entire subdivision would have involved a delay of completion of the project, causing the deadline on the original construction financing to be exceeded. The actual cost of restaking all the lots would have been $80,000.

Ultimately, the prime contractor and the developer agreed to pay Jack the market value his property would have had, had the land not contained the soil settlement problem. Jack's house was then removed, and, because of the continuing soil problem, no other structure could be erected on the property.

Analysis. Strong preproject planning might have avoided these problems. Frequent meetings at the project, along with inspection and quality control, might also have decreased the likelihood that the project would have defects. After the lawsuit was filed, the developer did not want to settle with the homeowners without knowing what amount was recoverable from the contractors. The developer could have settled with the homeowners and then pursued the contractor for indemnity. This failure to settle early was probably a tactical error, because the homeowners' resistance to a low settlement increased with the passage of time. In addition, as more time passed, new construction defects came to light. The cost of repairing these other defects had to be included in the overall settlement.

An earlier settlement with the homeowners might have precluded claims for other defects, depending on the wording of the release and on whether the subsequently discovered defects were related to the original problems.

§ 11.4 Concerns of Prime or General Contractor Working with Faltering Subcontractors

If a critical subcontractor is faltering in its performance, the prime contractor may wish to terminate the subcontractor before the subcontractor files bankruptcy. The question to be answered by a prime contractor is whether the problems are severe enough to constitute a default by the subcontractor under the terms of the contract, which would permit the prime contractor to terminate the subcontractor. If the prime contractor acts prematurely, his improper termination of the subcontractor makes the prime the party who has breached the contract. The greater the clarity of the subcontract termination provision, the less the prime contractor's risk in terminating the subcontractor, as long as the prime contractor follows the provisions of the termination clause.

For the prime contractor, an alternative to termination of the faltering subcontractor is negotiation of a voluntary termination agreement with the subcontractor. Surety companies for the subcontractor and the prime contractor should be involved in the negotiation process, to avoid having the parties lose coverage under the relevant bonds for the principals.

Another alternative for the prime contractor is to enter into an agreement with the subcontractor to modify the contract. In exchange for the subcontractor's being allowed to continue to perform, the prime contractor will want to obtain the right to remove the faltering subcontractor from the project expeditiously, if the subcontractor files bankruptcy. An agreement by the subcontractor not to file bankruptcy would be unenforceable. However, the prime contractor can (1) negotiate specific protection for creditors, which may eliminate litigation, or (2) simplify the bankruptcy process by reducing the number of creditors who present claims. The subcontractor can agree to have certain debts not discharged by the bankruptcy and to pay those debts after the bankruptcy process is completed.

Modifications to the original subcontract might also include provisions that make it unappealing for the subcontractor to file bankruptcy. The subcontractor could agree in advance to grant the prime contractor immediate relief from the automatic stay upon filing of the bankruptcy. Enforceability of this type of agreement varies. This type of clause has been specifically enforced in some bankruptcies following a workout, but other courts have refused to enforce such an agreement.[2]

§ 11.5 —Financial Data

One way to avoid the problems of a subcontractor filing bankruptcy is to obtain an advance financial statement, executed under oath by the principals of the organization. The prime might even request a recent audited financial statement, as well as written confirmation that the subcontractor has not recently defaulted on any of his loans. Taking these steps may avoid involvement in a project with a subcontractor who is not financially stable.

If a prime contractor wishes to impose such requirements on bidders, he should specify these requirements in the bid notice or solicitation for proposals. The bidders must know what will be required before making bids. Failure to advise bidders of financial requirements before the project is released for bids could provide a bidder who did not meet the requirements with an argument that he had been induced to spend time and effort preparing his bid while he was left ignorant of "unreasonable conditions."

---

[2] *In re* Citadel Properties, Inc., 86 Bankr. 275 (Bankr. M.D. Fla. 1988). For a contrary holding, *see In re* Sky Group International, Inc., 108 Bankr. 86 (Bankr. W.D. Pa. 1989).

One of the reasons for obtaining a financial statement from a subcontractor who is not incorporated is that a bankruptcy court may refuse to discharge any debts that are incurred by inducing reliance on a false financial statement.[3]

§ 11.6 —Performance Bonds

The prime contractor should require performance bonds from the critical subcontractors on the project. The added insurance cost is well worth the peace of mind obtained by this assurance that the subcontractor will perform, or that the bond will stand as security for performance or payment for the cost of performance. Because the language used by sureties may vary, it is important to review the bond. The financial strength of the surety should also be verified, if it is not a company that is familiar.

An added benefit of requiring the performance bond is that, if the subcontractor does file bankruptcy, the automatic stay of bankruptcy does not prohibit recovery by the prime contractor from the surety.

§ 11.7 —Individual Guarantees

The prime contractor may also wish to obtain personal guarantees from the principal or principals of the subcontracting entity. The Bankruptcy Code prohibits terminating a subcontractor solely because the subcontractor files for bankruptcy. However, the bankruptcy laws do not affect the validity of a clause in a third-party guarantee (the personal guarantee) whereby all amounts stated in the guarantee are due and payable to the prime contractor in the event of a subcontractor's bankruptcy.

The personal guarantee also provides leverage against the principals of the subcontractor, if the subcontractor files bankruptcy. With this type of third-party guarantee, the subcontractor's bankruptcy plan cannot provide for the release of guarantors. From a psychological perspective, if the subcontractor has provided a personal guarantee, he may be more inclined to complete the project successfully. If the subcontractor is facing various overwhelming financial obligations, those obligations for which he has given a personal guarantee are more likely to receive immediate attention.

When obtaining the personal guarantee, the prime contractor should also obtain a financial statement from the principal providing the guarantee. The guarantee should explicitly state that the prime contractor is relying on the accuracy of the financial information provided by the principal. This will

---

[3] 11 U.S.C.A. § 523(a)(2)(b) (West 1991).

bolster a claim for nondischargeability of the debt, in the event the subcontractor files bankruptcy. If the financial statement is misleading, the debt under the guarantee may be nondischargeable, even if the principal of the subcontractor files an individual bankruptcy.

§ 11.8 —Retainages; Entitlement of Subcontractor to Payment

It is essential to carefully draft and review the provisions in the subcontract that detail the subcontractor's right to payments. Payment clauses may provide that each payment request from a subcontractor shall contain a written representation, signed under penalty of perjury, that the subcontractor (1) has paid all of its sub-subcontractors and suppliers on a timely basis according to normal business terms, and (2) has no invoices standing against its account for more than a fixed period of time (for example, 60 days) except for specifically listed invoices.

One way to protect funds intended for payment of sub-subcontractors or suppliers is to state within the subcontract that all such funds are held in trust for the benefit of those lower-tier subcontractors and suppliers. Joint checks should be used wherever possible, in conjunction with this method.

The above protective steps regarding contract funds will help to identify those subcontractors who may present problems. The financial reporting requirements may allow a prime contractor to avoid using a subcontractor who is in financial trouble. If the prime contractor advances funds to the subcontractor on the basis of false financial representations that lower-tier subcontractors and suppliers have been paid, the prime contractor can argue that the entire resulting debt to the prime contractor, or a portion of that debt, should not be discharged by bankruptcy.[4]

When funds are disbursed to the subcontractor for payment of lower-tier subcontractors and suppliers, designating those funds as "trust funds" increases the likelihood that funds held by the subcontractor at the time of filing for bankruptcy will not be found to be property of the bankruptcy estate. Instead, the funds are more likely to be considered as funds held in trust.

Another benefit of designating the funds as trust funds and of using joint checks is that, if those funds are paid to lower-tier subcontractors or suppliers shortly before the filing for bankruptcy, those receiving the funds have a viable defense to a claim that those funds were paid as a **preference**. A preferential transfer—a transfer made before the bankruptcy filing—ignores the rights of creditors who would normally have preference over those who were paid.[5]

---

[4] 11 U.S.C.A. § 523(a)(2)(a) (West 1991).

[5] *See* 11 U.S.C.A. §§ 541, 547(b); Bergier v. Internal Revenue Service, 83 Bank. Rptr. 324, 110 S. Ct. 2258 (1990).

Under bankruptcy law, there is an order of preference for creditors. For example, creditors who have secured their debt with a trust deed against real property, or with a UCC-1 form (the document that secures personal property for debt), are entitled to have their debts satisfied ahead of unsecured creditors. The prime contractor has a strong incentive to see that payments made to lower-tier suppliers and subcontractors are not considered preferential, and that their payments are not included in the subcontractor's bankruptcy estate. If the trust funds for lower-tier subcontractors and suppliers are set aside as preferences by the bankruptcy court, the lower-tier subcontractors and the suppliers will make claims against the prime contractor, the prime's payment bond, or the project (by filing a mechanic's lien on the land).

Subcontracts should state that contract balances otherwise owing are not property of the subcontractor, or actually earned, until all lower-tier subcontractors and suppliers have been paid for all current and outstanding invoices. When this language is included, the contract balances that are otherwise owing are not considered officially "earned" and are therefore not part of the bankruptcy estate, unless the subcontractor has already satisfied all prior obligations to lower-tier subcontractors and suppliers. The prime should also provide in the subcontract that, if lower-tier subcontractors or suppliers are not paid by the subcontractor, the prime, may, at his option, pay those parties directly. If this language is lacking from the subcontract, the bankruptcy debtor or his trustee may have a valid action against the prime contractor on the subcontract.

The subcontract should also state that any retainage is not earned until the project is successfully completed by the subcontractor. This provision should avoid the retainage's being considered part of the bankruptcy estate. The existence of a substantial retainage creates an incentive for the subcontractor to complete the job. The retainage also will help fund completion by an alternate subcontractor, in the event of default by the first subcontractor.

Another problem may arise for a prime contractor if he advances funds to a faltering subcontractor. If the prime provides money to the subcontractor to meet payroll and the subcontractor files for bankruptcy, the prime contractor can be held liable for unpaid withholding taxes.[6] If the prime contractor pays the funds directly to the employees of the subcontractor, liability for unpaid withholding taxes is automatic.[7] Even if the money is paid to the subcontractor, the prime may be liable to the Internal Revenue Service if the advances are for the specific purpose of paying wages, and if the prime contractor has actual notice or knowledge that the subcontractor does not intend to make timely payments or deposits of employee withholding taxes.

---

[6] 26 U.S.C.A. § 3505 (West 1991).

[7] 26 U.S.C.A. § 3505(a) (West 1991).

§ 11.9 —Lower-Tier Subcontractors and Suppliers

The contracts between the subcontractor and the lower-tier subcontractors and suppliers should state that, in the event of a default by the subcontractor, the prime contractor has the right to assume and to hold the lower-tier contractors and suppliers to their contracts. The provision should explicitly state that this assumption is an option of the prime contractor, not an obligation; the prime can then select which lower-tier contractors and suppliers it wishes to retain. The master subcontract should require that all lower-tier subcontracts and supplier contracts contain this optional assumption provision. To ensure that these assumptions are included in the subcontracts, the prime contractor should review all the subcontracts prior to beginning work on the project.

§ 11.10 —Interaction of Contract Provisions and Bankruptcy

Construction contracts often contain arbitration provisions, attorney fee provisions, clauses that establish where suit may be brought, and designation of the jurisdiction whose laws the parties agree will be applied to disputes under the contract. Many of these clauses have not been drafted with sufficient care to ensure that they will survive a bankruptcy filing.

The contract should include a "selection of legal forum clause," explicitly stating the geographical selection of forum that is intended to be applied, even if one of the parties files bankruptcy. This clause may or may not be enforced, depending on the individual bankruptcy court. The question of forum will be especially critical if the subcontractor files bankruptcy in a location distant from the prime contractor. A New York contractor might be required to go to Texas to litigate a claim against a bankrupt subcontractor who filed in Texas for his own convenience.

An attorney's fees clause in a subcontract may also be drafted with attention to the possibility of bankruptcy. If a subcontractor on the job files for bankruptcy, the prime contractor may have to take a number of actions that require the aid of an attorney but do not make the prime contractor a "prevailing party" entitled to attorney's fees. Bankruptcy laws require that, if a debtor assumes or assigns a construction contract, the debtor must compensate the other party for all monetary losses caused by any breaches of the contract. By specifying that attorney's fees are to be considered monetary losses that must be compensated for in the event of an attempt of a party to assume or assign the subcontract, the prime contractor may be able to recover attorney's fees even though he is not considered the prevailing party.

Federal policy favors arbitration of disputes. Bankruptcy courts also may favor arbitration of disputes, even after bankruptcy has been filed. Specifying within the arbitration clauses that the parties intend arbitration to occur even if one of the parties is in bankruptcy will increase the likelihood that arbitration will continue to be available to resolve the disputes.

§ 11.11 —Subcontractor Termination

In accord with contract provisions, subcontractors may be terminated before bankruptcy is filed. After the subcontractor files bankruptcy, the automatic stay on any action against the subcontractor will prevent the prime contractor from terminating the subcontractor. The subcontract should specify that time is of the essence, describe the magnitude of the cost of delay, and state the grounds for termination of the subcontractor. If the termination provision is automatic, based on the occurrence or nonoccurrence of certain events, the automatic stay of the bankruptcy court may not prevent the subcontractor's termination. The subcontract may state that the prime contractor is to send notice that termination will be effective unless the subcontractor takes certain steps set out in the notice. If the subcontractor fails to comply, the subcontractor may be considered terminated. If the subcontractor fails to file bankruptcy before the deadline set out in the notice, the prime contractor may argue that the subcontractor was validly terminated under state law prior to the bankruptcy filing.

If the construction contract with a subcontractor was properly terminated before the subcontractor filed for bankruptcy, the bankrupt subcontractor has no right to assume the contract as executory.[8] The contract will not be considered subject to assumption, rejection, or assignment by the bankruptcy court or trustee in bankruptcy. If the debtor responds before the deadline set out in the notice, but fails to meet all the terms set out in the notice within the extension of time allowed under bankruptcy law, the prime contractor may then claim the contract was terminated. The prime can argue that the automatic stay does not affect the termination, because the prime contractor took no action against the debtor.

§ 11.12 —Specific Steps after Subcontractor's Filing of Bankruptcy

As soon as the prime contractor learns of a bankruptcy filing by a subcontractor, the prime should work with the subcontractor to agree on a date by which the debtor will have performed all obligations under the contract. If the subcontractor will not stipulate a date, the prime contractor can make a motion to the bankruptcy court to compel assumption or rejection of the subcontract. If

[8] For cases regarding actions against a debtor after bankruptcy filed, *see In re* Windmill Farms, Inc., 841 F.2d 1467 (9th Cir. 1988); *see also In re* B-K of Kansas, Inc., 69 Bankr. 812 (Bankr. D. Kan. 1987); *In re* Rosswog, 48 Bankr. 689 (Bankr. M.D. Pa. 1985).

the subcontractor's filing occurs at a critical point in the project, delay can be costly.

The prime contractor must file a proof of claim in the bankruptcy, to provide the court with evidence that the prime contractor has a claim against the subcontractor. The prime contractor will then be entitled to payment from the available assets of the subcontractor, depending on (1) the amount available to pay claims and (2) the priority of the prime contractor's claim in relation to other claims filed. The prime's failure to file a claim will result in the liabilities of the subcontractor to the prime contractor being discharged by the operation of bankruptcy laws.

§ 11.13 Effect of Bankruptcy Filing on Pending Litigation

If a party to a construction contract files bankruptcy after litigation has begun in the civil court system, adjudication of the dispute may be shifted to the bankruptcy courts. The bankruptcy courts have jurisdiction to hear all disputes that are "related to" the bankruptcy proceeding and involve the debtor. Bankruptcy courts broadly interpret this power as including the power to resolve disputes over whether the debtor owes money to, or can recover money from, other parties.[9]

In addition, the bankruptcy court has the power to serve parties or witnesses nationwide, rather than having jurisdiction restricted to a particular state, as it is in state court systems. A bankruptcy action may be filed wherever the domicile, residence, principal place of business, or principal assets of the debtor are located, or in any district in which there is a pending bankruptcy case concerning a related company.[10]

§ 11.14 Conclusion

As outlined in this chapter, good business practices, careful selection of the other parties to construction contracts, and intelligent preproject analysis will help to avoid construction disputes.

Retaining a written record of events as they occur may avoid some disputes. If disputes do occur and lawsuits are necessary to resolve them, written evidence as to what occurred will carry more force than testimony by principals involved in the dispute.

A negotiating strategy that is flexible and does not emphasize blame, but is instead pragmatically oriented, will aid in reducing the cost and magnitude of

---

[9] 28 U.S.C.A. §§ 157(a), 1334(b) (West 1991).

[10] 28 U.S.C.A. § 1408 (West 1991).

many disputes. It is important for all parties to recognize that litigation is rarely profitable.

As should be apparent, attorneys rarely create the problems that cause disputes on construction projects. Often, if an attorney had become involved in a project at an earlier stage, he or she could have prevented or minimized the problem or dispute.

The analyses of the cases demonstrate the type of approach a good attorney, mediator, or settlement judge brings to construction disputes: an ability to separate relevant facts from those that are irrelevant, a balancing of the respective proportions of fault or legal liability for damages, and a proposed solution to the dispute after all necessary factors have been taken into consideration.

CHAPTER 12

EFFECTIVE CONSTRUCTION DISPUTE RESOLUTION: A SUMMARY

§ 12.1 Using the Legal System and Avoiding Construction Litigation

The nature of the construction process makes conflicts and disputes almost inevitable. A number of different contracting entities, with different needs, are expected to cooperate and coordinate their efforts. Efficient, timely performance is critical. One party's failure to perform, for any reason, is likely to cost one or more of the parties involved substantial amounts of money.

Traditionally, conflicts, disputes, and losses have been handled by litigation. However, the court system has proven to be neither cost-effective nor timely in resolving construction issues. Attorneys' fees are costly. Courts are congested; criminal matters take priority over civil cases. Delays in reaching trial are likely to consume years. Finally, when trial is had, verdicts may be inconsistent with the parties' understanding and unpredictable in result.

You may find it difficult to locate attorneys in your area who are familiar with the construction industry. This text has offered you suggestions in locating competent counsel familiar with construction problems. Having retained such an attorney, you should receive prompt, responsive, and efficient service. Do not hesitate to change attorneys if you do not feel you are receiving good service.

Avoiding or minimizing disputes is the responsibility of both yourself and your attorney. You should consistently apply a methodical, thorough approach to negotiating or bidding on new projects. Identifying future problems, and specifying how they will be resolved in the course of a project, is essential.

If, prior to signing your contract, you have retained the services of an attorney experienced in construction matters, his or her review of the contract terms may avoid many problems that typically cause construction disputes. In addition, a substantial number of problems can be solved if you consult your attorney when disputes arise. Pragmatic determination of the cause of construction delays, and identification of responsibility for defective performance, can save many dollars in legal fees.

Following are some of the factors that should be taken into account when deciding whether to litigate a construction dispute.

Lawsuits must be brought before the relevant statutes of limitations have expired. The statutes of limitations for construction cases vary, depending on the type of legal theory used as a basis for a suit. The statutes range from a minimum of six months, under some government claims statutes, to a maximum of ten years in some jurisdictions, for latent construction defects.

If your case involves parties from different states, you may have to file a lawsuit in another state. Participating in a trial in a remote location can cause disruption to your business, if you are required to be absent for extended periods of time.

Where litigation is involved, cooperating with your attorney and responding to requests for documents from the other side, conducting site inspections, hiring experts, and responding to interrogatories, can be costly in direct, out-of-pocket costs, and in depriving you and your employees of large amounts of otherwise productive time.

Often, your clear *legal rights to recover* money from a party responsible for your losses are worthless. The party may lack the assets or funds from which to pay a judgment. Should that party file for bankruptcy, his liability to you will be discharged, unless the debt was procured by fraud.

If a construction dispute involves a large sum of money, you may be able to afford to hire *attorneys* who will invest substantial amounts of time in securing a recovery. However, your ability to obtain performance bonds can be hindered by delays in obtaining payment on this prior claim. The delays in recovery may cause you to miss opportunities to perform future work, or may destroy your ability to bid on new projects.

State and federal courts have instituted *procedures* to expedite the resolution of disputes before trial. These procedures include mediation and court-ordered arbitration. "Fast-track" guidelines are used by the courts to move cases to trial quickly, and to discourage the parties from filing suit unless they are planning to move the case to trial within a brief period of time.

Finally, even after trial of a case and a verdict in your favor, the other side may have a basis for an *appeal* if the judge or jury committed errors in deciding the case. An appeal may result in further delays of years, again delaying payment to you, even if the other parties are financially responsible.

§ 12.2 Construction Systems

Traditional interactions among owner, architect, general contractor, and subcontractors are giving way to more innovative approaches. As the economy continues to negatively affect those in construction, new approaches are being sought that will result in the production of higher quality construction at lower prices.

One alternative construction system involves using construction managers (CMs) to avoid having to pay a general or prime contractor. Construction managers are employees of the owner and are concerned only with the interests of the owner. Their sole interest is completion of the project at the lowest possible cost. Another alternative system is the use of multiple prime contractors, often in conjunction with fast-tracking. Using these systems, the entire design need not be complete prior to the commencement of construction.

These alternative systems are being used to avoid costly delays in completion of projects, and to attempt to effectively plan and complete projects for an amount reasonably close to that of the original contract. Change orders, extras, and increases that can occur in the cost of labor or supplies during the course of construction will result in completion of the project at a higher cost to the owners. These problems increase the likelihood of litigation, as the owner attempts to distribute the increased cost among those the owner deems responsible.

Design-build is another system used to avoid some of the faults perceived in the traditional construction process. By eliminating one of the parties used in the traditional system (usually either the architect or the prime contractor), costs may be reduced, which results in completion of the project at a higher profit.

Use of alternative construction systems entails some risks. Legal doctrines and statutes used to decide construction disputes usually are based on the traditional construction process. Most form construction contracts also assume the use of traditional construction systems. The use of innovative systems, without modifying form contracts, may result in unfair allocation of responsibilities or liabilities. Where alternative construction systems have been employed, arbitration or court decisions may be illogical, given the understandings of the parties. Consulting an attorney before agreeing to the use of alternative construction systems should help avoid some of these pitfalls.

§ 12.3 Contract Creation and Performance

Complex written contracts form the basis for most construction projects. Although an owner, design professional, or prime contractor need not always use a written contract to enforce performance or payment, a written contract is certainly preferable. The written contract will be the primary source of reference for determining responsibilities, rights, and liabilities that arise on construction projects. The more detailed the contract provisions for handling various eventualities, the more likely that disputes will be resolved without resort to litigation.

Keeping documentation of each step that may be disputed will help justify the reasonableness of your actions if problems arise.

If a contract dispute arises and a party can prove that a relevant contractual clause is ambiguous, that party may be able to have that clause effectively removed from the contract. If a court determines that a clause is ambiguous, the parties may be able to offer the court evidence to establish their intent regarding the issue addressed by the disputed clause.

Under certain circumstances, parties may be relieved of the duty of complying with written contractual obligations. For example, contracts commonly require that all change orders be provided in writing, and that written approval be received for extra work before such work is performed. If following these steps is not practical, the requirement may be waived.

Contractual clauses that work a hardship on a particular party may be disregarded by an arbitrator or court if it appears one party of greater economic strength imposed the clause on a weaker party.

Many construction contracts, particularly public contracts, are created by competitive bidding. Private contracts are more often created by negotiation between parties. With competitively bid contracts, it is far more likely that bidders will have no option regarding the inclusion or exclusion of a particular clause. In negotiated contracts, the existence of, or lack of, a particular clause will be significant in the event of a dispute. Important clauses include, but are not limited to, indemnity clauses, attorneys' fees clauses, liquidated damages provisions, pay-when-paid clauses, clauses regarding differing site conditions, arbitration clauses, and clauses limiting damages for delay.

You should understand the significance of these clauses before agreeing to sign a contract that contains such clauses. With negotiated contracts, you may wish to request inclusion or exclusion of some clauses before agreeing to perform the work covered by the contract.

In the event of a breach of contract, the most common remedy available is that of money damages. Under exceptional circumstances, courts may grant an injunction to order a party to do, or to abstain from doing, a particular act.

In the event of a dispute, the contracts executed between the parties will be the first source to consult regarding your duties. If you cannot resolve the dispute by discussion and negotiation with the other parties, you should consult an attorney regarding your options for resolution of the dispute.

§ 12.4 Subcontracts

Subcontracting portions of the work in construction projects serves a valuable function. Subcontracting allows entities to develop specialities in particular areas of work such as mechanical work, plumbing, masonry, framing, electrical, and painting. Development of the necessary expertise can result in more efficient, less costly performance of the work.

Trades whose work requires the investment of large sums of money for equipment can amortize the cost of that equipment by specializing in performing subcontract work that requires that equipment. General contractors can avoid having to purchase expensive equipment and can perform contracts at a lower amount than if they had had to invest in that equipment.

Contract and legal relationships between subcontractors, and with other parties to construction projects, are complex. Typically, the subcontractor has a contract only with the prime contractor, although all or part of the prime contract may be incorporated into the subcontract. Duties and responsibilities among different subcontractors are primarily governed by laws pertaining to negligence and other noncontractual torts.

§ 12.5 Torts

Legal theories for recovery of damages that do not involve contract responsibilities include negligence, fraud, and bad faith, also called the "breach of the implied covenant of good faith and fair dealing." Generally, to prove a liability under a tort theory, you must prove each of the three necessary elements: (1) the existence of a duty, (2) the breach of that duty, (3) the breach's role in causing your damage or injury.

Most lawsuits will assert rights to recovery under a variety of legal theories. Even where a suit alleges breach of a contract obligation, it may also allege that the defendants were negligent. Most lawsuits allege that the actions of defendants were both deliberate and accidental. Pleading rules allow plaintiffs to make claims under alternative theories, even though in many cases both allegations cannot be true. A common example of an allegation presenting alternate theories will state that the defendant "knew, or should have known" that his acts would cause damage.

Plaintiffs in a construction dispute may name as defendants everyone who was involved in the project, even though not every party was responsible for the plaintiff's damages. Facts will be developed during the litigation to determine who is actually responsible for the damages. The traditional rule of "joint and several liability" is that, if a particular defendant is liable for any percentage of the plaintiff's damage, the plaintiff may collect 100 per cent of the value of his damage from that one defendant. That defendant may claim the right to contribution from other defendants, based on respective percentages of fault.

For cases involving personal injury, such as construction accidents, there are some variations to this general rule. Some states, such as California and Washington, have enacted laws that alter the rule of joint and several liability for noneconomic damages, commonly called "pain and suffering." Payment by defendants for the elements of economic damages of plaintiffs (wage loss, property damage, and medical bills) are still based on the rule of joint and several liability.

In circumstances where the behavior of a defendant demonstrates gross negligence, or where a high level of disregard for the rights of others is shown, courts may award "punitive damages" to plaintiffs. These damages, often largely unrelated to the amount of injury suffered by the plaintiff, are calculated on the basis of the net worth of the defendant, or on the amount of money that would deter that defendant, or other defendants, from performing such socially undesirable conduct in the future. In the past five years, courts have attempted to place limitations on the award of punitive damages.

§ 12.6 Loss Prevention or Risk-Shifting: Insurance and Bonds

Insurance is widely used in construction as a risk allocation mechanism. To avoid a large financial risk, a party to a construction contract may obtain insurance or bonds by payment of a premium. Although indemnity clauses are also used for this purpose, the protection afforded by an indemnity clause depends on whether the party providing the indemnity has the financial ability to pay for losses. Insurance or bonds are intended to be guaranteed sources of payment for losses, as long as the loss is covered by the insurance policy or bond.

It is not uncommon for a party to believe insurance will pay for a loss, and later to discover that certain exclusions contained in the language of the contract of insurance negate coverage for the particular loss. In the past, courts would extend rules of contract construction to find in favor of a policyholder seeking insurance coverage. Courts are now less likely to favor policyholders.

Bonding contracts are similar to insurance policies. Under normal circumstances, large projects use payment and performance bonds to shift to the bonding companies the risk of loss due to nonperformance by the principals under the bonds. In exchange for accepting this risk, the bonding companies receive the premium payments and the right to pursue the principals for reimbursement.

In many states, licensing bonds are required for contractors to do business. A failure to comply with licensing procedures and licensing bonds may prevent a contractor from being able to collect for work performed.

§ 12.7 Government Involvement in Construction

Parties working in construction must work with regulatory government agencies such as the EPA, OSHA, and numerous federal, state, and local agencies. Attempts by those in construction to comply with rules and regulations of these organizations consume massive amounts of time, energy, and money. A failure to comply can result in the imposition, by government agencies, of large fines, and/or losses of necessary licenses or permits.

Public entities also act as owners paying for construction work. Complex bidding procedures control those competing for the award of public contracts. In many cases, contractors must be on approved lists of responsible bidders.

Many construction lawsuits arise from failures to comply with government bidding procedures, or from allegations by frustrated bidders of unfairness or lack of compliance by government agencies with their own rules. In many public contracts, the large amount of dollars involved may justify bringing lawsuits to attempt to recover perceived damages.

Many public contracts require that stated percentages of the work subcontracted by the bidder must be awarded to small businesses and to minority- (MBE) or women- (WBE) owned enterprises. These mechanisms have been instituted as a way of addressing social concerns about the discrimination toward these groups in the past, or, in the case of small businesses, to stimulate the growth and multiplication of small businesses.

It is common for public agencies to require that those wishing to be considered as qualifying under one of the above categories must comply with various certification procedures. Disputes regarding minority certification, or alleging bias or unfairness in the award of public contracts, occur frequently and often result in litigation.

§ 12.8 Conclusion

Almost every construction project has the potential for generating disputes and litigation. The majority of parties to construction appear to have been involved in lawsuits at some point. This fact represents a huge loss of productivity and profit for those in construction. In turn, these higher costs are passed on to the public in the form of higher costs of construction, higher taxes, and larger premiums for insurance of all types.

Use of foresight regarding the problems likely to arise, and a methodical approach to performing work on projects will reduce the risk of suffering large financial reverses because of construction disputes.

Consulting with an attorney who is experienced in construction matters, before, during, and after completion of projects, may eliminate many problems.

Maintaining a good relationship with other parties to projects should improve the chances of avoiding litigation. In the event litigation is unavoidable, approaching litigation with a rational and economically based perspective will improve the likelihood of resolving the litigation with a satisfactory result.

Absorbing the concepts described in this book will help you to understand the legal process and to be aware of the options available for resolution of construction disputes: litigation within the legal system, and arbitration or mediation through private organizations and private trials.

APPENDICES

APPENDIX A

AIA DOCUMENT A101, STANDARD FORM OF AGREEMENT BETWEEN OWNER AND CONTRACTOR (1987)

THE AMERICAN INSTITUTE OF ARCHITECTS

AIA Document A101

Standard Form of Agreement Between Owner and Contractor

where the basis of payment is a

STIPULATED SUM

1987 EDITION

THIS DOCUMENT HAS IMPORTANT LEGAL CONSEQUENCES; CONSULTATION WITH AN ATTORNEY IS ENCOURAGED WITH RESPECT TO ITS COMPLETION OR MODIFICATION.

The 1987 Edition of AIA Document A201, General Conditions of the Contract for Construction, is adopted in this document by reference. Do not use with other general conditions unless this document is modified.

This document has been approved and endorsed by The Associated General Contractors of America.

AGREEMENT

made as of the day of in the year of Nineteen Hundred and

BETWEEN the Owner:
(Name and address)

and the Contractor:
(Name and address)

The Project is:
(Name and location)

The Architect is:
(Name and address)

The Owner and Contractor agree as set forth below.

ARTICLE 1
THE CONTRACT DOCUMENTS

The Contract Documents consist of this Agreement, Conditions of the Contract (General, Supplementary and other Conditions), Drawings, Specifications, addenda issued prior to execution of this Agreement, other documents listed in this Agreement and Modifications issued after execution of this Agreement; these form the Contract, and are as fully a part of the Contract as if attached to this Agreement or repeated herein. The Contract represents the entire and integrated agreement between the parties hereto and supersedes prior negotiations, representations or agreements, either written or oral. An enumeration of the Contract Documents, other than Modifications, appears in Article 9.

ARTICLE 2
THE WORK OF THIS CONTRACT

The Contractor shall execute the entire Work described in the Contract Documents, except to the extent specifically indicated in the Contract Documents to be the responsibility of others, or as follows:

ARTICLE 3
DATE OF COMMENCEMENT AND SUBSTANTIAL COMPLETION

3.1 The date of commencement is the date from which the Contract Time of Paragraph 3.2 is measured, and shall be the date of this Agreement, as first written above, unless a different date is stated below or provision is made for the date to be fixed in a notice to proceed issued by the Owner.

(Insert the date of commencement, if it differs from the date of this Agreement or, if applicable, state that the date will be fixed in a notice to proceed.)

Unless the date of commencement is established by a notice to proceed issued by the Owner, the Contractor shall notify the Owner in writing not less than five days before commencing the Work to permit the timely filing of mortgages, mechanic's liens and other security interests.

3.2 The Contractor shall achieve Substantial Completion of the entire Work not later than

(Insert the calendar date or number of calendar days after the date of commencement. Also insert any requirements for earlier Substantial Completion of certain portions of the Work, if not stated elsewhere in the Contract Documents.)

, subject to adjustments of this Contract Time as provided in the Contract Documents.

(Insert provisions, if any, for liquidated damages relating to failure to complete on time.)

ARTICLE 4
CONTRACT SUM

4.1 The Owner shall pay the Contractor in current funds for the Contractor's performance of the Contract the Contract Sum of Dollars ($), subject to additions and deductions as provided in the Contract Documents.

4.2 The Contract Sum is based upon the following alternates, if any, which are described in the Contract Documents and are hereby accepted by the Owner:

(State the numbers or other identification of accepted alternates. If decisions on other alternates are to be made by the Owner subsequent to the execution of this Agreement, attach a schedule of such other alternates showing the amount for each and the date until which that amount is valid.)

4.3 Unit prices, if any, are as follows:

ARTICLE 5
PROGRESS PAYMENTS

5.1 Based upon Applications for Payment submitted to the Architect by the Contractor and Certificates for Payment issued by the Architect, the Owner shall make progress payments on account of the Contract Sum to the Contractor as provided below and elsewhere in the Contract Documents.

5.2 The period covered by each Application for Payment shall be one calendar month ending on the last day of the month, or as follows:

5.3 Provided an Application for Payment is received by the Architect not later than the day of a month, the Owner shall make payment to the Contractor not later than the day of the month. If an Application for Payment is received by the Architect after the application date fixed above, payment shall be made by the Owner not later than days after the Architect receives the Application for Payment.

5.4 Each Application for Payment shall be based upon the schedule of values submitted by the Contractor in accordance with the Contract Documents. The schedule of values shall allocate the entire Contract Sum among the various portions of the Work and be prepared in such form and supported by such data to substantiate its accuracy as the Architect may require. This schedule, unless objected to by the Architect, shall be used as a basis for reviewing the Contractor's Applications for Payment.

5.5 Applications for Payment shall indicate the percentage of completion of each portion of the Work as of the end of the period covered by the Application for Payment.

5.6 Subject to the provisions of the Contract Documents, the amount of each progress payment shall be computed as follows:

5.6.1. Take that portion of the Contract Sum properly allocable to completed Work as determined by multiplying the percentage completion of each portion of the Work by the share of the total Contract Sum allocated to that portion of the Work in the schedule of values, less retainage of percent (%). Pending final determination of cost to the Owner of changes in the Work, amounts not in the dispute may be included as provided in Subparagraph [illegible] of the General Conditions even though the Contract Sum has not yet been adjusted by Change Order;

5.6.2 Add that portion of the Contract Sum properly allocable to materials and equipment delivered and suitably stored at the site for subsequent incorporation in the completed construction (or, if approved in advance by the Owner, suitably stored off the site at a location agreed upon in writing), less retainage of percent (%);

5.6.3 Subtract the aggregate of previous payments made by the Owner; and

5.6.4 Subtract amounts, if any, for which the Architect has withheld or nullified a Certificate for Payment as provided in Paragraph 9.5 of the General Conditions.

5.7 The progress payment amount determined in accordance with Paragraph 5.6 shall be further modified under the following circumstances:

5.7.1 Add, upon Substantial Completion of the Work, a sum sufficient to increase the total payments to percent (%) of the Contract Sum, less such amounts as the Architect shall determine for incomplete Work and unsettled claims; and

5.7.2 Add, if final completion of the Work is thereafter materially delayed through no fault of the Contractor, any additional amounts payable in accordance with Subparagraph 9.10.3 of the General Conditions.

5.8 Reduction or limitation of retainage, if any, shall be as follows:

(If it is intended, prior to Substantial Completion of the entire Work, to reduce or limit the retainage resulting from the percentages inserted in Subparagraphs 5.6.1 and 5.6.2 above, and this is not explained elsewhere in the Contract Documents, insert here provisions for such reduction or limitation.)

ARTICLE 6
FINAL PAYMENT

Final payment, constituting the entire unpaid balance of the Contract Sum, shall be made by the Owner to the Contractor when (1) the Contract has been fully performed by the Contractor except for the Contractor's responsibility to correct nonconforming Work as provided in Subparagraph 12.2.2 of the General Conditions and to satisfy other requirements, if any, which necessarily survive final payment; and (2) a final Certificate for Payment has been issued by the Architect; such final payment shall be made by the Owner not more than 30 days after the issuance of the Architect's final Certificate for Payment, or as follows:

ARTICLE 7
MISCELLANEOUS PROVISIONS

7.1 Where reference is made in this Agreement to a provision of the General Conditions or another Contract Document, the reference refers to that provision as amended or supplemented by other provisions of the Contract Documents.

7.2 Payments due and unpaid under the Contract shall bear interest from the date payment is due at the rate stated below, or in the absence thereof, at the legal rate prevailing from time to time at the place where the Project is located.

(Insert rate of interest agreed upon, if any.)

(Usury laws and requirements under the Federal Truth in Lending Act, similar state and local consumer credit laws and other regulations at the Owner's and Contractor's principal places of business, the location of the Project and elsewhere may affect the validity of this provision. Legal advice should be obtained with respect to deletions or modifications, and also regarding requirements such as written disclosures or waivers.)

7.3 Other provisions:

ARTICLE 8
TERMINATION OR SUSPENSION

8.1 The Contract may be terminated by the Owner or the Contractor as provided in Article 14 of the General Conditions.

8.2 The Work may be suspended by the Owner as provided in Article 14 of the General Conditions.

ARTICLE 9
ENUMERATION OF CONTRACT DOCUMENTS

9.1 The Contract Documents, except for Modifications issued after execution of this Agreement, are enumerated as follows:

9.1.1 The Agreement is this executed Standard Form of Agreement Between Owner and Contractor, AIA Document A101, 1987 Edition.

9.1.2 The General Conditions are the General Conditions of the Contract for Construction, AIA Document A201, 1987 Edition.

9.1.3 The Supplementary and other Conditions of the Contract are those contained in the Project Manual dated , and are as follows:

Document	Title	Pages

9.1.4 The Specifications are those contained in the Project Manual dated as in Subparagraph 9.1.3, and are as follows:
(Either list the Specifications here or refer to an exhibit attached to this Agreement.)

Section	Title	Pages

9.1.5 The Drawings are as follows, and are dated unless a different date is shown below:
(Either list the Drawings here or refer to an exhibit attached to this Agreement.)

Number	Title	Date

9.1.6 The addenda, if any, are as follows:

Number	Date	Pages

Portions of addenda relating to bidding requirements are not part of the Contract Documents unless the bidding requirements are also enumerated in this Article 9.

9.1.7 Other documents, if any, forming part of the Contract Documents are as follows:

(List here any additional documents which are intended to form part of the Contract Documents. The General Conditions provide that bidding requirements such as advertisement or invitation to bid, Instructions to Bidders, sample forms and the Contractor's bid are not part of the Contract Documents unless enumerated in this Agreement. They should be listed here only if intended to be part of the Contract Documents.)

This Agreement is entered into as of the day and year first written above and is executed in at least three original copies of which one is to be delivered to the Contractor, one to the Architect for use in the administration of the Contract, and the remainder to the Owner.

OWNER	CONTRACTOR
(Signature)	*(Signature)*
(Printed name and title)	*(Printed name and title)*

CAUTION: You should sign an original AIA document which has this caution printed in red. An original assures that changes will not be obscured as may occur when documents are reproduced.

APPENDIX B

AIA DOCUMENT A201, GENERAL CONDITIONS OF THE CONTRACT FOR CONSTRUCTION (1987)

THE AMERICAN INSTITUTE OF ARCHITECTS

Because AIA Documents are revised from time to time, users should ascertain from the AIA the current edition of this document. Copies of the current edition of this AIA document may be purchased from The American Institute of Architects or its local distributors.

The text of this document is not "model language" and is not intended for use in other documents without permission of the AIA.

AIA Document A201

General Conditions of the Contract for Construction

THIS DOCUMENT HAS IMPORTANT LEGAL CONSEQUENCES; CONSULTATION WITH AN ATTORNEY IS ENCOURAGED WITH RESPECT TO ITS MODIFICATION

1987 EDITION

TABLE OF ARTICLES

This document has been approved and endorsed by the Associated General Contractors of America.

AIA **CAUTION: You should use an original AIA document which has this caution printed in red. An original assures that changes will not be obscured as may occur when documents are reproduced.**

INDEX

AIA DOCUMENT A201 • GENERAL CONDITIONS OF THE CONTRACT FOR CONSTRUCTION • FOURTEENTH EDITION

GENERAL CONDITIONS OF THE CONTRACT FOR CONSTRUCTION

ARTICLE 1

GENERAL PROVISIONS

1.1 BASIC DEFINITIONS

1.1.1 THE CONTRACT DOCUMENTS

The Contract Documents consist of the Agreement between Owner and Contractor (hereinafter the Agreement), Conditions of the Contract (General, Supplementary and other Conditions), Drawings, Specifications, addenda issued prior to execution of the Contract, other documents listed in the Agreement and Modifications issued after execution of the Contract. A Modification is (1) a written amendment to the Contract signed by both parties, (2) a Change Order, (3) a Construction Change Directive or (4) a written order for a minor change in the Work issued by the Architect. Unless specifically enumerated in the Agreement, the Contract Documents do not include other documents such as bidding requirements (advertisement or invitation to bid, Instructions to Bidders, sample forms, the Contractor's bid or portions of addenda relating to bidding requirements).

1.1.2 THE CONTRACT

The Contract Documents form the Contract for Construction. The Contract represents the entire and integrated agreement between the parties hereto and supersedes prior negotiations, representations or agreements, either written or oral. The Contract may be amended or modified only by a Modification. The Contract Documents shall not be construed to create a contractual relationship of any kind (1) between the Architect and Contractor, (2) between the Owner and a Subcontractor or Sub-subcontractor or (3) between any persons or entities other than the Owner and Contractor. The Architect shall, however, be entitled to performance and enforcement of obligations under the Contract intended to facilitate performance of the Architect's duties.

1.1.3 THE WORK

The term "Work" means the construction and services required by the Contract Documents, whether completed or partially completed, and includes all other labor, materials, equipment and services provided or to be provided by the Contractor to fulfill the Contractor's obligations. The Work may constitute the whole or a part of the Project.

1.1.4 THE PROJECT

The Project is the total construction of which the Work performed under the Contract Documents may be the whole or a part and which may include construction by the Owner or by separate contractors.

1.1.5 THE DRAWINGS

The Drawings are the graphic and pictorial portions of the Contract Documents, wherever located and whenever issued, showing the design, location and dimensions of the Work, generally including plans, elevations, sections, details, schedules and diagrams.

1.1.6 THE SPECIFICATIONS

The Specifications are that portion of the Contract Documents consisting of the written requirements for materials, equipment, construction systems, standards and workmanship for the Work, and performance of related services.

1.1.7 THE PROJECT MANUAL

The Project Manual is the volume usually assembled for the Work which may include the bidding requirements, sample forms, Conditions of the Contract and Specifications.

1.2 EXECUTION, CORRELATION AND INTENT

1.2.1 The Contract Documents shall be signed by the Owner and Contractor as provided in the Agreement. If either the Owner or Contractor or both do not sign all the Contract Documents, the Architect shall identify such unsigned Documents upon request.

1.2.2 Execution of the Contract by the Contractor is a representation that the Contractor has visited the site, become familiar with local conditions under which the Work is to be performed and correlated personal observations with requirements of the Contract Documents.

1.2.3 The intent of the Contract Documents is to include all items necessary for the proper execution and completion of the Work by the Contractor. The Contract Documents are complementary, and what is required by one shall be as binding as if required by all; performance by the Contractor shall be required only to the extent consistent with the Contract Documents and reasonably inferable from them as being necessary to produce the intended results.

1.2.4 Organization of the Specifications into divisions, sections and articles, and arrangement of Drawings shall not control the Contractor in dividing the Work among Subcontractors or in establishing the extent of Work to be performed by any trade.

1.2.5 Unless otherwise stated in the Contract Documents, words which have well-known technical or construction industry meanings are used in the Contract Documents in accordance with such recognized meanings.

1.3 OWNERSHIP AND USE OF ARCHITECT'S DRAWINGS, SPECIFICATIONS AND OTHER DOCUMENTS

1.3.1 The Drawings, Specifications and other documents prepared by the Architect are instruments of the Architect's service through which the Work to be executed by the Contractor is described. The Contractor may retain one contract record set. Neither the Contractor nor any Subcontractor, Sub-subcontractor or material or equipment supplier shall own or claim a copyright in the Drawings, Specifications and other documents prepared by the Architect, and unless otherwise indicated the Architect shall be deemed the author of them and will retain all common law, statutory and other reserved rights, in addition to the copyright. All copies of them, except the Contractor's record set, shall be returned or suitably accounted for to the Architect, on request, upon completion of the Work. The Drawings, Specifications and other documents prepared by the Architect, and copies thereof furnished to the Contractor, are for use solely with respect to this Project. They are not to be used by the Contractor or any Subcontractor, Sub-subcontractor or material or equipment supplier on other projects or for additions to this Project outside the scope of the

Work without the specific written consent of the Owner and Architect. The Contractor, Subcontractors, Sub-subcontractors and material or equipment suppliers are granted a limited license to use and reproduce applicable portions of the Drawings, Specifications and other documents prepared by the Architect appropriate to and for use in the execution of their Work under the Contract Documents. All copies made under this license shall bear the statutory copyright notice, if any, shown on the Drawings, Specifications and other documents prepared by the Architect. Submittal or distribution to meet official regulatory requirements or for other purposes in connection with this Project is not to be construed as publication in derogation of the Architect's copyright or other reserved rights.

1.4 CAPITALIZATION

1.4.1 Terms capitalized in these General Conditions include those which are (1) specifically defined, (2) the titles of numbered articles and identified references to Paragraphs, Subparagraphs and Clauses in the document or (3) the titles of other documents published by the American Institute of Architects.

1.5 INTERPRETATION

1.5.1 In the interest of brevity the Contract Documents frequently omit modifying words such as "all" and "any" and articles such as "the" and "an," but the fact that a modifier or an article is absent from one statement and appears in another is not intended to affect the interpretation of either statement.

ARTICLE 2

OWNER

2.1 DEFINITION

2.1.1 The Owner is the person or entity identified as such in the Agreement and is referred to throughout the Contract Documents as if singular in number. The term "Owner" means the Owner or the Owner's authorized representative.

2.1.2 The Owner upon reasonable written request shall furnish to the Contractor in writing information which is necessary and relevant for the Contractor to evaluate, give notice of or enforce mechanic's lien rights. Such information shall include a correct statement of the record legal title to the property on which the Project is located, usually referred to as the site, and the Owner's interest therein at the time of execution of the Agreement and, within five days after any change, information of such change in title, recorded or unrecorded.

2.2 INFORMATION AND SERVICES REQUIRED OF THE OWNER

2.2.1 The Owner shall, at the request of the Contractor, prior to execution of the Agreement and promptly from time to time thereafter, furnish to the Contractor reasonable evidence that financial arrangements have been made to fulfill the Owner's obligations under the Contract. *[Note: Unless such reasonable evidence were furnished on request prior to the execution of the Agreement, the prospective contractor would not be required to execute the Agreement or to commence the Work.]*

2.2.2 The Owner shall furnish surveys describing physical characteristics, legal limitations and utility locations for the site of the Project, and a legal description of the site.

2.2.3 Except for permits and fees which are the responsibility of the Contractor under the Contract Documents, the Owner shall secure and pay for necessary approvals, easements, assessments and charges required for construction, use or occupancy of permanent structures or for permanent changes in existing facilities.

2.2.4 Information or services under the Owner's control shall be furnished by the Owner with reasonable promptness to avoid delay in orderly progress of the Work.

2.2.5 Unless otherwise provided in the Contract Documents, the Contractor will be furnished, free of charge, such copies of Drawings and Project Manuals as are reasonably necessary for execution of the Work.

2.2.6 The foregoing are in addition to other duties and responsibilities of the Owner enumerated herein and especially those in respect to Article 6 (Construction by Owner or by Separate Contractors), Article 9 (Payments and Completion) and Article 11 (Insurance and Bonds).

2.3 OWNER'S RIGHT TO STOP THE WORK

2.3.1 If the Contractor fails to correct Work which is not in accordance with the requirements of the Contract Documents as required by Paragraph 12.2 or persistently fails to carry out Work in accordance with the Contract Documents, the Owner, by written order signed personally or by an agent specifically so empowered by the Owner in writing, may order the Contractor to stop the Work, or any portion thereof, until the cause for such order has been eliminated; however, the right of the Owner to stop the Work shall not give rise to a duty on the part of the Owner to exercise this right for the benefit of the Contractor or any other person or entity, except to the extent required by Subparagraph 6.1.3.

2.4 OWNER'S RIGHT TO CARRY OUT THE WORK

2.4.1 If the Contractor defaults or neglects to carry out the Work in accordance with the Contract Documents and fails within a seven-day period after receipt of written notice from the Owner to commence and continue correction of such default or neglect with diligence and promptness, the Owner may after such seven-day period give the Contractor a second written notice to correct such deficiencies within a second seven-day period. If the Contractor within such second seven-day period after receipt of such second notice fails to commence and continue to correct any deficiencies, the Owner may, without prejudice to other remedies the Owner may have, correct such deficiencies. In such case an appropriate Change Order shall be issued deducting from payments then or thereafter due the Contractor the cost of correcting such deficiencies, including compensation for the Architect's additional services and expenses made necessary by such default, neglect or failure. Such action by the Owner and amounts charged to the Contractor are both subject to prior approval of the Architect. If payments then or thereafter due the Contractor are not sufficient to cover such amounts, the Contractor shall pay the difference to the Owner.

ARTICLE 3

CONTRACTOR

3.1 DEFINITION

3.1.1 The Contractor is the person or entity identified as such in the Agreement and is referred to throughout the Contract Documents as if singular in number. The term "Contractor" means the Contractor or the Contractor's authorized representative.

3.2 REVIEW OF CONTRACT DOCUMENTS AND FIELD CONDITIONS BY CONTRACTOR

3.2.1 The Contractor shall carefully study and compare the Contract Documents with each other and with information furnished by the Owner pursuant to Subparagraph 2.2.2 and shall at once report to the Architect errors, inconsistencies or omissions discovered. The Contractor shall not be liable to the Owner or Architect for damage resulting from errors, inconsistencies or omissions in the Contract Documents unless the Contractor recognized such error, inconsistency or omission and knowingly failed to report it to the Architect. If the Contractor performs any construction activity knowing it involves a recognized error, inconsistency or omission in the Contract Documents without such notice to the Architect, the Contractor shall assume appropriate responsibility for such performance and shall bear an appropriate amount of the attributable costs for correction.

3.2.2 The Contractor shall take field measurements and verify field conditions and shall carefully compare such field measurements and conditions and other information known to the Contractor with the Contract Documents before commencing activities. Errors, inconsistencies or omissions discovered shall be reported to the Architect at once.

3.2.3 The Contractor shall perform the Work in accordance with the Contract Documents and submittals approved pursuant to Paragraph 3.12.

3.3 SUPERVISION AND CONSTRUCTION PROCEDURES

3.3.1 The Contractor shall supervise and direct the Work, using the Contractor's best skill and attention. The Contractor shall be solely responsible for and have control over construction means, methods, techniques, sequences and procedures and for coordinating all portions of the Work under the Contract, unless Contract Documents give other specific instructions concerning these matters.

3.3.2 The Contractor shall be responsible to the Owner for acts and omissions of the Contractor's employees, subcontractors and their agents and employees, and other persons performing portions of the Work under a contract with the Contractor.

3.3.3 The Contractor shall not be relieved of obligations to perform the Work in accordance with the Contract Documents either by activities or duties of the Architect in the Architect's administration of the Contract, or by tests, inspections or approvals required or performed by persons other than the Contractor.

3.3.4 The Contractor shall be responsible for inspection of portions of Work already performed under this Contract to determine that such portions are in proper condition to receive subsequent Work.

3.4 LABOR AND MATERIALS

3.4.1 Unless otherwise provided in the Contract Documents, the Contractor shall provide and pay for labor, materials, equipment, tools, construction equipment and machinery, water, heat, utilities, transportation, and other facilities and services necessary for proper execution and completion of the Work, whether temporary or permanent and whether or not incorporated or to be incorporated in the Work.

3.4.2 The Contractor shall enforce strict discipline and good order among the Contractor's employees and other persons carrying out the Contract. The Contractor shall not permit employment of unfit persons or persons not skilled in tasks assigned to them.

3.5 WARRANTY

3.5.1 The Contractor warrants to the Owner and Architect that materials and equipment furnished under the Contract will be of good quality and new unless otherwise required or permitted by the Contract Documents, that the Work will be free from defects not inherent in the quality required or permitted, and that the Work will conform with the requirements of the Contract Documents. Work not conforming to these requirements, including substitutions not properly approved and authorized, may be considered defective. The Contractor's warranty excludes remedy for damage or defect caused by abuse, modifications not executed by the Contractor, improper or insufficient maintenance, improper operation, or normal wear and tear under normal usage. If required by the Architect, the Contractor shall furnish satisfactory evidence as to the kind and quality of materials and equipment.

3.6 TAXES

3.6.1 The Contractor shall pay sales, consumer, use and similar taxes for the Work or portions thereof provided by the Contractor which are legally enacted when bids are received or negotiations concluded, whether or not yet effective or merely scheduled to go into effect.

3.7 PERMITS, FEES AND NOTICES

3.7.1 Unless otherwise provided in the Contract Documents, the Contractor shall secure and pay for the building permit and other permits and governmental fees, licenses and inspections necessary for proper execution and completion of the Work which are customarily secured after execution of the Contract and which are legally required when bids are received or negotiations concluded.

3.7.2 The Contractor shall comply with and give notices required by laws, ordinances, rules, regulations and lawful orders of public authorities bearing on performance of the Work.

3.7.3 It is not the Contractor's responsibility to ascertain that the Contract Documents are in accordance with applicable laws, statutes, ordinances, building codes, and rules and regulations. However, if the Contractor observes that portions of the Contract Documents are at variance therewith, the Contractor shall promptly notify the Architect and Owner in writing, and necessary changes shall be accomplished by appropriate Modification.

3.7.4 If the Contractor performs Work knowing it to be contrary to laws, statutes, ordinances, building codes, and rules and regulations without such notice to the Architect and Owner, the Contractor shall assume full responsibility for such Work and shall bear the attributable costs.

3.8 ALLOWANCES

3.8.1 The Contractor shall include in the Contract Sum all allowances stated in the Contract Documents. Items covered by allowances shall be supplied for such amounts and by such persons or entities as the Owner may direct, but the Contractor shall not be required to employ persons or entities against which the Contractor makes reasonable objection.

3.8.2 Unless otherwise provided in the Contract Documents:

.1 materials and equipment under an allowance shall be selected promptly by the Owner to avoid delay in the Work;

.2 allowances shall cover the cost to the Contractor of materials and equipment delivered at the site and all required taxes, less applicable trade discounts;

.3 Contractor's costs for unloading and handling at the site, labor, installation costs, overhead, profit and other expenses contemplated for stated allowance amounts shall be included in the Contract Sum and not in the allowances;

.4 whenever costs are more than or less than allowances, the Contract Sum shall be adjusted accordingly by Change Order. The amount of the Change Order shall reflect (1) the difference between actual costs and the allowances under Clause 3.8.2.2 and (2) changes in Contractor's costs under Clause 3.8.2.3.

3.9 SUPERINTENDENT

3.9.1 The Contractor shall employ a competent superintendent and necessary assistants who shall be in attendance at the Project site during performance of the Work. The superintendent shall represent the Contractor, and communications given to the superintendent shall be as binding as if given to the Contractor. Important communications shall be confirmed in writing. Other communications shall be similarly confirmed on written request in each case.

3.10 CONTRACTOR'S CONSTRUCTION SCHEDULES

3.10.1 The Contractor, promptly after being awarded the Contract, shall prepare and submit for the Owner's and Architect's information a Contractor's construction schedule for the Work. The schedule shall not exceed time limits current under the Contract Documents, shall be revised at appropriate intervals as required by the conditions of the Work and Project, shall be related to the entire Project to the extent required by the Contract Documents, and shall provide for expeditious and practicable execution of the Work.

3.10.2 The Contractor shall prepare and keep current, for the Architect's approval, a schedule of submittals which is coordinated with the Contractor's construction schedule and allows the Architect reasonable time to review submittals.

3.10.3 The Contractor shall conform to the most recent schedules.

3.11 DOCUMENTS AND SAMPLES AT THE SITE

3.11.1 The Contractor shall maintain at the site for the Owner one record copy of the Drawings, Specifications, addenda, Change Orders and other Modifications, in good order and marked currently to record changes and selections made during construction, and in addition approved Shop Drawings, Product Data, Samples and similar required submittals. These shall be available to the Architect and shall be delivered to the Architect for submittal to the Owner upon completion of the Work.

3.12 SHOP DRAWINGS, PRODUCT DATA AND SAMPLES

3.12.1 Shop Drawings are drawings, diagrams, schedules and other data specially prepared for the Work by the Contractor or a Subcontractor, Sub-subcontractor, manufacturer, supplier or distributor to illustrate some portion of the Work.

3.12.2 Product Data are illustrations, standard schedules, performance charts, instructions, brochures, diagrams and other information furnished by the Contractor to illustrate materials or equipment for some portion of the Work.

3.12.3 Samples are physical examples which illustrate materials, equipment or workmanship and establish standards by which the Work will be judged.

3.12.4 Shop Drawings, Product Data, Samples and similar submittals are not Contract Documents. The purpose of their submittal is to demonstrate for those portions of the Work for which submittals are required the way the Contractor proposes to conform to the information given and the design concept expressed in the Contract Documents. Review by the Architect is subject to the limitations of Subparagraph 4.2.7.

3.12.5 The Contractor shall review, approve and submit to the Architect Shop Drawings, Product Data, Samples and similar submittals required by the Contract Documents with reasonable promptness and in such sequence as to cause no delay in the Work or in the activities of the Owner or of separate contractors. Submittals made by the Contractor which are not required by the Contract Documents may be returned without action.

3.12.6 The Contractor shall perform no portion of the Work requiring submittal and review of Shop Drawings, Product Data, Samples or similar submittals until the respective submittal has been approved by the Architect. Such Work shall be in accordance with approved submittals.

3.12.7 By approving and submitting Shop Drawings, Product Data, Samples and similar submittals, the Contractor represents that the Contractor has determined and verified materials, field measurements and field construction criteria related thereto, or will do so, and has checked and coordinated the information contained within such submittals with the requirements of the Work and of the Contract Documents.

3.12.8 The Contractor shall not be relieved of responsibility for deviations from requirements of the Contract Documents by the Architect's approval of Shop Drawings, Product Data, Samples or similar submittals unless the Contractor has specifically informed the Architect in writing of such deviation at the time of submittal and the Architect has given written approval to the specific deviation. The Contractor shall not be relieved of responsibility for errors or omissions in Shop Drawings, Product Data, Samples or similar submittals by the Architect's approval thereof.

3.12.9 The Contractor shall direct specific attention, in writing or on resubmitted Shop Drawings, Product Data, Samples or similar submittals, to revisions other than those requested by the Architect on previous submittals.

3.12.10 Informational submittals upon which the Architect is not expected to take responsive action may be so identified in the Contract Documents.

3.12.11 When professional certification of performance criteria of materials, systems or equipment is required by the Contract Documents, the Architect shall be entitled to rely upon the accuracy and completeness of such calculations and certifications.

3.13 USE OF SITE

3.13.1 The Contractor shall confine operations at the site to areas permitted by law, ordinances, permits and the Contract Documents and shall not unreasonably encumber the site with materials or equipment.

3.14 CUTTING AND PATCHING

3.14.1 The Contractor shall be responsible for cutting, fitting or patching required to complete the Work or to make its parts fit together properly.

3.14.2 The Contractor shall not damage or endanger a portion of the Work or fully or partially completed construction of the Owner or separate contractors by cutting, patching or otherwise altering such construction, or by excavation. The Contractor shall not cut or otherwise alter such construction by the

Owner or a separate contractor except with written consent of the Owner and of such separate contractor; such consent shall not be unreasonably withheld. The Contractor shall not unreasonably withhold from the Owner or a separate contractor the Contractor's consent to cutting or otherwise altering the Work.

3.15 CLEANING UP

3.15.1 The Contractor shall keep the premises and surrounding area free from accumulation of waste materials or rubbish caused by operations under the Contract. At completion of the Work the Contractor shall remove from and about the Project waste materials, rubbish, the Contractor's tools, construction equipment, machinery and surplus materials.

3.15.2 If the Contractor fails to clean up as provided in the Contract Documents, the Owner may do so and the cost thereof shall be charged to the Contractor.

3.16 ACCESS TO WORK

3.16.1 The Contractor shall provide the Owner and Architect access to the Work in preparation and progress wherever located.

3.17 ROYALTIES AND PATENTS

3.17.1 The Contractor shall pay all royalties and license fees. The Contractor shall defend suits or claims for infringement of patent rights and shall hold the Owner and Architect harmless from loss on account thereof, but shall not be responsible for such defense or loss when a particular design, process or product of a particular manufacturer or manufacturers is required by the Contract Documents. However, if the Contractor has reason to believe that the required design, process or product is an infringement of a patent, the Contractor shall be responsible for such loss unless such information is promptly furnished to the Architect.

3.18 INDEMNIFICATION

3.18.1 To the fullest extent permitted by law, the Contractor shall indemnify and hold harmless the Owner, Architect, Architect's consultants, and agents and employees of any of them from and against claims, damages, losses and expenses, including but not limited to attorneys' fees, arising out of or resulting from performance of the Work, provided that such claim, damage, loss or expense is attributable to bodily injury, sickness, disease or death, or to injury to or destruction of tangible property (other than the Work itself) including loss of use resulting therefrom, but only to the extent caused in whole or in part by negligent acts or omissions of the Contractor, a Subcontractor, anyone directly or indirectly employed by them or anyone for whose acts they may be liable, regardless of whether or not such claim, damage, loss or expense is caused in part by a party indemnified hereunder. Such obligation shall not be construed to negate, abridge, or reduce other rights or obligations of indemnity which would otherwise exist as to a party or person described in this Paragraph 3.18.

3.18.2 In claims against any person or entity indemnified under this Paragraph 3.18 by an employee of the Contractor, a Subcontractor, anyone directly or indirectly employed by them or anyone for whose acts they may be liable, the indemnification obligation under this Paragraph 3.18 shall not be limited by a limitation on amount or type of damages, compensation or benefits payable by or for the Contractor or a Subcontractor under workers' or workmen's compensation acts, disability benefit acts or other employee benefit acts.

3.18.3 The obligations of the Contractor under this Paragraph 3.18 shall not extend to the liability of the Architect, the Architect's consultants, and agents and employees of any of them arising out of (1) the preparation or approval of maps, drawings, opinions, reports, surveys, Change Orders, designs or specifications, or (2) the giving of or the failure to give directions or instructions by the Architect, the Architect's consultants, and agents and employees of any of them provided such giving or failure to give is the primary cause of the injury or damage.

ARTICLE 4

ADMINISTRATION OF THE CONTRACT

4.1 ARCHITECT

4.1.1 The Architect is the person lawfully licensed to practice architecture or an entity lawfully practicing architecture identified as such in the Agreement and is referred to throughout the Contract Documents as if singular in number. The term "Architect" means the Architect or the Architect's authorized representative.

4.1.2 Duties, responsibilities and limitations of authority of the Architect as set forth in the Contract Documents shall not be restricted, modified or extended without written consent of the Owner, Contractor and Architect. Consent shall not be unreasonably withheld.

4.1.3 In case of termination of employment of the Architect, the Owner shall appoint an architect against whom the Contractor makes no reasonable objection and whose status under the Contract Documents shall be that of the former architect.

4.1.4 Disputes arising under Subparagraphs 4.1.2 and 4.1.3 shall be subject to arbitration.

4.2 ARCHITECT'S ADMINISTRATION OF THE CONTRACT

4.2.1 The Architect will provide administration of the Contract as described in the Contract Documents, and will be the Owner's representative (1) during construction, (2) until final payment is due and (3) with the Owner's concurrence, from time to time during the correction period described in Paragraph 12.2. The Architect will advise and consult with the Owner. The Architect will have authority to act on behalf of the Owner only to the extent provided in the Contract Documents, unless otherwise modified by written instrument in accordance with other provisions of the Contract.

4.2.2 The Architect will visit the site at intervals appropriate to the stage of construction to become generally familiar with the progress and quality of the completed Work and to determine in general if the Work is being performed in a manner indicating that the Work, when completed, will be in accordance with the Contract Documents. However, the Architect will not be required to make exhaustive or continuous on-site inspections to check quality or quantity of the Work. On the basis of on-site observations as an architect, the Architect will keep the Owner informed of progress of the Work, and will endeavor to guard the Owner against defects and deficiencies in the Work.

4.2.3 The Architect will not have control over or charge of and will not be responsible for construction means, methods, techniques, sequences or procedures, or for safety precautions and programs in connection with the Work, since these are solely the Contractor's responsibility as provided in Paragraph 3.3. The Architect will not be responsible for the Contractor's failure to carry out the Work in accordance with the Contract Documents. The Architect will not have control over or charge of and will not be responsible for acts or omissions of the Contractor, Subcontractors, or their agents or employees, or of any other persons performing portions of the Work.

4.2.4 Communications Facilitating Contract Administration. Except as otherwise provided in the Contract Documents or when direct communications have been specially authorized, the Owner and Contractor shall endeavor to communicate through the Architect. Communications by and with the Architect's consultants shall be through the Architect. Communications by and with Subcontractors and material suppliers shall be through the Contractor. Communications by and with separate contractors shall be through the Owner.

4.2.5 Based on the Architect's observations and evaluations of the Contractor's Applications for Payment, the Architect will review and certify the amounts due the Contractor and will issue Certificates for Payment in such amounts.

4.2.6 The Architect will have authority to reject Work which does not conform to the Contract Documents. Whenever the Architect considers it necessary or advisable for implementation of the intent of the Contract Documents, the Architect will have authority to require additional inspection or testing of the Work in accordance with Subparagraphs 13.5.2 and 13.5.3, whether or not such Work is fabricated, installed or completed. However, neither this authority of the Architect nor a decision made in good faith either to exercise or not to exercise such authority shall give rise to a duty or responsibility of the Architect to the Contractor, Subcontractors, material and equipment suppliers, their agents or employees, or other persons performing portions of the Work.

4.2.7 The Architect will review and approve or take other appropriate action upon the Contractor's submittals such as Shop Drawings, Product Data and Samples, but only for the limited purpose of checking for conformance with information given and the design concept expressed in the Contract Documents. The Architect's action will be taken with such reasonable promptness as to cause no delay in the Work or in the activities of the Owner, Contractor or separate contractors, while allowing sufficient time in the Architect's professional judgment to permit adequate review. Review of such submittals is not conducted for the purpose of determining the accuracy and completeness of other details such as dimensions and quantities, or for substantiating instructions for installation or performance of equipment or systems, all of which remain the responsibility of the Contractor as required by the Contract Documents. The Architect's review of the Contractor's submittals shall not relieve the Contractor of the obligations under Paragraphs 3.3, 3.5 and 3.12. The Architect's review shall not constitute approval of safety precautions or, unless otherwise specifically stated by the Architect, of any construction means, methods, techniques, sequences or procedures. The Architect's approval of a specific item shall not indicate approval of an assembly of which the item is a component.

4.2.8 The Architect will prepare Change Orders and Construction Change Directives, and may authorize minor changes in the Work as provided in Paragraph 7.4.

4.2.9 The Architect will conduct inspections to determine the date or dates of Substantial Completion and the date of final completion, will receive and forward to the Owner for the Owner's review and records written warranties and related documents required by the Contract and assembled by the Contractor, and will issue a final Certificate for Payment upon compliance with the requirements of the Contract Documents.

4.2.10 If the Owner and Architect agree, the Architect will provide one or more project representatives to assist in carrying out the Architect's responsibilities at the site. The duties, responsibilities and limitations of authority of such project representatives shall be as set forth in an exhibit to be incorporated in the Contract Documents.

4.2.11 The Architect will interpret and decide matters concerning performance under and requirements of the Contract Documents on written request of either the Owner or Contractor. The Architect's response to such requests will be made with reasonable promptness and within any time limits agreed upon. If no agreement is made concerning the time within which interpretations required of the Architect shall be furnished in compliance with this Paragraph 4.2, then delay shall not be recognized on account of failure by the Architect to furnish such interpretations until 15 days after written request is made for them.

4.2.12 Interpretations and decisions of the Architect will be consistent with the intent of and reasonably inferable from the Contract Documents and will be in writing or in the form of drawings. When making such interpretations and decisions, the Architect will endeavor to secure faithful performance by both Owner and Contractor, will not show partiality to either and will not be liable for results of interpretations or decisions so rendered in good faith.

4.2.13 The Architect's decisions on matters relating to aesthetic effect will be final if consistent with the intent expressed in the Contract Documents.

4.3 CLAIMS AND DISPUTES

4.3.1 Definition. A Claim is a demand or assertion by one of the parties seeking, as a matter of right, adjustment or interpretation of Contract terms, payment of money, extension of time or other relief with respect to the terms of the Contract. The term "Claim" also includes other disputes and matters in question between the Owner and Contractor arising out of or relating to the Contract. Claims must be made by written notice. The responsibility to substantiate Claims shall rest with the party making the Claim.

4.3.2 Decision of Architect. Claims, including those alleging an error or omission by the Architect, shall be referred initially to the Architect for action as provided in Paragraph 4.4. A decision by the Architect, as provided in Subparagraph 4.4.4, shall be required as a condition precedent to arbitration or litigation of a Claim between the Contractor and Owner as to all such matters arising prior to the date final payment is due, regardless of (1) whether such matters relate to execution and progress of the Work or (2) the extent to which the Work has been completed. The decision by the Architect in response to a Claim shall not be a condition precedent to arbitration or litigation in the event (1) the position of Architect is vacant, (2) the Architect has not received evidence or has failed to render a decision within agreed time limits, (3) the Architect has failed to take action required under Subparagraph 4.4.4 within 30 days after the Claim is made, (4) 45 days have passed after the Claim has been referred to the Architect or (5) the Claim relates to a mechanic's lien.

4.3.3 Time Limits on Claims. Claims by either party must be made within 21 days after occurrence of the event giving rise to such Claim or within 21 days after the claimant first recognizes the condition giving rise to the Claim, whichever is later. Claims must be made by written notice. An additional Claim made after the initial Claim has been implemented by Change Order will not be considered unless submitted in a timely manner.

4.3.4 Continuing Contract Performance. Pending final resolution of a Claim including arbitration, unless otherwise agreed in writing the Contractor shall proceed diligently with performance of the Contract and the Owner shall continue to make payments in accordance with the Contract Documents.

4.3.5 Waiver of Claims: Final Payment. The making of final payment shall constitute a waiver of Claims by the Owner except those arising from:

.1 liens, Claims, security interests or encumbrances arising out of the Contract and unsettled;

.2 failure of the Work to comply with the requirements of the Contract Documents; or

.3 terms of special warranties required by the Contract Documents.

4.3.6 Claims for Concealed or Unknown Conditions. If conditions are encountered at the site which are (1) subsurface or otherwise concealed physical conditions which differ materially from those indicated in the Contract Documents or (2) unknown physical conditions of an unusual nature, which differ materially from those ordinarily found to exist and generally recognized as inherent in construction activities of the character provided for in the Contract Documents, then notice by the observing party shall be given to the other party promptly before conditions are disturbed and in no event later than 21 days after first observance of the conditions. The Architect will promptly investigate such conditions and, if they differ materially and cause an increase or decrease in the Contractor's cost of, or time required for, performance of any part of the Work, will recommend an equitable adjustment in the Contract Sum or Contract Time, or both. If the Architect determines that the conditions at the site are not materially different from those indicated in the Contract Documents and that no change in the terms of the Contract is justified, the Architect shall so notify the Owner and Contractor in writing, stating the reasons. Claims by either party in opposition to such determination must be made within 21 days after the Architect has given notice of the decision. If the Owner and Contractor cannot agree on an adjustment in the Contract Sum or Contract Time, the adjustment shall be referred to the Architect for initial determination, subject to further proceedings pursuant to Paragraph 4.4.

4.3.7 Claims for Additional Cost. If the Contractor wishes to make Claim for an increase in the Contract Sum, written notice as provided herein shall be given before proceeding to execute the Work. Prior notice is not required for Claims relating to an emergency endangering life or property arising under Paragraph 10.3. If the Contractor believes additional cost is involved for reasons including but not limited to (1) a written interpretation from the Architect, (2) an order by the Owner to stop the Work where the Contractor was not at fault, (3) a written order for a minor change in the Work issued by the Architect, (4) failure of payment by the Owner, (5) termination of the Contract by the Owner, (6) Owner's suspension or (7) other reasonable grounds, Claim shall be filed in accordance with the procedure established herein.

4.3.8 Claims for Additional Time

4.3.8.1 If the Contractor wishes to make Claim for an increase in the Contract Time, written notice as provided herein shall be given. The Contractor's Claim shall include an estimate of cost and of probable effect of delay on progress of the Work. In the case of a continuing delay only one Claim is necessary.

4.3.8.2 If adverse weather conditions are the basis for a Claim for additional time, such Claim shall be documented by data substantiating that weather conditions were abnormal for the period of time and could not have been reasonably anticipated, and that weather conditions had an adverse effect on the scheduled construction.

4.3.9 Injury or Damage to Person or Property. If either party to the Contract suffers injury or damage to person or property because of an act or omission of the other party, of any of the other party's employees or agents, or of others for whose acts such party is legally liable, written notice of such injury or damage, whether or not insured, shall be given to the other party within a reasonable time not exceeding 21 days after first observance. The notice shall provide sufficient detail to enable the other party to investigate the matter. If a Claim for additional cost or time related to this Claim is to be asserted, it shall be filed as provided in Subparagraphs 4.3.7 or 4.3.8.

4.4 RESOLUTION OF CLAIMS AND DISPUTES

4.4.1 The Architect will review Claims and take one or more of the following preliminary actions within ten days of receipt of a Claim: (1) request additional supporting data from the claimant, (2) submit a schedule to the parties indicating when the Architect expects to take action, (3) reject the Claim in whole or in part, stating reasons for rejection, (4) recommend approval of the Claim by the other party or (5) suggest a compromise. The Architect may also, but is not obligated to, notify the surety, if any, of the nature and amount of the Claim.

4.4.2 If a Claim has been resolved, the Architect will prepare or obtain appropriate documentation.

4.4.3 If a Claim has not been resolved, the party making the Claim shall, within ten days after the Architect's preliminary response, take one or more of the following actions: (1) submit additional supporting data requested by the Architect, (2) modify the initial Claim or (3) notify the Architect that the initial Claim stands.

4.4.4 If a Claim has not been resolved after consideration of the foregoing and of further evidence presented by the parties or requested by the Architect, the Architect will notify the parties in writing that the Architect's decision will be made within seven days, which decision shall be final and binding on the parties but subject to arbitration. Upon expiration of such time period, the Architect will render to the parties the Architect's written decision relative to the Claim, including any change in the Contract Sum or Contract Time or both. If there is a surety and there appears to be a possibility of a Contractor's default, the Architect may, but is not obligated to, notify the surety and request the surety's assistance in resolving the controversy.

4.5 ARBITRATION

4.5.1 Controversies and Claims Subject to Arbitration. Any controversy or Claim arising out of or related to the Contract, or the breach thereof, shall be settled by arbitration in accordance with the Construction Industry Arbitration Rules of the American Arbitration Association, and judgment upon the award rendered by the arbitrator or arbitrators may be entered in any court having jurisdiction thereof, except controversies or Claims relating to aesthetic effect and except those waived as provided for in Subparagraph 4.3.5. Such controversies or Claims upon which the Architect has given notice and rendered a decision as provided in Subparagraph 4.4.4 shall be subject to arbitration upon written demand of either party. Arbitration may be commenced when 45 days have passed after a Claim has been referred to the Architect as provided in Paragraph 4.3 and no decision has been rendered.

4.5.2 Rules and Notices for Arbitration. Claims between the Owner and Contractor not resolved under Paragraph 4.4 shall, if subject to arbitration under Subparagraph 4.5.1, be decided by arbitration in accordance with the Construction Industry Arbitration Rules of the American Arbitration Association currently in effect, unless the parties mutually agree otherwise. Notice of demand for arbitration shall be filed in writing with the other party to the Agreement between the Owner and Contractor and with the American Arbitration Association, and a copy shall be filed with the Architect.

4.5.3 Contract Performance During Arbitration. During arbitration proceedings, the Owner and Contractor shall comply with Subparagraph 4.3.4.

4.5.4 When Arbitration May Be Demanded. Demand for arbitration of any Claim may not be made until the earlier of (1) the date on which the Architect has rendered a final written decision on the Claim, (2) the tenth day after the parties have presented evidence to the Architect or have been given reasonable opportunity to do so, if the Architect has not rendered a final written decision by that date, or (3) any of the five events described in Subparagraph 4.3.2.

4.5.4.1 When a written decision of the Architect states that (1) the decision is final but subject to arbitration and (2) a demand for arbitration of a Claim covered by such decision must be made within 30 days after the date on which the party making the demand receives the final written decision, then failure to demand arbitration within said 30 days' period shall result in the Architect's decision becoming final and binding upon the Owner and Contractor. If the Architect renders a decision after arbitration proceedings have been initiated, such decision may be entered as evidence, but shall not supersede arbitration proceedings unless the decision is acceptable to all parties concerned.

4.5.4.2 A demand for arbitration shall be made within the time limits specified in Subparagraphs 4.5.1 and 4.5.4 and Clause 4.5.4.1 as applicable, and in other cases within a reasonable time after the Claim has arisen, and in no event shall it be made after the date when institution of legal or equitable proceedings based on such Claim would be barred by the applicable statute of limitations as determined pursuant to Paragraph 13.7.

4.5.5 Limitation on Consolidation or Joinder. No arbitration arising out of or relating to the Contract Documents shall include, by consolidation or joinder or in any other manner, the Architect, the Architect's employees or consultants, except by written consent containing specific reference to the Agreement and signed by the Architect, Owner, Contractor and any other person or entity sought to be joined. No arbitration shall include, by consolidation or joinder or in any other manner, parties other than the Owner, Contractor, a separate contractor as described in Article 6 and other persons substantially involved in a common question of fact or law whose presence is required if complete relief is to be accorded in arbitration. No person or entity other than the Owner, Contractor or a separate contractor as described in Article 6 shall be included as an original third party or additional third party to an arbitration whose interest or responsibility is insubstantial. Consent to arbitration involving an additional person or entity shall not constitute consent to arbitration of a dispute not described therein or with a person or entity not named or described therein. The foregoing agreement to arbitrate and other agreements to arbitrate with an additional person or entity duly consented to by parties to the Agreement shall be specifically enforceable under applicable law in any court having jurisdiction thereof.

4.5.6 Claims and Timely Assertion of Claims. A party who files a notice of demand for arbitration must assert in the demand all Claims then known to that party on which arbitration is permitted to be demanded. When a party fails to include a Claim through oversight, inadvertence or excusable neglect, or when a Claim has matured or been acquired subsequently, the arbitrator or arbitrators may permit amendment.

4.5.7 Judgment on Final Award. The award rendered by the arbitrator or arbitrators shall be final, and judgment may be entered upon it in accordance with applicable law in any court having jurisdiction thereof.

ARTICLE 5

SUBCONTRACTORS

5.1 DEFINITIONS

5.1.1 A Subcontractor is a person or entity who has a direct contract with the Contractor to perform a portion of the Work at the site. The term "Subcontractor" is referred to throughout the Contract Documents as if singular in number and means a Subcontractor or an authorized representative of the Subcontractor. The term "Subcontractor" does not include a separate contractor or subcontractors of a separate contractor.

5.1.2 A Sub-subcontractor is a person or entity who has a direct or indirect contract with a Subcontractor to perform a portion of the Work at the site. The term "Sub-subcontractor" is referred to throughout the Contract Documents as if singular in number and means a Sub-subcontractor or an authorized representative of the Sub-subcontractor.

5.2 AWARD OF SUBCONTRACTS AND OTHER CONTRACTS FOR PORTIONS OF THE WORK

5.2.1 Unless otherwise stated in the Contract Documents or the bidding requirements, the Contractor, as soon as practicable after award of the Contract, shall furnish in writing to the Owner through the Architect the names of persons or entities (including those who are to furnish materials or equipment fabricated to a special design) proposed for each principal portion of the Work. The Architect will promptly reply to the Contractor in writing stating whether or not the Owner or the Architect, after due investigation, has reasonable objection to any such proposed person or entity. Failure of the Owner or Architect to reply promptly shall constitute notice of no reasonable objection.

5.2.2 The Contractor shall not contract with a proposed person or entity to whom the Owner or Architect has made reasonable and timely objection. The Contractor shall not be required to contract with anyone to whom the Contractor has made reasonable objection.

5.2.3 If the Owner or Architect has reasonable objection to a person or entity proposed by the Contractor, the Contractor shall propose another to whom the Owner or Architect has no reasonable objection. The Contract Sum shall be increased or decreased by the difference in cost occasioned by such change and an appropriate Change Order shall be issued. However, no increase in the Contract Sum shall be allowed for such change unless the Contractor has acted promptly and responsively in submitting names as required.

5.2.4 The Contractor shall not change a Subcontractor, person or entity previously selected if the Owner or Architect makes reasonable objection to such change.

5.3 SUBCONTRACTUAL RELATIONS

5.3.1 By appropriate agreement, written where legally required for validity, the Contractor shall require each Subcontractor, to the extent of the Work to be performed by the Subcontractor, to be bound to the Contractor by terms of the Contract Documents, and to assume toward the Contractor all the obligations and responsibilities which the Contractor, by these Documents, assumes toward the Owner and Architect. Each subcontract agreement shall preserve and protect the rights of the Owner and Architect under the Contract Documents with respect to the Work to be performed by the Subcontractor so that subcontracting thereof will not prejudice such rights, and shall allow to the Subcontractor, unless specifically provided otherwise in the subcontract agreement, the benefit of all rights, remedies and redress against the Contractor that the Contractor, by the Contract Documents, has against the Owner. Where appropriate, the Contractor shall require each Subcontractor to enter into similar agreements with Sub-subcontractors. The Contractor shall make available to each proposed Subcontractor, prior to the execution of the subcontract agreement, copies of the Contract Documents to which the Subcontractor will be bound, and, upon written request of the Subcontractor, identify to the Subcontractor terms and conditions of the proposed subcontract agreement which may be at variance with the Contract Documents. Subcontractors shall similarly make copies of applicable portions of such documents available to their respective proposed Sub-subcontractors.

5.4 CONTINGENT ASSIGNMENT OF SUBCONTRACTS

5.4.1 Each subcontract agreement for a portion of the Work is assigned by the Contractor to the Owner provided that:

.1 assignment is effective only after termination of the Contract by the Owner for cause pursuant to Paragraph 14.2 and only for those subcontract agreements which the Owner accepts by notifying the Subcontractor in writing; and

.2 assignment is subject to the prior rights of the surety, if any, obligated under bond relating to the Contract.

5.4.2 If the Work has been suspended for more than 30 days, the Subcontractor's compensation shall be equitably adjusted.

ARTICLE 6

CONSTRUCTION BY OWNER OR BY SEPARATE CONTRACTORS

6.1 OWNER'S RIGHT TO PERFORM CONSTRUCTION AND TO AWARD SEPARATE CONTRACTS

6.1.1 The Owner reserves the right to perform construction or operations related to the Project with the Owner's own forces, and to award separate contracts in connection with other portions of the Project or other construction or operations on the site under Conditions of the Contract identical or substantially similar to these including those portions related to insurance and waiver of subrogation. If the Contractor claims that delay or additional cost is involved because of such action by the Owner, the Contractor shall make such Claim as provided elsewhere in the Contract Documents.

6.1.2 When separate contracts are awarded for different portions of the Project or other construction or operations on the site, the term "Contractor" in the Contract Documents in each case shall mean the Contractor who executes each separate Owner-Contractor Agreement.

6.1.3 The Owner shall provide for coordination of the activities of the Owner's own forces and of each separate contractor with the Work of the Contractor, who shall cooperate with them. The Contractor shall participate with other separate contractors and the Owner in reviewing their construction schedules when directed to do so. The Contractor shall make any revisions to the construction schedule and Contract Sum deemed necessary after a joint review and mutual agreement. The construction schedules shall then constitute the schedules to be used by the Contractor, separate contractors and the Owner until subsequently revised.

6.1.4 Unless otherwise provided in the Contract Documents, when the Owner performs construction or operations related to the Project with the Owner's own forces, the Owner shall be deemed to be subject to the same obligations and to have the same rights which apply to the Contractor under the Conditions of the Contract, including, without excluding others, those stated in Article 3, this Article 6 and Articles 10, 11 and 12.

6.2 MUTUAL RESPONSIBILITY

6.2.1 The Contractor shall afford the Owner and separate contractors reasonable opportunity for introduction and storage of their materials and equipment and performance of their activities and shall connect and coordinate the Contractor's construction and operations with theirs as required by the Contract Documents.

6.2.2 If part of the Contractor's Work depends for proper execution or results upon construction or operations by the Owner or a separate contractor, the Contractor shall, prior to proceeding with that portion of the Work, promptly report to the Architect apparent discrepancies or defects in such other construction that would render it unsuitable for such proper execution and results. Failure of the Contractor so to report shall constitute an acknowledgment that the Owner's or separate contractors' completed or partially completed construction is fit and proper to receive the Contractor's Work, except as to defects not then reasonably discoverable.

6.2.3 Costs caused by delays or by improperly timed activities or defective construction shall be borne by the party responsible therefor.

6.2.4 The Contractor shall promptly remedy damage wrongfully caused by the Contractor to completed or partially completed construction or to property of the Owner or separate contractors as provided in Subparagraph 10.2.5.

6.2.5 Claims and other disputes and matters in question between the Contractor and a separate contractor shall be subject to the provisions of Paragraph 4.3 provided the separate contractor has reciprocal obligations.

6.2.6 The Owner and each separate contractor shall have the same responsibilities for cutting and patching as are described for the Contractor in Paragraph 3.14.

6.3 OWNER'S RIGHT TO CLEAN UP

6.3.1 If a dispute arises among the Contractor, separate contractors and the Owner as to the responsibility under their respective contracts for maintaining the premises and surrounding area free from waste materials and rubbish as described in Paragraph 3.15, the Owner may clean up and allocate the cost among those responsible as the Architect determines to be just.

ARTICLE 7

CHANGES IN THE WORK

7.1 CHANGES

7.1.1 Changes in the Work may be accomplished after execution of the Contract, and without invalidating the Contract, by Change Order, Construction Change Directive or order for a minor change in the Work, subject to the limitations stated in this Article 7 and elsewhere in the Contract Documents.

7.1.2 A Change Order shall be based upon agreement among the Owner, Contractor and Architect; a Construction Change Directive requires agreement by the Owner and Architect and may or may not be agreed to by the Contractor; an order for a minor change in the Work may be issued by the Architect alone.

7.1.3 Changes in the Work shall be performed under applicable provisions of the Contract Documents, and the Contractor shall proceed promptly, unless otherwise provided in the Change Order, Construction Change Directive or order for a minor change in the Work.

7.1.4 If unit prices are stated in the Contract Documents or subsequently agreed upon, and if quantities originally contemplated are so changed in a proposed Change Order or Construction Change Directive that application of such unit prices to quantities of Work proposed will cause substantial inequity to the Owner or Contractor, the applicable unit prices shall be equitably adjusted.

7.2 CHANGE ORDERS

7.2.1 A Change Order is a written instrument prepared by the Architect and signed by the Owner, Contractor and Architect, stating their agreement upon all of the following:

- **.1** a change in the Work;
- **.2** the amount of the adjustment in the Contract Sum, if any; and
- **.3** the extent of the adjustment in the Contract Time, if any.

7.2.2 Methods used in determining adjustments to the Contract Sum may include those listed in Subparagraph 7.3.3.

7.3 CONSTRUCTION CHANGE DIRECTIVES

7.3.1 A Construction Change Directive is a written order prepared by the Architect and signed by the Owner and Architect, directing a change in the Work and stating a proposed basis for adjustment, if any, in the Contract Sum or Contract Time, or both. The Owner may by Construction Change Directive, without invalidating the Contract, order changes in the Work within the general scope of the Contract consisting of additions, deletions or other revisions, the Contract Sum and Contract Time being adjusted accordingly.

7.3.2 A Construction Change Directive shall be used in the absence of total agreement on the terms of a Change Order.

7.3.3 If the Construction Change Directive provides for an adjustment to the Contract Sum, the adjustment shall be based on one of the following methods:

- **.1** mutual acceptance of a lump sum properly itemized and supported by sufficient substantiating data to permit evaluation;
- **.2** unit prices stated in the Contract Documents or subsequently agreed upon;
- **.3** cost to be determined in a manner agreed upon by the parties and a mutually acceptable fixed or percentage fee; or
- **.4** as provided in Subparagraph 7.3.6.

7.3.4 Upon receipt of a Construction Change Directive, the Contractor shall promptly proceed with the change in the Work involved and advise the Architect of the Contractor's agreement or disagreement with the method, if any, provided in the Construction Change Directive for determining the proposed adjustment in the Contract Sum or Contract Time.

7.3.5 A Construction Change Directive signed by the Contractor indicates the agreement of the Contractor therewith, including adjustment in Contract Sum and Contract Time or the method for determining them. Such agreement shall be effective immediately and shall be recorded as a Change Order.

7.3.6 If the Contractor does not respond promptly or disagrees with the method for adjustment in the Contract Sum, the method and the adjustment shall be determined by the Architect on the basis of reasonable expenditures and savings of those performing the Work attributable to the change, including, in case of an increase in the Contract Sum, a reasonable allowance for overhead and profit. In such case, and also under Clause 7.3.3.3, the Contractor shall keep and present, in such form as the Architect may prescribe, an itemized accounting together with appropriate supporting data. Unless otherwise provided in the Contract Documents, costs for the purposes of this Subparagraph 7.3.6 shall be limited to the following:

- **.1** costs of labor, including social security, old age and unemployment insurance, fringe benefits required by agreement or custom, and workers' or workmen's compensation insurance;
- **.2** costs of materials, supplies and equipment, including cost of transportation, whether incorporated or consumed;
- **.3** rental costs of machinery and equipment, exclusive of hand tools, whether rented from the Contractor or others;
- **.4** costs of premiums for all bonds and insurance, permit fees, and sales, use or similar taxes related to the Work; and
- **.5** additional costs of supervision and field office personnel directly attributable to the change.

7.3.7 Pending final determination of cost to the Owner, amounts not in dispute may be included in Applications for Payment. The amount of credit to be allowed by the Contractor to the Owner for a deletion or change which results in a net decrease in the Contract Sum shall be actual net cost as confirmed by the Architect. When both additions and credits covering related Work or substitutions are involved in a change, the allowance for overhead and profit shall be figured on the basis of net increase, if any, with respect to that change.

7.3.8 If the Owner and Contractor do not agree with the adjustment in Contract Time or the method for determining it, the adjustment or the method shall be referred to the Architect for determination.

7.3.9 When the Owner and Contractor agree with the determination made by the Architect concerning the adjustments in the Contract Sum and Contract Time, or otherwise reach agreement upon the adjustments, such agreement shall be effective immediately and shall be recorded by preparation and execution of an appropriate Change Order.

7.4 MINOR CHANGES IN THE WORK

7.4.1 The Architect will have authority to order minor changes in the Work not involving adjustment in the Contract Sum or extension of the Contract Time and not inconsistent with the intent of the Contract Documents. Such changes shall be effected by written order and shall be binding on the Owner and Contractor. The Contractor shall carry out such written orders promptly.

ARTICLE 8

TIME

8.1 DEFINITIONS

8.1.1 Unless otherwise provided, Contract Time is the period of time, including authorized adjustments, allotted in the Contract Documents for Substantial Completion of the Work.

8.1.2 The date of commencement of the Work is the date established in the Agreement. The date shall not be postponed by the failure to act of the Contractor or of persons or entities for whom the Contractor is responsible.

8.1.3 The date of Substantial Completion is the date certified by the Architect in accordance with Paragraph 9.8.

8.1.4 The term "day" as used in the Contract Documents shall mean calendar day unless otherwise specifically defined.

8.2 PROGRESS AND COMPLETION

8.2.1 Time limits stated in the Contract Documents are of the essence of the Contract. By executing the Agreement the Contractor confirms that the Contract Time is a reasonable period for performing the Work.

8.2.2 The Contractor shall not knowingly, except by agreement or instruction of the Owner in writing, prematurely commence operations on the site or elsewhere prior to the effective date of insurance required by Article 11 to be furnished by the Contractor. The date of commencement of the Work shall not be changed by the effective date of such insurance. Unless the date of commencement is established by a notice to proceed given by the Owner, the Contractor shall notify the Owner in writing not less than five days or other agreed period before commencing the Work to permit the timely filing of mortgages, mechanic's liens and other security interests.

8.2.3 The Contractor shall proceed expeditiously with adequate forces and shall achieve Substantial Completion within the Contract Time.

8.3 DELAYS AND EXTENSIONS OF TIME

8.3.1 If the Contractor is delayed at any time in progress of the Work by an act or neglect of the Owner or Architect, or of an employee of either, or of a separate contractor employed by the Owner, or by changes ordered in the Work, or by labor disputes, fire, unusual delay in deliveries, unavoidable casualties or other causes beyond the Contractor's control, or by delay authorized by the Owner pending arbitration, or by other causes which the Architect determines may justify delay, then the Contract Time shall be extended by Change Order for such reasonable time as the Architect may determine.

8.3.2 Claims relating to time shall be made in accordance with applicable provisions of Paragraph 4.3.

8.3.3 This Paragraph 8.3 does not preclude recovery of damages for delay by either party under other provisions of the Contract Documents.

ARTICLE 9

PAYMENTS AND COMPLETION

9.1 CONTRACT SUM

9.1.1 The Contract Sum is stated in the Agreement and, including authorized adjustments, is the total amount payable by the Owner to the Contractor for performance of the Work under the Contract Documents.

9.2 SCHEDULE OF VALUES

9.2.1 Before the first Application for Payment, the Contractor shall submit to the Architect a schedule of values allocated to various portions of the Work, prepared in such form and supported by such data to substantiate its accuracy as the Architect may require. This schedule, unless objected to by the Architect, shall be used as a basis for reviewing the Contractor's Applications for Payment.

9.3 APPLICATIONS FOR PAYMENT

9.3.1 At least ten days before the date established for each progress payment, the Contractor shall submit to the Architect an itemized Application for Payment for operations completed in accordance with the schedule of values. Such application shall be notarized, if required, and supported by such data substantiating the Contractor's right to payment as the Owner or Architect may require, such as copies of requisitions from Subcontractors and material suppliers, and reflecting retainage if provided for elsewhere in the Contract Documents.

9.3.1.1 Such applications may include requests for payment on account of changes in the Work which have been properly authorized by Construction Change Directives but not yet included in Change Orders.

9.3.1.2 Such applications may not include requests for payment of amounts the Contractor does not intend to pay to a Subcontractor or material supplier because of a dispute or other reason.

9.3.2 Unless otherwise provided in the Contract Documents, payments shall be made on account of materials and equipment delivered and suitably stored at the site for subsequent incorporation in the Work. If approved in advance by the Owner, payment may similarly be made for materials and equipment suitably stored off the site at a location agreed upon in writing. Payment for materials and equipment stored on or off the site shall be conditioned upon compliance by the Contractor with procedures satisfactory to the Owner to establish the Owner's title to such materials and equipment or otherwise protect the Owner's interest, and shall include applicable insurance, storage and transportation to the site for such materials and equipment stored off the site.

9.3.3 The Contractor warrants that title to all Work covered by an Application for Payment will pass to the Owner no later than the time of payment. The Contractor further warrants that upon submittal of an Application for Payment all Work for which Certificates for Payment have been previously issued and payments received from the Owner shall, to the best of the Contractor's knowledge, information and belief, be free and clear of liens, claims, security interests or encumbrances in favor of the Contractor, Subcontractors, material suppliers, or other persons or entities making a claim by reason of having provided labor, materials and equipment relating to the Work.

9.4 CERTIFICATES FOR PAYMENT

9.4.1 The Architect will, within seven days after receipt of the Contractor's Application for Payment, either issue to the

Owner a Certificate for Payment, with a copy to the Contractor, for such amount as the Architect determines is properly due, or notify the Contractor and Owner in writing of the Architect's reasons for withholding certification in whole or in part as provided in Subparagraph 9.5.1.

9.4.2 The issuance of a Certificate for Payment will constitute a representation by the Architect to the Owner, based on the Architect's observations at the site and the data comprising the Application for Payment, that the Work has progressed to the point indicated and that, to the best of the Architect's knowledge, information and belief, quality of the Work is in accordance with the Contract Documents. The foregoing representations are subject to an evaluation of the Work for conformance with the Contract Documents upon Substantial Completion, to results of subsequent tests and inspections, to minor deviations from the Contract Documents correctable prior to completion and to specific qualifications expressed by the Architect. The issuance of a Certificate for Payment will further constitute a representation that the Contractor is entitled to payment in the amount certified. However, the issuance of a Certificate for Payment will not be a representation that the Architect has (1) made exhaustive or continuous on-site inspections to check the quality or quantity of the Work, (2) reviewed construction means, methods, techniques, sequences or procedures, (3) reviewed copies of requisitions received from Subcontractors and material suppliers and other data requested by the Owner to substantiate the Contractor's right to payment or (4) made examination to ascertain how or for what purpose the Contractor has used money previously paid on account of the Contract Sum.

9.5 DECISIONS TO WITHHOLD CERTIFICATION

9.5.1 The Architect may decide not to certify payment and may withhold a Certificate for Payment in whole or in part, to the extent reasonably necessary to protect the Owner, if in the Architect's opinion the representations to the Owner required by Subparagraph 9.4.2 cannot be made. If the Architect is unable to certify payment in the amount of the Application, the Architect will notify the Contractor and Owner as provided in Subparagraph 9.4.1. If the Contractor and Architect cannot agree on a revised amount, the Architect will promptly issue a Certificate for Payment for the amount for which the Architect is able to make such representations to the Owner. The Architect may also decide not to certify payment or, because of subsequently discovered evidence or subsequent observations, may nullify the whole or a part of a Certificate for Payment previously issued, to such extent as may be necessary in the Architect's opinion to protect the Owner from loss because of:

- **.1** defective Work not remedied;
- **.2** third party claims filed or reasonable evidence indicating probable filing of such claims;
- **.3** failure of the Contractor to make payments properly to Subcontractors or for labor, materials or equipment;
- **.4** reasonable evidence that the Work cannot be completed for the unpaid balance of the Contract Sum;
- **.5** damage to the Owner or another contractor;
- **.6** reasonable evidence that the Work will not be completed within the Contract Time, and that the unpaid balance would not be adequate to cover actual or liquidated damages for the anticipated delay; or
- **.7** persistent failure to carry out the Work in accordance with the Contract Documents.

9.5.2 When the above reasons for withholding certification are removed, certification will be made for amounts previously withheld.

9.6 PROGRESS PAYMENTS

9.6.1 After the Architect has issued a Certificate for Payment, the Owner shall make payment in the manner and within the time provided in the Contract Documents, and shall so notify the Architect.

9.6.2 The Contractor shall promptly pay each Subcontractor, upon receipt of payment from the Owner, out of the amount paid to the Contractor on account of such Subcontractor's portion of the Work, the amount to which said Subcontractor is entitled, reflecting percentages actually retained from payments to the Contractor on account of such Subcontractor's portion of the Work. The Contractor shall, by appropriate agreement with each Subcontractor, require each Subcontractor to make payments to Sub-subcontractors in similar manner.

9.6.3 The Architect will, on request, furnish to a Subcontractor, if practicable, information regarding percentages of completion or amounts applied for by the Contractor and action taken thereon by the Architect and Owner on account of portions of the Work done by such Subcontractor.

9.6.4 Neither the Owner nor Architect shall have an obligation to pay or to see to the payment of money to a Subcontractor except as may otherwise be required by law.

9.6.5 Payment to material suppliers shall be treated in a manner similar to that provided in Subparagraphs 9.6.2, 9.6.3 and 9.6.4.

9.6.6 A Certificate for Payment, a progress payment, or partial or entire use or occupancy of the Project by the Owner shall not constitute acceptance of Work not in accordance with the Contract Documents.

9.7 FAILURE OF PAYMENT

9.7.1 If the Architect does not issue a Certificate for Payment, through no fault of the Contractor, within seven days after receipt of the Contractor's Application for Payment, or if the Owner does not pay the Contractor within seven days after the date established in the Contract Documents the amount certified by the Architect or awarded by arbitration, then the Contractor may, upon seven additional days' written notice to the Owner and Architect, stop the Work until payment of the amount owing has been received. The Contract Time shall be extended appropriately and the Contract Sum shall be increased by the amount of the Contractor's reasonable costs of shut-down, delay and start-up, which shall be accomplished as provided in Article 7.

9.8 SUBSTANTIAL COMPLETION

9.8.1 Substantial Completion is the stage in the progress of the Work when the Work or designated portion thereof is sufficiently complete in accordance with the Contract Documents so the Owner can occupy or utilize the Work for its intended use.

9.8.2 When the Contractor considers that the Work, or a portion thereof which the Owner agrees to accept separately, is substantially complete, the Contractor shall prepare and submit to the Architect a comprehensive list of items to be completed or corrected. The Contractor shall proceed promptly to complete and correct items on the list. Failure to include an item on such list does not alter the responsibility of the Contractor to complete all Work in accordance with the Contract Documents. Upon receipt of the Contractor's list, the Architect will make an inspection to determine whether the Work or designated portion thereof is substantially complete. If the Architect's inspection discloses any item, whether or not included on the Contractor's list, which is not in accordance with the requirements of the Contract Documents, the Contractor shall, before issuance of the Certificate of Substantial Completion, complete or correct such item upon notification by the Architect. The Contractor shall then submit a request for another inspection by the Architect to determine Substantial Completion. When the Work or designated portion thereof is substantially complete, the Architect will prepare a Certificate of Substantial Completion which shall establish the date of Substantial Completion, shall establish responsibilities of the Owner and Contractor for security, maintenance, heat, utilities, damage to the Work and insurance, and shall fix the time within which the Contractor shall finish all items on the list accompanying the Certificate. Warranties required by the Contract Documents shall commence on the date of Substantial Completion of the Work or designated portion thereof unless otherwise provided in the Certificate of Substantial Completion. The Certificate of Substantial Completion shall be submitted to the Owner and Contractor for their written acceptance of responsibilities assigned to them in such Certificate.

9.8.3 Upon Substantial Completion of the Work or designated portion thereof and upon application by the Contractor and certification by the Architect, the Owner shall make payment, reflecting adjustment in retainage, if any, for such Work or portion thereof as provided in the Contract Documents.

9.9 PARTIAL OCCUPANCY OR USE

9.9.1 The Owner may occupy or use any completed or partially completed portion of the Work at any stage when such portion is designated by separate agreement with the Contractor, provided such occupancy or use is consented to by the insurer as required under Subparagraph 11.3.11 and authorized by public authorities having jurisdiction over the Work. Such partial occupancy or use may commence whether or not the portion is substantially complete, provided the Owner and Contractor have accepted in writing the responsibilities assigned to each of them for payments, retainage if any, security, maintenance, heat, utilities, damage to the Work and insurance, and have agreed in writing concerning the period for correction of the Work and commencement of warranties required by the Contract Documents. When the Contractor considers a portion substantially complete, the Contractor shall prepare and submit a list to the Architect as provided under Subparagraph 9.8.2. Consent of the Contractor to partial occupancy or use shall not be unreasonably withheld. The stage of the progress of the Work shall be determined by written agreement between the Owner and Contractor or, if no agreement is reached, by decision of the Architect.

9.9.2 Immediately prior to such partial occupancy or use, the Owner, Contractor and Architect shall jointly inspect the area to be occupied or portion of the Work to be used in order to determine and record the condition of the Work.

9.9.3 Unless otherwise agreed upon, partial occupancy or use of a portion or portions of the Work shall not constitute acceptance of Work not complying with the requirements of the Contract Documents.

9.10 FINAL COMPLETION AND FINAL PAYMENT

9.10.1 Upon receipt of written notice that the Work is ready for final inspection and acceptance and upon receipt of a final Application for Payment, the Architect will promptly make such inspection and, when the Architect finds the Work acceptable under the Contract Documents and the Contract fully performed, the Architect will promptly issue a final Certificate for Payment stating that to the best of the Architect's knowledge, information and belief, and on the basis of the Architect's observations and inspections, the Work has been completed in accordance with terms and conditions of the Contract Documents and that the entire balance found to be due the Contractor and noted in said final Certificate is due and payable. The Architect's final Certificate for Payment will constitute a further representation that conditions listed in Subparagraph 9.10.2 as precedent to the Contractor's being entitled to final payment have been fulfilled.

9.10.2 Neither final payment nor any remaining retained percentage shall become due until the Contractor submits to the Architect (1) an affidavit that payrolls, bills for materials and equipment, and other indebtedness connected with the Work for which the Owner or the Owner's property might be responsible or encumbered (less amounts withheld by Owner) have been paid or otherwise satisfied, (2) a certificate evidencing that insurance required by the Contract Documents to remain in force after final payment is currently in effect and will not be cancelled or allowed to expire until at least 30 days' prior written notice has been given to the Owner, (3) a written statement that the Contractor knows of no substantial reason that the insurance will not be renewable to cover the period required by the Contract Documents, (4) consent of surety, if any, to final payment and (5), if required by the Owner, other data establishing payment or satisfaction of obligations, such as receipts, releases and waivers of liens, claims, security interests or encumbrances arising out of the Contract, to the extent and in such form as may be designated by the Owner. If a Subcontractor refuses to furnish a release or waiver required by the Owner, the Contractor may furnish a bond satisfactory to the Owner to indemnify the Owner against such lien. If such lien remains unsatisfied after payments are made, the Contractor shall refund to the Owner all money that the Owner may be compelled to pay in discharging such lien, including all costs and reasonable attorneys' fees.

9.10.3 If, after Substantial Completion of the Work, final completion thereof is materially delayed through no fault of the Contractor or by issuance of Change Orders affecting final completion, and the Architect so confirms, the Owner shall, upon application by the Contractor and certification by the Architect, and without terminating the Contract, make payment of the balance due for that portion of the Work fully completed and accepted. If the remaining balance for Work not fully completed or corrected is less than retainage stipulated in the Contract Documents, and if bonds have been furnished, the written consent of surety to payment of the balance due for that portion of the Work fully completed and accepted shall be submitted by the Contractor to the Architect prior to certification of such payment. Such payment shall be made under terms and conditions governing final payment, except that it shall not constitute a waiver of claims. The making of final payment shall constitute a waiver of claims by the Owner as provided in Subparagraph 4.3.5.

9.10.4 Acceptance of final payment by the Contractor, a Subcontractor or material supplier shall constitute a waiver of claims by that payee except those previously made in writing and identified by that payee as unsettled at the time of final Application for Payment. Such waivers shall be in addition to the waiver described in Subparagraph 4.3.5.

ARTICLE 10

PROTECTION OF PERSONS AND PROPERTY

10.1 SAFETY PRECAUTIONS AND PROGRAMS

10.1.1 The Contractor shall be responsible for initiating, maintaining and supervising all safety precautions and programs in connection with the performance of the Contract.

10.1.2 In the event the Contractor encounters on the site material reasonably believed to be asbestos or polychlorinated biphenyl (PCB) which has not been rendered harmless, the Contractor shall immediately stop Work in the area affected and report the condition to the Owner and Architect in writing. The Work in the affected area shall not thereafter be resumed except by written agreement of the Owner and Contractor if in fact the material is asbestos or polychlorinated biphenyl (PCB) and has not been rendered harmless. The Work in the affected area shall be resumed in the absence of asbestos or polychlorinated biphenyl (PCB), or when it has been rendered harmless, by written agreement of the Owner and Contractor, or in accordance with final determination by the Architect on which arbitration has not been demanded, or by arbitration under Article 4.

10.1.3 The Contractor shall not be required pursuant to Article 7 to perform without consent any Work relating to asbestos or polychlorinated biphenyl (PCB).

10.1.4 To the fullest extent permitted by law, the Owner shall indemnify and hold harmless the Contractor, Architect, Architect's consultants and agents and employees of any of them from and against claims, damages, losses and expenses, including but not limited to attorneys' fees, arising out of or resulting from performance of the Work in the affected area if in fact the material is asbestos or polychlorinated biphenyl (PCB) and has not been rendered harmless, provided that such claim, damage, loss or expense is attributable to bodily injury, sickness, disease or death, or to injury to or destruction of tangible property (other than the Work itself) including loss of use resulting therefrom, but only to the extent caused in whole or in part by negligent acts or omissions of the Owner, anyone directly or indirectly employed by the Owner or anyone for whose acts the Owner may be liable, regardless of whether or not such claim, damage, loss or expense is caused in part by a party indemnified hereunder. Such obligation shall not be construed to negate, abridge, or reduce other rights or obligations of indemnity which would otherwise exist as to a party or person described in this Subparagraph 10.1.4.

10.2 SAFETY OF PERSONS AND PROPERTY

10.2.1 The Contractor shall take reasonable precautions for safety of, and shall provide reasonable protection to prevent damage, injury or loss to:

- **.1** employees on the Work and other persons who may be affected thereby;
- **.2** the Work and materials and equipment to be incorporated therein, whether in storage on or off the site, under care, custody or control of the Contractor or the Contractor's Subcontractors or Sub-subcontractors; and
- **.3** other property at the site or adjacent thereto, such as trees, shrubs, lawns, walks, pavements, roadways, structures and utilities not designated for removal, relocation or replacement in the course of construction.

10.2.2 The Contractor shall give notices and comply with applicable laws, ordinances, rules, regulations and lawful orders of public authorities bearing on safety of persons or property or their protection from damage, injury or loss.

10.2.3 The Contractor shall erect and maintain, as required by existing conditions and performance of the Contract, reasonable safeguards for safety and protection, including posting danger signs and other warnings against hazards, promulgating safety regulations and notifying owners and users of adjacent sites and utilities.

10.2.4 When use or storage of explosives or other hazardous materials or equipment or unusual methods are necessary for execution of the Work, the Contractor shall exercise utmost care and carry on such activities under supervision of properly qualified personnel.

10.2.5 The Contractor shall promptly remedy damage and loss (other than damage or loss insured under property insurance required by the Contract Documents) to property referred to in Clauses 10.2.1.2 and 10.2.1.3 caused in whole or in part by the Contractor, a Subcontractor, a Sub-subcontractor, or anyone directly or indirectly employed by any of them, or by anyone for whose acts they may be liable and for which the Contractor is responsible under Clauses 10.2.1.2 and 10.2.1.3, except damage or loss attributable to acts or omissions of the Owner or Architect or anyone directly or indirectly employed by either of them, or by anyone for whose acts either of them may be liable, and not attributable to the fault or negligence of the Contractor. The foregoing obligations of the Contractor are in addition to the Contractor's obligations under Paragraph 3.18.

10.2.6 The Contractor shall designate a responsible member of the Contractor's organization at the site whose duty shall be the prevention of accidents. This person shall be the Contractor's superintendent unless otherwise designated by the Contractor in writing to the Owner and Architect.

10.2.7 The Contractor shall not load or permit any part of the construction or site to be loaded so as to endanger its safety.

10.3 EMERGENCIES

10.3.1 In an emergency affecting safety of persons or property, the Contractor shall act, at the Contractor's discretion, to prevent threatened damage, injury or loss. Additional compensation or extension of time claimed by the Contractor on account of an emergency shall be determined as provided in Paragraph 4.3 and Article 7.

ARTICLE 11

INSURANCE AND BONDS

11.1 CONTRACTOR'S LIABILITY INSURANCE

11.1.1 The Contractor shall purchase from and maintain in a company or companies lawfully authorized to do business in the jurisdiction in which the Project is located such insurance as will protect the Contractor from claims set forth below which may arise out of or result from the Contractor's operations under the Contract and for which the Contractor may be legally liable, whether such operations be by the Contractor or by a Subcontractor or by anyone directly or indirectly employed by any of them, or by anyone for whose acts any of them may be liable:

- **.1** claims under workers' or workmen's compensation, disability benefit and other similar employee benefit acts which are applicable to the Work to be performed;

.2 claims for damages because of bodily injury, occupational sickness or disease, or death of the Contractor's employees;

.3 claims for damages because of bodily injury, sickness or disease, or death of any person other than the Contractor's employees;

.4 claims for damages insured by usual personal injury liability coverage which are sustained (1) by a person as a result of an offense directly or indirectly related to employment of such person by the Contractor, or (2) by another person;

.5 claims for damages, other than to the Work itself, because of injury to or destruction of tangible property, including loss of use resulting therefrom;

.6 claims for damages because of bodily injury, death of a person or property damage arising out of ownership, maintenance or use of a motor vehicle; and

.7 claims involving contractual liability insurance applicable to the Contractor's obligations under Paragraph 3.18.

11.1.2 The insurance required by Subparagraph 11.1.1 shall be written for not less than limits of liability specified in the Contract Documents or required by law, whichever coverage is greater. Coverages, whether written on an occurrence or claims-made basis, shall be maintained without interruption from date of commencement of the Work until date of final payment and termination of any coverage required to be maintained after final payment.

11.1.3 Certificates of Insurance acceptable to the Owner shall be filed with the Owner prior to commencement of the Work. These Certificates and the insurance policies required by this Paragraph 11.1 shall contain a provision that coverages afforded under the policies will not be cancelled or allowed to expire until at least 30 days' prior written notice has been given to the Owner. If any of the foregoing insurance coverages are required to remain in force after final payment and are reasonably available, an additional certificate evidencing continuation of such coverage shall be submitted with the final Application for Payment as required by Subparagraph 9.10.2. Information concerning reduction of coverage shall be furnished by the Contractor with reasonable promptness in accordance with the Contractor's information and belief.

11.2 OWNER'S LIABILITY INSURANCE

11.2.1 The Owner shall be responsible for purchasing and maintaining the Owner's usual liability insurance. Optionally, the Owner may purchase and maintain other insurance for self-protection against claims which may arise from operations under the Contract. The Contractor shall not be responsible for purchasing and maintaining this optional Owner's liability insurance unless specifically required by the Contract Documents.

11.3 PROPERTY INSURANCE

11.3.1 Unless otherwise provided, the Owner shall purchase and maintain, in a company or companies lawfully authorized to do business in the jurisdiction in which the Project is located, property insurance in the amount of the initial Contract Sum as well as subsequent modifications thereto for the entire Work at the site on a replacement cost basis without voluntary deductibles. Such property insurance shall be maintained, unless otherwise provided in the Contract Documents or otherwise agreed in writing by all persons and entities who are beneficiaries of such insurance, until final payment has been made as provided in Paragraph 9.10 or until no person or entity other than the Owner has an insurable interest in the property required by this Paragraph 11.3 to be covered, whichever is earlier. This insurance shall include interests of the Owner, the Contractor, Subcontractors and Sub-subcontractors in the Work.

11.3.1.1 Property insurance shall be on an all-risk policy form and shall insure against the perils of fire and extended coverage and physical loss or damage including, without duplication of coverage, theft, vandalism, malicious mischief, collapse, falsework, temporary buildings and debris removal including demolition occasioned by enforcement of any applicable legal requirements, and shall cover reasonable compensation for Architect's services and expenses required as a result of such insured loss. Coverage for other perils shall not be required unless otherwise provided in the Contract Documents.

11.3.1.2 If the Owner does not intend to purchase such property insurance required by the Contract and with all of the coverages in the amount described above, the Owner shall so inform the Contractor in writing prior to commencement of the Work. The Contractor may then effect insurance which will protect the interests of the Contractor, Subcontractors and Sub-subcontractors in the Work, and by appropriate Change Order the cost thereof shall be charged to the Owner. If the Contractor is damaged by the failure or neglect of the Owner to purchase or maintain insurance as described above, without so notifying the Contractor, then the Owner shall bear all reasonable costs properly attributable thereto.

11.3.1.3 If the property insurance requires minimum deductibles and such deductibles are identified in the Contract Documents, the Contractor shall pay costs not covered because of such deductibles. If the Owner or insurer increases the required minimum deductibles above the amounts so identified or if the Owner elects to purchase this insurance with voluntary deductible amounts, the Owner shall be responsible for payment of the additional costs not covered because of such increased or voluntary deductibles. If deductibles are not identified in the Contract Documents, the Owner shall pay costs not covered because of deductibles.

11.3.1.4 Unless otherwise provided in the Contract Documents, this property insurance shall cover portions of the Work stored off the site after written approval of the Owner at the value established in the approval, and also portions of the Work in transit.

11.3.2 Boiler and Machinery Insurance. The Owner shall purchase and maintain boiler and machinery insurance required by the Contract Documents or by law, which shall specifically cover such insured objects during installation and until final acceptance by the Owner; this insurance shall include interests of the Owner, Contractor, Subcontractors and Sub-subcontractors in the Work, and the Owner and Contractor shall be named insureds.

11.3.3 Loss of Use Insurance. The Owner, at the Owner's option, may purchase and maintain such insurance as will insure the Owner against loss of use of the Owner's property due to fire or other hazards, however caused. The Owner waives all rights of action against the Contractor for loss of use of the Owner's property, including consequential losses due to fire or other hazards however caused.

11.3.4 If the Contractor requests in writing that insurance for risks other than those described herein or for other special hazards be included in the property insurance policy, the Owner shall, if possible, include such insurance, and the cost thereof shall be charged to the Contractor by appropriate Change Order.

11.3.5 If during the Project construction period the Owner insures properties, real or personal or both, adjoining or adjacent to the site by property insurance under policies separate from those insuring the Project, or if after final payment property insurance is to be provided on the completed Project through a policy or policies other than those insuring the Project during the construction period, the Owner shall waive all rights in accordance with the terms of Subparagraph 11.3.7 for damages caused by fire or other perils covered by this separate property insurance. All separate policies shall provide this waiver of subrogation by endorsement or otherwise.

11.3.6 Before an exposure to loss may occur, the Owner shall file with the Contractor a copy of each policy that includes insurance coverages required by this Paragraph 11.3. Each policy shall contain all generally applicable conditions, definitions, exclusions and endorsements related to this Project. Each policy shall contain a provision that the policy will not be cancelled or allowed to expire until at least 30 days' prior written notice has been given to the Contractor.

11.3.7 Waivers of Subrogation. The Owner and Contractor waive all rights against (1) each other and any of their subcontractors, sub-subcontractors, agents and employees, each of the other, and (2) the Architect, Architect's consultants, separate contractors described in Article 6, if any, and any of their subcontractors, sub-subcontractors, agents and employees, for damages caused by fire or other perils to the extent covered by property insurance obtained pursuant to this Paragraph 11.3 or other property insurance applicable to the Work, except such rights as they have to proceeds of such insurance held by the Owner as fiduciary. The Owner or Contractor, as appropriate, shall require of the Architect, Architect's consultants, separate contractors described in Article 6, if any, and the subcontractors, sub-subcontractors, agents and employees of any of them, by appropriate agreements, written where legally required for validity, similar waivers each in favor of other parties enumerated herein. The policies shall provide such waivers of subrogation by endorsement or otherwise. A waiver of subrogation shall be effective as to a person or entity even though that person or entity would otherwise have a duty of indemnification, contractual or otherwise, did not pay the insurance premium directly or indirectly, and whether or not the person or entity had an insurable interest in the property damaged.

11.3.8 A loss insured under Owner's property insurance shall be adjusted by the Owner as fiduciary and made payable to the Owner as fiduciary for the insureds, as their interests may appear, subject to requirements of any applicable mortgagee clause and of Subparagraph 11.3.10. The Contractor shall pay Subcontractors their just shares of insurance proceeds received by the Contractor, and by appropriate agreements, written where legally required for validity, shall require Subcontractors to make payments to their Sub-subcontractors in similar manner.

11.3.9 If required in writing by a party in interest, the Owner as fiduciary shall, upon occurrence of an insured loss, give bond for proper performance of the Owner's duties. The cost of required bonds shall be charged against proceeds received as fiduciary. The Owner shall deposit in a separate account proceeds so received, which the Owner shall distribute in accordance with such agreement as the parties in interest may reach, or in accordance with an arbitration award in which case the procedure shall be as provided in Paragraph 4.5. If after such loss no other special agreement is made, replacement of damaged property shall be covered by appropriate Change Order.

11.3.10 The Owner as fiduciary shall have power to adjust and settle a loss with insurers unless one of the parties in interest shall object in writing within five days after occurrence of loss to the Owner's exercise of this power; if such objection be made, arbitrators shall be chosen as provided in Paragraph 4.5. The Owner as fiduciary shall, in that case, make settlement with insurers in accordance with directions of such arbitrators. If distribution of insurance proceeds by arbitration is required, the arbitrators will direct such distribution.

11.3.11 Partial occupancy or use in accordance with Paragraph 9.9 shall not commence until the insurance company or companies providing property insurance have consented to such partial occupancy or use by endorsement or otherwise. The Owner and the Contractor shall take reasonable steps to obtain consent of the insurance company or companies and shall, without mutual written consent, take no action with respect to partial occupancy or use that would cause cancellation, lapse or reduction of insurance.

11.4 PERFORMANCE BOND AND PAYMENT BOND

11.4.1 The Owner shall have the right to require the Contractor to furnish bonds covering faithful performance of the Contract and payment of obligations arising thereunder as stipulated in bidding requirements or specifically required in the Contract Documents on the date of execution of the Contract.

11.4.2 Upon the request of any person or entity appearing to be a potential beneficiary of bonds covering payment of obligations arising under the Contract, the Contractor shall promptly furnish a copy of the bonds or shall permit a copy to be made.

ARTICLE 12

UNCOVERING AND CORRECTION OF WORK

12.1 UNCOVERING OF WORK

12.1.1 If a portion of the Work is covered contrary to the Architect's request or to requirements specifically expressed in the Contract Documents, it must, if required in writing by the Architect, be uncovered for the Architect's observation and be replaced at the Contractor's expense without change in the Contract Time.

12.1.2 If a portion of the Work has been covered which the Architect has not specifically requested to observe prior to its being covered, the Architect may request to see such Work and it shall be uncovered by the Contractor. If such Work is in accordance with the Contract Documents, costs of uncovering and replacement shall, by appropriate Change Order, be charged to the Owner. If such Work is not in accordance with the Contract Documents, the Contractor shall pay such costs unless the condition was caused by the Owner or a separate contractor in which event the Owner shall be responsible for payment of such costs.

12.2 CORRECTION OF WORK

12.2.1 The Contractor shall promptly correct Work rejected by the Architect or failing to conform to the requirements of the Contract Documents, whether observed before or after Substantial Completion and whether or not fabricated, installed or completed. The Contractor shall bear costs of correcting such rejected Work, including additional testing and inspections and compensation for the Architect's services and expenses made necessary thereby.

12.2.2 If, within one year after the date of Substantial Completion of the Work or designated portion thereof, or after the date for commencement of warranties established under Subparagraph 9.9.1, or by terms of an applicable special warranty required by the Contract Documents, any of the Work is found to be not in accordance with the requirements of the Contract Documents, the Contractor shall correct it promptly after receipt of written notice from the Owner to do so unless the Owner has previously given the Contractor a written acceptance of such condition. This period of one year shall be extended with respect to portions of Work first performed after Substantial Completion by the period of time between Substantial Completion and the actual performance of the Work. This obligation under this Subparagraph 12.2.2 shall survive acceptance of the Work under the Contract and termination of the Contract. The Owner shall give such notice promptly after discovery of the condition.

12.2.3 The Contractor shall remove from the site portions of the Work which are not in accordance with the requirements of the Contract Documents and are neither corrected by the Contractor nor accepted by the Owner.

12.2.4 If the Contractor fails to correct nonconforming Work within a reasonable time, the Owner may correct it in accordance with Paragraph 2.4. If the Contractor does not proceed with correction of such nonconforming Work within a reasonable time fixed by written notice from the Architect, the Owner may remove it and store the salvable materials or equipment at the Contractor's expense. If the Contractor does not pay costs of such removal and storage within ten days after written notice, the Owner may upon ten additional days' written notice sell such materials and equipment at auction or at private sale and shall account for the proceeds thereof, after deducting costs and damages that should have been borne by the Contractor, including compensation for the Architect's services and expenses made necessary thereby. If such proceeds of sale do not cover costs which the Contractor should have borne, the Contract Sum shall be reduced by the deficiency. If payments then or thereafter due the Contractor are not sufficient to cover such amount, the Contractor shall pay the difference to the Owner.

12.2.5 The Contractor shall bear the cost of correcting destroyed or damaged construction, whether completed or partially completed, of the Owner or separate contractors caused by the Contractor's correction or removal of Work which is not in accordance with the requirements of the Contract Documents.

12.2.6 Nothing contained in this Paragraph 12.2 shall be construed to establish a period of limitation with respect to other obligations which the Contractor might have under the Contract Documents. Establishment of the time period of one year as described in Subparagraph 12.2.2 relates only to the specific obligation of the Contractor to correct the Work, and has no relationship to the time within which the obligation to comply with the Contract Documents may be sought to be enforced, nor to the time within which proceedings may be commenced to establish the Contractor's liability with respect to the Contractor's obligations other than specifically to correct the Work.

12.3 ACCEPTANCE OF NONCONFORMING WORK

12.3.1 If the Owner prefers to accept Work which is not in accordance with the requirements of the Contract Documents, the Owner may do so instead of requiring its removal and correction, in which case the Contract Sum will be reduced as appropriate and equitable. Such adjustment shall be effected whether or not final payment has been made.

ARTICLE 13

MISCELLANEOUS PROVISIONS

13.1 GOVERNING LAW

13.1.1 The Contract shall be governed by the law of the place where the Project is located.

13.2 SUCCESSORS AND ASSIGNS

13.2.1 The Owner and Contractor respectively bind themselves, their partners, successors, assigns and legal representatives to the other party hereto and to partners, successors, assigns and legal representatives of such other party in respect to covenants, agreements and obligations contained in the Contract Documents. Neither party to the Contract shall assign the Contract as a whole without written consent of the other. If either party attempts to make such an assignment without such consent, that party shall nevertheless remain legally responsible for all obligations under the Contract.

13.3 WRITTEN NOTICE

13.3.1 Written notice shall be deemed to have been duly served if delivered in person to the individual or a member of the firm or entity or to an officer of the corporation for which it was intended, or if delivered at or sent by registered or certified mail to the last business address known to the party giving notice.

13.4 RIGHTS AND REMEDIES

13.4.1 Duties and obligations imposed by the Contract Documents and rights and remedies available thereunder shall be in addition to and not a limitation of duties, obligations, rights and remedies otherwise imposed or available by law.

13.4.2 No action or failure to act by the Owner, Architect or Contractor shall constitute a waiver of a right or duty afforded them under the Contract, nor shall such action or failure to act constitute approval of or acquiescence in a breach thereunder, except as may be specifically agreed in writing.

13.5 TESTS AND INSPECTIONS

13.5.1 Tests, inspections and approvals of portions of the Work required by the Contract Documents or by laws, ordinances, rules, regulations or orders of public authorities having jurisdiction shall be made at an appropriate time. Unless otherwise provided, the Contractor shall make arrangements for such tests, inspections and approvals with an independent testing laboratory or entity acceptable to the Owner, or with the appropriate public authority, and shall bear all related costs of tests, inspections and approvals. The Contractor shall give the Architect timely notice of when and where tests and inspections are to be made so the Architect may observe such procedures. The Owner shall bear costs of tests, inspections or approvals which do not become requirements until after bids are received or negotiations concluded.

13.5.2 If the Architect, Owner or public authorities having jurisdiction determine that portions of the Work require additional testing, inspection or approval not included under Subparagraph 13.5.1, the Architect will, upon written authorization from the Owner, instruct the Contractor to make arrangements for such additional testing, inspection or approval by an entity acceptable to the Owner, and the Contractor shall give timely notice to the Architect of when and where tests and inspections are to be made so the Architect may observe such procedures.

The Owner shall bear such costs except as provided in Subparagraph 13.5.3.

13.5.3 If such procedures for testing, inspection or approval under Subparagraphs 13.5.1 and 13.5.2 reveal failure of the portions of the Work to comply with requirements established by the Contract Documents, the Contractor shall bear all costs made necessary by such failure including those of repeated procedures and compensation for the Architect's services and expenses.

13.5.4 Required certificates of testing, inspection or approval shall, unless otherwise required by the Contract Documents, be secured by the Contractor and promptly delivered to the Architect.

13.5.5 If the Architect is to observe tests, inspections or approvals required by the Contract Documents, the Architect will do so promptly and, where practicable, at the normal place of testing.

13.5.6 Tests or inspections conducted pursuant to the Contract Documents shall be made promptly to avoid unreasonable delay in the Work.

13.6 INTEREST

13.6.1 Payments due and unpaid under the Contract Documents shall bear interest from the date payment is due at such rate as the parties may agree upon in writing or, in the absence thereof, at the legal rate prevailing from time to time at the place where the Project is located.

13.7 COMMENCEMENT OF STATUTORY LIMITATION PERIOD

13.7.1 As between the Owner and Contractor:

.1 Before Substantial Completion. As to acts or failures to act occurring prior to the relevant date of Substantial Completion, any applicable statute of limitations shall commence to run and any alleged cause of action shall be deemed to have accrued in any and all events not later than such date of Substantial Completion;

.2 Between Substantial Completion and Final Certificate for Payment. As to acts or failures to act occurring subsequent to the relevant date of Substantial Completion and prior to issuance of the final Certificate for Payment, any applicable statute of limitations shall commence to run and any alleged cause of action shall be deemed to have accrued in any and all events not later than the date of issuance of the final Certificate for Payment; and

.3 After Final Certificate for Payment. As to acts or failures to act occurring after the relevant date of issuance of the final Certificate for Payment, any applicable statute of limitations shall commence to run and any alleged cause of action shall be deemed to have accrued in any and all events not later than the date of any act or failure to act by the Contractor pursuant to any warranty provided under Paragraph 3.5, the date of any correction of the Work or failure to correct the Work by the Contractor under Paragraph 12.2, or the date of actual commission of any other act or failure to perform any duty or obligation by the Contractor or Owner, whichever occurs last.

ARTICLE 14

TERMINATION OR SUSPENSION OF THE CONTRACT

14.1 TERMINATION BY THE CONTRACTOR

14.1.1 The Contractor may terminate the Contract if the Work is stopped for a period of 30 days through no act or fault of the Contractor or a Subcontractor, Sub-subcontractor or their agents or employees or any other persons performing portions of the Work under contract with the Contractor, for any of the following reasons:

.1 issuance of an order of a court or other public authority having jurisdiction;

.2 an act of government, such as a declaration of national emergency, making material unavailable;

.3 because the Architect has not issued a Certificate for Payment and has not notified the Contractor of the reason for withholding certification as provided in Subparagraph 9.4.1, or because the Owner has not made payment on a Certificate for Payment within the time stated in the Contract Documents;

.4 if repeated suspensions, delays or interruptions by the Owner as described in Paragraph 14.3 constitute in the aggregate more than 100 percent of the total number of days scheduled for completion, or 120 days in any 365-day period, whichever is less; or

.5 the Owner has failed to furnish to the Contractor promptly, upon the Contractor's request, reasonable evidence as required by Subparagraph 2.2.1.

14.1.2 If one of the above reasons exists, the Contractor may, upon seven additional days' written notice to the Owner and Architect, terminate the Contract and recover from the Owner payment for Work executed and for proven loss with respect to materials, equipment, tools, and construction equipment and machinery, including reasonable overhead, profit and damages.

14.1.3 If the Work is stopped for a period of 60 days through no act or fault of the Contractor or a Subcontractor or their agents or employees or any other persons performing portions of the Work under contract with the Contractor because the Owner has persistently failed to fulfill the Owner's obligations under the Contract Documents with respect to matters important to the progress of the Work, the Contractor may, upon seven additional days' written notice to the Owner and the Architect, terminate the Contract and recover from the Owner as provided in Subparagraph 14.1.2.

14.2 TERMINATION BY THE OWNER FOR CAUSE

14.2.1 The Owner may terminate the Contract if the Contractor:

.1 persistently or repeatedly refuses or fails to supply enough properly skilled workers or proper materials;

.2 fails to make payment to Subcontractors for materials or labor in accordance with the respective agreements between the Contractor and the Subcontractors;

.3 persistently disregards laws, ordinances, or rules, regulations or orders of a public authority having jurisdiction; or

.4 otherwise is guilty of substantial breach of a provision of the Contract Documents.

14.2.2 When any of the above reasons exist, the Owner, upon certification by the Architect that sufficient cause exists to justify such action, may without prejudice to any other rights or remedies of the Owner and after giving the Contractor and the Contractor's surety, if any, seven days' written notice, terminate employment of the Contractor and may, subject to any prior rights of the surety:

.1 take possession of the site and of all materials, equipment, tools, and construction equipment and machinery thereon owned by the Contractor;

.2 accept assignment of subcontracts pursuant to Paragraph 5.4; and

.3 finish the Work by whatever reasonable method the Owner may deem expedient.

14.2.3 When the Owner terminates the Contract for one of the reasons stated in Subparagraph 14.2.1, the Contractor shall not be entitled to receive further payment until the Work is finished.

14.2.4 If the unpaid balance of the Contract Sum exceeds costs of finishing the Work, including compensation for the Architect's services and expenses made necessary thereby, such excess shall be paid to the Contractor. If such costs exceed the unpaid balance, the Contractor shall pay the difference to the Owner. The amount to be paid to the Contractor or Owner, as the case may be, shall be certified by the Architect, upon application, and this obligation for payment shall survive termination of the Contract.

14.3 SUSPENSION BY THE OWNER FOR CONVENIENCE

14.3.1 The Owner may, without cause, order the Contractor in writing to suspend, delay or interrupt the Work in whole or in part for such period of time as the Owner may determine.

14.3.2 An adjustment shall be made for increases in the cost of performance of the Contract, including profit on the increased cost of performance, caused by suspension, delay or interruption. No adjustment shall be made to the extent:

.1 that performance is, was or would have been so suspended, delayed or interrupted by another cause for which the Contractor is responsible; or

.2 that an equitable adjustment is made or denied under another provision of this Contract.

14.3.3 Adjustments made in the cost of performance may have a mutually agreed fixed or percentage fee.

APPENDIX C

AIA DOCUMENT A311, PERFORMANCE BOND (FEBRUARY 1970)

THE AMERICAN INSTITUTE OF ARCHITECTS

AIA Document A311

Performance Bond

KNOW ALL MEN BY THESE PRESENTS: that

(Here insert full name and address or legal title of Contractor)

as Principal, hereinafter called Contractor, and,

(Here insert full name and address or legal title of Surety)

as Surety, hereinafter called Surety, are held and firmly bound unto

(Here insert full name and address or legal title of Owner)

as Obligee, hereinafter called Owner, in the amount of

Dollars ($),

for the payment whereof Contractor and Surety bind themselves, their heirs, executors, administrators, successors and assigns, jointly and severally, firmly by these presents.

WHEREAS,

Contractor has by written agreement dated 19 , entered into a contract with Owner for

(Here insert full name, address and description of project)

in accordance with Drawings and Specifications prepared by

(Here insert full name and address or legal title of Architect)

which contract is by reference made a part hereof, and is hereinafter referred to as the Contract.

PERFORMANCE BOND

NOW, THEREFORE, THE CONDITION OF THIS OBLIGATION is such that, if Contractor shall promptly and faithfully perform said Contract, then this obligation shall be null and void; otherwise it shall remain in full force and effect.

The Surety hereby waives notice of any alteration or extension of time made by the Owner.

Whenever Contractor shall be, and declared by Owner to be in default under the Contract, the Owner having performed Owner's obligations thereunder, the Surety may promptly remedy the default, or shall promptly

1) Complete the Contract in accordance with its terms and conditions, or

2) Obtain a bid or bids for completing the Contract in accordance with its terms and conditions, and upon determination by Surety of the lowest responsible bidder, or, if the Owner elects, upon determination by the Owner and the Surety jointly of the lowest responsible bidder, arrange for a contract between such bidder and Owner, and make available as Work progresses (even though there should be a default or a succession of defaults under the contract or contracts of completion arranged under this paragraph) sufficient funds to pay the cost of completion less the balance of the contract price; but not exceeding, including other costs and damages for which the Surety may be liable hereunder, the amount set forth in the first paragraph hereof. The term "balance of the contract price," as used in this paragraph, shall mean the total amount payable by Owner to Contractor under the Contract and any amendments thereto, less the amount properly paid by Owner to Contractor.

Any suit under this bond must be instituted before the expiration of two (2) years from the date on which final payment under the Contract falls due.

No right of action shall accrue on this bond to or for the use of any person or corporation other than the Owner named herein or the heirs, executors, administrators or successors of the Owner.

Signed and sealed this day of 19

_______________ (Principal) (Seal)

_______________ (Witness)

_______________ (Title)

_______________ (Surety) (Seal)

_______________ (Witness)

_______________ (Title)

AIA CAUTION: You should sign an original AIA document which has this caution printed in red. An original assures that changes will not be obscured as may occur when documents are reproduced.

APPENDIX D

AIA DOCUMENT A311, LABOR AND MATERIAL PAYMENT BOND (FEBRUARY 1970)

THE AMERICAN INSTITUTE OF ARCHITECTS

AIA Document A311

Labor and Material Payment Bond

THIS BOND IS ISSUED SIMULTANEOUSLY WITH PERFORMANCE BOND IN FAVOR OF THE OWNER CONDITIONED ON THE FULL AND FAITHFUL PERFORMANCE OF THE CONTRACT

KNOW ALL MEN BY THESE PRESENTS: that (Here insert full name and address or legal title of Contractor)

as Principal, hereinafter called Principal, and, (Here insert full name and address or legal title of Surety)

as Surety, hereinafter called Surety, are held and firmly bound unto (Here insert full name and address or legal title of Owner)

as Obligee, hereinafter called Owner, for the use and benefit of claimants as hereinbelow defined, in the amount of (Here insert a sum equal to at least one-half of the contract price) Dollars ($), for the payment whereof Principal and Surety bind themselves, their heirs, executors, administrators, successors and assigns, jointly and severally, firmly by these presents.

WHEREAS,

Principal has by written agreement dated 19 , entered into a contract with Owner for (Here insert full name, address and description of project)

in accordance with Drawings and Specifications prepared by (Here insert full name and address or legal title of Architect)

which contract is by reference made a part hereof, and is hereinafter referred to as the Contract.

LABOR AND MATERIAL PAYMENT BOND

NOW, THEREFORE, THE CONDITION OF THIS OBLIGATION is such that, if Principal shall promptly make payment to all claimants as hereinafter defined, for all labor and material used or reasonably required for use in the performance of the Contract, then this obligation shall be void, otherwise it shall remain in full force and effect, subject, however, to the following conditions:

1. A claimant is defined as one having a direct contract with the Principal or with a Subcontractor of the Principal for labor, material, or both, used or reasonably required for use in the performance of the Contract, labor and material being construed to include that part of water, gas, power, light, heat, oil, gasoline, telephone service or rental of equipment directly applicable to the Contract.

2. The above named Principal and Surety hereby jointly and severally agree with the Owner that every claimant as herein defined, who has not been paid in full before the expiration of a period of ninety (90) days after the date on which the last of such claimant's work or labor was done or performed, or materials were furnished by such claimant, may sue on this bond for the use of such claimant, prosecute the suit to final judgment for such sum or sums as may be justly due claimant, and have execution thereon. The Owner shall not be liable for the payment of any costs or expenses of any such suit.

3. No suit or action shall be commenced hereunder by any claimant:

a) Unless claimant, other than one having a direct contract with the Principal, shall have given written notice to any two of the following: the Principal, the Owner, or the Surety above named, within ninety (90) days after such claimant did or performed the last of the work or labor, or furnished the last of the materials for which said claim is made, stating with substantial accuracy the amount claimed and the name of the party to whom the materials were furnished, or for whom the work or labor was done or performed. Such notice shall be served by mailing the same by registered mail or certified mail, postage prepaid, in an envelope addressed to the Principal, Owner or Surety, at any place where an office is regularly maintained for the transaction of business, or served in any manner in which legal process may be served in the state in which the aforesaid project is located, save that such service need not be made by a public officer.

b) After the expiration of one (1) year following the date on which Principal ceased Work on said Contract, it being understood, however, that if any limitation embodied in this bond is prohibited by any law controlling the construction hereof such limitation shall be deemed to be amended so as to be equal to the minimum period of limitation permitted by such law.

c) Other than in a state court of competent jurisdiction in and for the county or other political subdivision of the state in which the Project, or any part thereof, is situated, or in the United States District Court for the district in which the Project, or any part thereof, is situated, and not elsewhere.

4. The amount of this bond shall be reduced by and to the extent of any payment or payments made in good faith hereunder, inclusive of the payment by Surety of mechanics' liens which may be filed of record against said improvement, whether or not claim for the amount of such lien be presented under and against this bond.

Signed and sealed this day of 19

(Witness)	(Principal) (Seal)
	(Title)

(Witness)	(Surety) (Seal)
	(Title)

CAUTION: You should sign an original AIA document which has this caution printed in red. An original assures that changes will not be obscured as may occur when documents are reproduced.

APPENDIX E

AIA DOCUMENT A312, PERFORMANCE BOND (1984)

THE AMERICAN INSTITUTE OF ARCHITECTS

Because AIA Documents are revised from time to time, users should ascertain from the AIA the current edition of this document. Copies of the current edition of this AIA document may be purchased from The American Institute of Architects or its local distributors.

The text of this document is not "model language" and is not intended for use in other documents without permission of the AIA.

AIA Document A312

Performance Bond

Any singular reference to Contractor, Surety, Owner or other party shall be considered plural where applicable.

CONTRACTOR (Name and Address):

SURETY (Name and Principal Place of Business):

OWNER (Name and Address):

SAMPLE

CONSTRUCTION CONTRACT
Date:
Amount:
Description (Name and Location):

BOND
Date (Not earlier than Construction Contract Date):
Amount:
Modifications to this Bond: ☐ None ☐ See Page 3

CONTRACTOR AS PRINCIPAL
Company: (Corporate Seal)

Signature: ____________________
Name and Title:

SURETY
Company: (Corporate Seal)

Signature: ____________________
Name and Title:

(Any additional signatures appear on page 3)

(FOR INFORMATION ONLY—Name, Address and Telephone)
AGENT or BROKER:

OWNER'S REPRESENTATIVE (Architect, Engineer or other party):

1 The Contractor and the Surety, jointly and severally, bind themselves, their heirs, executors, administrators, successors and assigns to the Owner for the performance of the Construction Contract, which is incorporated herein by reference.

2 If the Contractor performs the Construction Contract, the Surety and the Contractor shall have no obligation under this Bond, except to participate in conferences as provided in Subparagraph 3.1.

3 If there is no Owner Default, the Surety's obligation under this Bond shall arise after:

3.1 The Owner has notified the Contractor and the Surety at its address described in Paragraph 10 below that the Owner is considering declaring a Contractor Default and has requested and attempted to arrange a conference with the Contractor and the Surety to be held not later than fifteen days after receipt of such notice to discuss methods of performing the Construction Contract. If the Owner, the Contractor and the Surety agree, the Contractor shall be allowed a reasonable time to perform the Construction Contract, but such an agreement shall not waive the Owner's right, if any, subsequently to declare a Contractor Default; and

3.2 The Owner has declared a Contractor Default and formally terminated the Contractor's right to complete the contract. Such Contractor Default shall not be declared earlier than twenty days after the Contractor and the Surety have received notice as provided in Subparagraph 3.1; and

3.3 The Owner has agreed to pay the Balance of the Contract Price to the Surety in accordance with the terms of the Construction Contract or to a contractor selected to perform the Construction Contract in accordance with the terms of the contract with the Owner.

4 When the Owner has satisfied the conditions of Paragraph 3, the Surety shall promptly and at the Surety's expense take one of the following actions:

4.1 Arrange for the Contractor, with consent of the Owner, to perform and complete the Construction Contract; or

4.2 Undertake to perform and complete the Construction Contract itself, through its agents or through independent contractors; or

4.3 Obtain bids or negotiated proposals from qualified contractors acceptable to the Owner for a contract for performance and completion of the Construction Contract, arrange for a contract to be prepared for execution by the Owner and the contractor selected with the Owner's concurrence, to be secured with performance and payment bonds executed by a qualified surety equivalent to the bonds issued on the Construction Contract, and pay to the Owner the amount of damages as described in Paragraph 6 in excess of the Balance of the Contract Price incurred by the Owner resulting from the Contractor's default; or

4.4 Waive its right to perform and complete, arrange for completion, or obtain a new contractor and with reasonable promptness under the circumstances:

.1 After investigation, determine the amount for which it may be liable to the Owner and, as soon as practicable after the amount is determined, tender payment therefor to the Owner; or

.2 Deny liability in whole or in part and notify the Owner citing reasons therefor.

5 If the Surety does not proceed as provided in Paragraph 4 with reasonable promptness, the Surety shall be deemed to be in default on this Bond fifteen days after receipt of an additional written notice from the Owner to the Surety demanding that the Surety perform its obligations under this Bond, and the Owner shall be entitled to enforce any remedy available to the Owner. If the Surety proceeds as provided in Subparagraph 4.4, and the Owner refuses the payment tendered or the Surety has denied liability, in whole or in part, without further notice the Owner shall be entitled to enforce any remedy available to the Owner.

6 After the Owner has terminated the Contractor's right to complete the Construction Contract, and if the Surety elects to act under Subparagraph 4.1, 4.2, or 4.3 above, then the responsibilities of the Surety to the Owner shall not be greater than those of the Contractor under the Construction Contract, and the responsibilities of the Owner to the Surety shall not be greater than those of the Owner under the Construction Contract. To the limit of the amount of this Bond, but subject to commitment by the Owner of the Balance of the Contract Price to mitigation of costs and damages on the Construction Contract, the Surety is obligated without duplication for:

6.1 The responsibilities of the Contractor for correction of defective work and completion of the Construction Contract;

6.2 Additional legal, design professional and delay costs resulting from the Contractor's Default, and resulting from the actions or failure to act of the Surety under Paragraph 4; and

6.3 Liquidated damages, or if no liquidated damages are specified in the Construction Contract, actual damages caused by delayed performance or non-performance of the Contractor.

7 The Surety shall not be liable to the Owner or others for obligations of the Contractor that are unrelated to the Construction Contract, and the Balance of the Contract Price shall not be reduced or set off on account of any such unrelated obligations. No right of action shall accrue on this Bond to any person or entity other than the Owner or its heirs, executors, administrators or successors.

8 The Surety hereby waives notice of any change, including changes of time, to the Construction Contract or to related subcontracts, purchase orders and other obligations.

9 Any proceeding, legal or equitable, under this Bond may be instituted in any court of competent jurisdiction in the location in which the work or part of the work is located and shall be instituted within two years after Contractor Default or within two years after the Contractor ceased working or within two years after the Surety refuses or fails to perform its obligations under this Bond, whichever occurs first. If the provisions of this Paragraph are void or prohibited by law, the minimum period of limitation available to sureties as a defense in the jurisdiction of the suit shall be applicable.

10 Notice to the Surety, the Owner or the Contractor shall be mailed or delivered to the address shown on the signature page.

11 When this Bond has been furnished to comply with a statutory or other legal requirement in the location where the construction was to be performed, any provision in this Bond conflicting with said statutory or legal requirement shall be deemed deleted herefrom and provisions conforming to such statutory or other legal requirement shall be deemed incorporated herein. The intent is that this Bond shall be construed as a statutory bond and not as a common law bond.

12 DEFINITIONS

12.1 Balance of the Contract Price: The total amount payable by the Owner to the Contractor under the Construction Contract after all proper adjustments have been made, including allowance to the Contractor of any amounts received or to be received by the Owner in settlement of insurance or other claims for damages to which the Contractor is entitled, reduced by all valid and proper payments made to or on behalf of the Contractor under the Construction Contract.

12.2 Construction Contract: The agreement between the Owner and the Contractor identified on the signature page, including all Contract Documents and changes thereto.

12.3 Contractor Default: Failure of the Contractor, which has neither been remedied nor waived, to perform or otherwise to comply with the terms of the Construction Contract.

12.4 Owner Default: Failure of the Owner, which has neither been remedied nor waived, to pay the Contractor as required by the Construction Contract or to perform and complete or comply with the other terms thereof.

MODIFICATIONS TO THIS BOND ARE AS FOLLOWS:

(Space is provided below for additional signatures of added parties, other than those appearing on the cover page.)

CONTRACTOR AS PRINCIPAL		SURETY	
Company:	(Corporate Seal)	Company:	(Corporate Seal)
Signature: ______________________		Signature: ______________________	
Name and Title:		Name and Title:	
Address:		Address:	

APPENDIX F

AIA DOCUMENT A312, PAYMENT BOND (1984)

THE AMERICAN INSTITUTE OF ARCHITECTS

AIA Document A312

Payment Bond

Any singular reference to Contractor, Surety, Owner or other party shall be considered plural where applicable.

CONTRACTOR (Name and Address):

SURETY (Name and Principal Place of Business):

OWNER (Name and Address):

CONSTRUCTION CONTRACT
Date:
Amount:
Description (Name and Location):

BOND
Date (Not earlier than Construction Contract Date):
Amount:
Modifications to this Bond: ☐ None ☐ See Page 6

CONTRACTOR AS PRINCIPAL		SURETY	
Company:	(Corporate Seal)	Company:	(Corporate Seal)
Signature: ______________		Signature: ______________	
Name and Title:		Name and Title:	

(Any additional signatures appear on page 6)

(FOR INFORMATION ONLY—Name, Address and Telephone)

AGENT or BROKER:

OWNER'S REPRESENTATIVE (Architect, Engineer or other party):

1 The Contractor and the Surety, jointly and severally, bind themselves, their heirs, executors, administrators, successors and assigns to the Owner to pay for labor, materials and equipment furnished for use in the performance of the Construction Contract, which is incorporated herein by reference.

2 With respect to the Owner, this obligation shall be null and void if the Contractor:

2.1 Promptly makes payment, directly or indirectly, for all sums due Claimants, and

2.2 Defends, indemnifies and holds harmless the Owner from claims, demands, liens or suits by any person or entity whose claim, demand, lien or suit is for the payment for labor, materials or equipment furnished for use in the performance of the Construction Contract, provided the Owner has promptly notified the Contractor and the Surety (at the address described in Paragraph 12) of any claims, demands, liens or suits and tendered defense of such claims, demands, liens or suits to the Contractor and the Surety, and provided there is no Owner Default.

3 With respect to Claimants, this obligation shall be null and void if the Contractor promptly makes payment, directly or indirectly, for all sums due.

4 The Surety shall have no obligation to Claimants under this Bond until:

4.1 Claimants who are employed by or have a direct contract with the Contractor have given notice to the Surety (at the address described in Paragraph 12) and sent a copy, or notice thereof, to the Owner, stating that a claim is being made under this Bond and, with substantial accuracy, the amount of the claim.

4.2 Claimants who do not have a direct contract with the Contractor:

.1 Have furnished written notice to the Contractor and sent a copy, or notice thereof, to the Owner, within 90 days after having last performed labor or last furnished materials or equipment included in the claim stating, with substantial accuracy, the amount of the claim and the name of the party to whom the materials were furnished or supplied or for whom the labor was done or performed; and

.2 Have either received a rejection in whole or in part from the Contractor, or not received within 30 days of furnishing the above notice any communication from the Contractor by which the Contractor has indicated the claim will be paid directly or indirectly; and

.3 Not having been paid within the above 30 days, have sent a written notice to the Surety (at the address described in Paragraph 12) and sent a copy, or notice thereof, to the Owner, stating that a claim is being made under this Bond and enclosing a copy of the previous written notice furnished to the Contractor.

5 If a notice required by Paragraph 4 is given by the Owner to the Contractor or to the Surety, that is sufficient compliance.

6 When the Claimant has satisfied the conditions of Paragraph 4, the Surety shall promptly and at the Surety's expense take the following actions:

6.1 Send an answer to the Claimant, with a copy to the Owner, within 45 days after receipt of the claim, stating the amounts that are undisputed and the basis for challenging any amounts that are disputed.

6.2 Pay or arrange for payment of any undisputed amounts.

7 The Surety's total obligation shall not exceed the amount of this Bond, and the amount of this Bond shall be credited for any payments made in good faith by the Surety.

8 Amounts owed by the Owner to the Contractor under the Construction Contract shall be used for the performance of the Construction Contract and to satisfy claims, if any, under any Construction Performance Bond. By the Contractor furnishing and the Owner accepting this Bond, they agree that all funds earned by the Contractor in the performance of the Construction Contract are dedicated to satisfy obligations of the Contractor and the Surety under this Bond, subject to the Owner's priority to use the funds for the completion of the work.

9 The Surety shall not be liable to the Owner, Claimants or others for obligations of the Contractor that are unrelated to the Construction Contract. The Owner shall not be liable for payment of any costs or expenses of any Claimant under this Bond, and shall have under this Bond no obligations to make payments to, give notices on behalf of, or otherwise have obligations to Claimants under this Bond.

10 The Surety hereby waives notice of any change, including changes of time, to the Construction Contract or to related subcontracts, purchase orders and other obligations.

11 No suit or action shall be commenced by a Claimant under this Bond other than in a court of competent jurisdiction in the location in which the work or part of the work is located or after the expiration of one year from the date (1) on which the Claimant gave the notice required by Subparagraph 4.1 or Clause 4.2.3, or (2) on which the last labor or service was performed by anyone or the last materials or equipment were furnished by anyone under the Construction Contract, whichever of (1) or (2) first occurs. If the provisions of this Paragraph are void or prohibited by law, the minimum period of limitation available to sureties as a defense in the jurisdiction of the suit shall be applicable.

12 Notice to the Surety, the Owner or the Contractor shall be mailed or delivered to the address shown on the signature page. Actual receipt of notice by Surety, the Owner or the Contractor, however accomplished, shall be sufficient compliance as of the date received at the address shown on the signature page.

13 When this Bond has been furnished to comply with a statutory or other legal requirement in the location where the construction was to be performed, any provision in this Bond conflicting with said statutory or legal requirement shall be deemed deleted herefrom and provisions conforming to such statutory or other legal requirement shall be deemed incorporated herein. The intent is that this

Bond shall be construed as a statutory bond and not as a common law bond.

14 Upon request by any person or entity appearing to be a potential beneficiary of this Bond, the Contractor shall promptly furnish a copy of this Bond or shall permit a copy to be made.

15 DEFINITIONS

15.1 Claimant: An individual or entity having a direct contract with the Contractor or with a subcontractor of the Contractor to furnish labor, materials or equipment for use in the performance of the Contract. The intent of this Bond shall be to include without limitation in the terms "labor, materials or equipment" that part of water, gas, power, light, heat, oil, gasoline, telephone service or rental equipment used in the Construction Contract, architectural and engineering services required for performance of the work of the Contractor and the Contractor's subcontractors, and all other items for which a mechanic's lien may be asserted in the jurisdiction where the labor, materials or equipment were furnished.

15.2 Construction Contract: The agreement between the Owner and the Contractor identified on the signature page, including all Contract Documents and changes thereto.

15.3 Owner Default: Failure of the Owner, which has neither been remedied nor waived, to pay the Contractor as required by the Construction Contract or to perform and complete or comply with the other terms thereof.

MODIFICATIONS TO THIS BOND ARE AS FOLLOWS:

(Space is provided below for additional signatures of added parties, other than those appearing on the cover page.)

CONTRACTOR AS PRINCIPAL		SURETY	
Company:	(Corporate Seal)	Company:	(Corporate Seal)
Signature: ______________		Signature: ______________	
Name and Title:		Name and Title:	
Address:		Address:	

APPENDIX G

AIA DOCUMENT A401, STANDARD FORM OF AGREEMENT BETWEEN CONTRACTOR AND SUBCONTRACTOR (1987)

THE AMERICAN INSTITUTE OF ARCHITECTS

AIA Document A401

Standard Form of Agreement Between Contractor and Subcontractor

1987 EDITION

THIS DOCUMENT HAS IMPORTANT LEGAL CONSEQUENCES; CONSULTATION WITH AN ATTORNEY IS ENCOURAGED WITH RESPECT TO ITS COMPLETION OR MODIFICATION.

This document has been approved and endorsed by the American Subcontractors Association and the Associated Specialty Contractors, Inc.

AGREEMENT

made as of the day of in the year of Nineteen Hundred and

BETWEEN the Contractor:
(Name and Address)

and the Subcontractor:
(Name and Address)

The Contractor has made a contract for construction dated with The Owner:
(Name and Address)

For the following Project:
(Name and Location)

which Contract is hereinafter referred to as the Prime Contract and which provides for the furnishing of labor, materials, equipment and services in connection with the construction of the Project. A copy of the Prime Contract, consisting of the Agreement Between Owner and Contractor (from which compensation amounts may be deleted) and the other Contract Documents enumerated therein has been made available to the Subcontractor.

The Architect for the Project is:
(Name and Address)

The Contractor and the Subcontractor agree as set forth below.

TERMS AND CONDITIONS OF AGREEMENT BETWEEN CONTRACTOR AND SUBCONTRACTOR

ARTICLE 1
THE SUBCONTRACT DOCUMENTS

1.1. The Subcontract Documents consist of (1) this Agreement; (2) the Prime Contract, consisting of the Agreement between the Owner and Contractor and the other Contract Documents enumerated therein, including Conditions of the Contract (General, Supplementary and other Conditions), Drawings, Specifications, Addenda issued prior to execution of the Agreement between the Owner and Contractor and Modifications issued subsequent to the execution of the Agreement between the Owner and Contractor, whether before or after the execution of this Agreement, and other Contract Documents, if any, listed in the Owner-Contractor Agreement; (3) other documents listed in Article 16 of this Agreement; and (4) Modifications to this Subcontract issued after execution of this Agreement. These form the Subcontract, and are as fully a part of the Subcontract as if attached to this Agreement or repeated herein. The Subcontract represents the entire and integrated agreement between the parties hereto and supersedes prior negotiations, representations or agreements, either written or oral. An enumeration of the Subcontract Documents, other than Modifications issued subsequent to the execution of this Agreement, appears in Article 16.

1.2 The Subcontractor shall be furnished copies of the Subcontract Documents upon request, but the Contractor may charge the Subcontractor for the cost of reproduction.

ARTICLE 2
MUTUAL RIGHTS AND RESPONSIBILITIES

2.1 The Contractor and Subcontractor shall be mutually bound by the terms of this Agreement and, to the extent that provisions of the Prime Contract apply to the Work of the Subcontractor, the Contractor shall assume toward the Subcontractor all obligations and responsibilities that the Owner, under the Prime Contract, assumes toward the Contractor, and the Subcontractor shall assume toward the Contractor all obligations and responsibilities which the Contractor, under the Prime Contract, assumes toward the Owner and the Architect. The Contractor shall have the benefit of all rights, remedies and redress against the Subcontractor which the Owner, under the Prime Contract, has against the Contractor, and the Subcontractor shall have the benefit of all rights, remedies and redress against the Contractor which the Contractor, under the Prime Contract, has against the Owner, insofar as applicable to this Subcontract. Where a provision of the Prime Contract is inconsistent with a provision of this Agreement, this Agreement shall govern.

2.2 The Contractor may require the Subcontractor to enter into agreements with Sub-subcontractors performing portions of the Work of this Subcontract by which the Subcontractor and the Sub-subcontractor are mutually bound, to the extent of the Work to be performed by the Sub-subcontractor, assuming toward each other all obligations and responsibilities which the Contractor and Subcontractor assume toward each other and having the benefit of all rights, remedies and redress each against the other which the Contractor and Subcontractor have by virtue of the provisions of this Agreement.

ARTICLE 3
CONTRACTOR

3.1 SERVICES PROVIDED BY THE CONTRACTOR

3.1.1 The Contractor shall cooperate with the Subcontractor in scheduling and performing the Contractor's Work to avoid conflicts or interference in the Subcontractor's Work and shall expedite written responses to submittals made by the Subcontractor in accordance with Paragraph 4.1 and Article 5. As soon as practicable after execution of this Agreement, the Contractor shall provide the Subcontractor copies of the Contractor's construction schedule and schedule of submittals, together with such additional scheduling details as will enable the Subcontractor to plan and perform the Subcontractor's Work properly. The Subcontractor shall be notified promptly of subsequent changes in the construction and submittal schedules and additional scheduling details.

3.1.2 The Contractor shall provide suitable areas for storage of the Subcontractor's materials and equipment during the course of the Work. Additional costs to the Subcontractor resulting from relocation of such facilities at the direction of the Contractor, except as previously agreed upon, shall be reimbursed by the Contractor.

3.1.3 Except as provided in Article 14, the Contractor's equipment will be available to the Subcontractor only at the Contractor's discretion and on mutually satisfactory terms.

3.2 COMMUNICATIONS

3.2.1 The Contractor shall promptly make available to the Subcontractor information which affects this Subcontract and which becomes available to the Contractor subsequent to execution of this Subcontract.

3.2.2 The Contractor shall not give instructions or orders directly to employees or workmen of the Subcontractor, except to persons designated as authorized representatives of the Subcontractor.

3.2.3 The Contractor shall permit the Subcontractor to request directly from the Architect information regarding the percentages of completion and the amount certified on account of Work done by the Subcontractor.

3.2.4 If hazardous substances of a type of which an employer is required by law to notify its employees are being used on the site by the Contractor, a subcontractor or anyone directly or indirectly employed by them (other than the Subcontractor), the Contractor shall, prior to harmful exposure of the Subcontractor's employees to such substance, give written notice of the chemical composition thereof to the Subcontractor in sufficient detail and time to permit the Subcontractor's compliance with such laws.

3.3 CLAIMS BY THE CONTRACTOR

3.3.1 Liquidated damages for delay, if provided for in Paragraph 9.3 of this Agreement, shall be assessed against the Subcontractor only to the extent caused by the Subcontractor, the Subcontractor's employees and agents, Sub-subcontractors, suppliers or any person or entity for whose acts the Subcontractor may be liable, and in no case for delays or causes arising outside the scope of this Subcontract.

3.3.2 Except as may be indicated in this Agreement, the Contractor agrees that no claim for payment for services rendered or materials and equipment furnished by the Contractor to the Subcontractor shall be valid without prior notice to the Subcontractor and unless written notice thereof is given by the Contractor to the Subcontractor not later than the tenth day of the calendar month following that in which the claim originated.

3.4 CONTRACTOR'S REMEDIES

3.4.1 If the Subcontractor defaults or neglects to carry out the Work in accordance with this Agreement and fails within three working days after receipt of written notice from the Contractor to commence and continue correction of such default or neglect with diligence and promptness, the Contractor may, after three days following receipt by the Subcontractor of an additional written notice, and without prejudice to any other remedy the Contractor may have, make good such deficiencies and may deduct the cost thereof from the payments then or thereafter due the Subcontractor, provided, however, that if such action is based upon faulty workmanship or materials and equipment, the Architect shall first have determined that the workmanship or materials and equipment are not in accordance with requirements of the Prime Contract.

ARTICLE 4
SUBCONTRACTOR

4.1 EXECUTION AND PROGRESS OF THE WORK

4.1.1 The Subcontractor shall cooperate with the Contractor in scheduling and performing the Subcontractor's Work to avoid conflict, delay in or interference with the Work of the Contractor, other subcontractors or Owner's own forces.

4.1.2 The Subcontractor shall promptly submit Shop Drawings, Product Data, Samples and similar submittals required by the Subcontract Documents with reasonable promptness and in such sequence as to cause no delay in the Work or in the activities of the Contractor or other subcontractors.

4.1.3 The Subcontractor shall submit to the Contractor a schedule of values allocated to the various parts of the Work of this Subcontract, aggregating the Subcontract Sum, made out in such detail as the Contractor and Subcontractor may agree upon or as required by the Owner, and supported by such evidence as the Contractor may direct. In applying for payment, the Subcontractor shall submit statements based upon this schedule.

4.1.4 The Subcontractor shall furnish to the Contractor periodic progress reports on the Work of this Subcontract as mutually agreed, including information on the status of materials and equipment which may be in the course of preparation or manufacture.

4.1.5 The Subcontractor agrees that the Architect will have the authority to reject Work which does not conform to the Prime Contract. The Architect's decisions on matters relating to aesthetic effect shall be final if consistent with the intent expressed in the Prime Contract.

4.1.6 The Subcontractor shall pay for materials, equipment and labor used in connection with the performance of this Subcontract through the period covered by previous payments received from the Contractor, and shall furnish satisfactory evidence, when requested by the Contractor, to verify compliance with the above requirements.

4.1.7 The Subcontractor shall take necessary precautions to protect properly the Work of other subcontractors from damage caused by operations under this Subcontract.

4.1.8 The Subcontractor shall cooperate with the Contractor, other subcontractors and the Owner's own forces whose Work might interfere with the Subcontractor's Work. The Subcontractor shall participate in the preparation of coordinated drawings in areas of congestion, if required by the Prime Contract, specifically noting and advising the Contractor of potential conflicts between the Work of the Subcontractor and that of the Contractor, other subcontractors or the Owner's own forces.

4.2 LAWS, PERMITS, FEES AND NOTICES

4.2.1 The Subcontractor shall give notices and comply with laws, ordinances, rules, regulations and orders of public authorities bearing on performance of the Work of this Subcontract. The Subcontractor shall secure and pay for permits and governmental fees, licenses and inspections necessary for proper execution and completion of the Subcontractor's Work, the furnishing of which is required of the Contractor by the Prime Contract.

4.2.2 The Subcontractor shall comply with Federal, state and local tax laws, social security acts, unemployment compensation acts and workers' or workmen's compensation acts insofar as applicable to the performance of this Subcontract.

4.3 SAFETY PRECAUTIONS AND PROCEDURES

4.3.1 The Subcontractor shall take reasonable safety precautions with respect to performance of this Subcontract, shall comply with safety measures initiated by the Contractor and with applicable laws, ordinances, rules, regulations and orders of public authorities for the safety of persons or property in accordance with the requirements of the Prime Contract. The Subcontractor shall report to the Contractor within three days an injury to an employee or agent of the Subcontractor which occurred at the site.

4.3.2 If hazardous substances of a type of which an employer is required by law to notify its employees are being used on the site by the Subcontractor, the Subcontractor's Sub-subcontractors or anyone directly or indirectly employed by them, the Subcontractor shall, prior to harmful exposure of any employees on the site to such substance, give written notice of the chemical composition thereof to the Contractor in sufficient detail and time to permit compliance with such laws by the Contractor, other subcontractors and other employers on the site.

4.3.3 In the event the Subcontractor encounters on the site material reasonably believed to be asbestos or polychlorinated biphenyl (PCB) which has not been rendered harmless, the Subcontractor shall immediately stop Work in the area affected and report the condition to the Contractor in writing. The Work in the affected area shall resume in the absence of asbestos or polychlorinated biphenyl (PCB), or when it has been rendered harmless, by written agreement of the Contractor and Subcontractor, or in accordance with final determination by the Architect on which arbitration has not been demanded, or by arbitration as provided in this Agreement. The Subcontractor shall not be required pursuant to Article 5 to perform without consent any Work relating to asbestos or polychlorinated biphenyl (PCB).

4.3.4 To the fullest extent permitted by law, the Contractor shall indemnify and hold harmless the Subcontractor, the Subcontractor's Sub-subcontractors, and agents and employees of any of them from and against claims, damages, losses and expenses, including but not limited to attorneys' fees, arising out of or resulting from performance of the Work in the affected area if in fact the material is asbestos or polychlorinated biphenyl (PCB) and has not been rendered harmless, provided that such claim, damage, loss or expense is attributable to bodily injury, sickness, disease or death, or to injury to or destruction of tangible property (other than the Work itself) including loss of use resulting therefrom, but only to the extent caused in whole or in part by negligent acts or omissions of the Contractor, Architect, Owner, anyone directly or indirectly employed by any of them, or anyone for whose acts any of them may be liable, regardless of whether or not such claim, damage, loss or expense is caused in part by a party indemnified hereunder. Such obligation shall not be construed to negate, abridge, or reduce other rights or obligations of indemnity which would otherwise exist as to a party or person described in this Subparagraph 4.3.4.

4.4 CLEANING UP

4.4.1 The Subcontractor shall keep the premises and surrounding area free from accumulation of waste materials or rubbish caused by operations performed under this Subcontract. The Subcontractor shall not be held responsible for unclean conditions caused by other contractors or subcontractors.

4.5 WARRANTY

4.5.1 The Subcontractor warrants to the Owner, Architect and Contractor that materials and equipment furnished under this Subcontract will be of good quality and new unless otherwise required or permitted by the Subcontract Documents, that the Work of this Subcontract will be free from defects not inherent in the quality required or permitted, and that the Work will conform with the requirements of the Subcontract Documents. Work not conforming to these requirements, including substitutions not properly approved and authorized, may be considered defective. The Subcontractor's warranty excludes remedy for damage or defect caused by abuse, modifications not executed by the Subcontractor, improper or insufficient maintenance, improper operation, or normal wear and tear under normal usage. This warranty shall be in addition to and not in limitation of any other warranty or remedy required by law or by the Subcontract Documents.

4.6 INDEMNIFICATION

4.6.1 To the fullest extent permitted by law, the Subcontractor shall indemnify and hold harmless the Owner, Contractor, Architect, Architect's consultants, and agents and employees of any of them from and against claims, damages, losses and expenses, including but not limited to attorney's fees, arising out of or resulting from performance of the Subcontractor's Work under this Subcontract, provided that such claim, damage, loss or expense is attributable to bodily injury, sickness, disease or death, or to injury to or destruction of tangible property (other than the Work itself) including loss of use resulting therefrom, but only to the extent caused in whole or in part by negligent acts or omissions of the Subcontractor, the Subcontractor's Sub-subcontractors, anyone directly or indirectly employed by them or anyone for whose acts they may be liable, regardless of whether or not such claim, damage, loss or expense is caused in part by a party indemnified hereunder. Such obligation shall not be construed to negate, abridge, or otherwise reduce other rights or obligations of indemnity which would otherwise exist as to a party or person described in this Paragraph 4.6.

4.6.2 In claims against any person or entity idemnified under this Paragraph 4.6 by an employee of the Subcontractor, the Subcontractor's Sub-subcontractors, anyone directly or indirectly employed by them or anyone for whose acts they may be liable, the indemnification obligation under this Paragraph 4.6 shall not be limited by a limitation on amount or type of damages, compensation or benefits payable by or for the Subcontractor or the Subcontractor's Sub-subcontractors under workers' or workmen's compensation acts, disability benefit acts or other employee benefit acts.

4.6.3 The obligations of the Subcontractor under this Paragraph 4.6 shall not extend to the liability of the Architect, the Architect's consultants, and agents and employees of any of them arising out of (1) the preparation or approval of maps, drawings, opinions, reports, surveys, Change Orders, designs or specifications, or (2) the giving of or the failure to give directions or instructions by the Architect, the Architect's consultants, and agents and employees of any of them, provided such giving or failure to give is the primary cause of the injury or damage.

4.7 REMEDIES FOR NONPAYMENT

4.7.1 If the Contractor does not pay the Subcontractor through no fault of the Subcontractor, within seven days from the time payment should be made as provided in this Agreement, the Subcontractor may, without prejudice to other available remedies, upon seven additional days' written notice to the Contractor, stop the Work of this Subcontract until payment of the amount owing has been received. The Subcontract Sum shall, by appropriate adjustment, be increased by the amount of the Subcontractor's reasonable costs of shutdown, delay and start-up.

ARTICLE 5

CHANGES IN THE WORK

5.1 The Owner may make changes in the Work by issuing Modifications to the Prime Contract. Upon receipt of such a Modification issued subsequent to the execution of the Subcontract Agreement, the Contractor shall promptly notify the Subcontractor of the Modification. Unless otherwise directed by the Contractor, the Subcontractor shall not thereafter order materials or perform Work which would be inconsistent with the changes made by the Modifications to the Prime Contract.

5.2 The Subcontractor may be ordered in writing by the Contractor, without invalidating this Subcontract, to make changes in the Work within the general scope of this Subcontract consisting of additions, deletions or other revisions, including those required by Modifications to the Prime Contract issued subsequent to the execution of this Agreement, the Subcontract Sum and the Subcontract Time being adjusted accordingly. The Subcontractor, prior to the commencement of such changed or revised Work, shall submit promptly to the Contractor written copies of a claim for adjustment to the Subcontract Sum and Subcontract Time for such revised Work in a manner consistent with requirements of the Subcontract Documents.

5.3 The Subcontractor shall make claims promptly to the Contractor for additional cost, extensions of time and damages for delays or other causes in accordance with the Subcontract

Documents. A claim which will affect or become part of a claim which the Contractor is required to make under the Prime Contract within a specified time period or in a specified manner shall be made in sufficient time to permit the Contractor to satisfy the requirements of the Prime Contract. Such claims shall be received by the Contractor not less than two working days preceding the time by which the Contractor's claim must be made. Failure of the Subcontractor to make such a timely claim shall bind the Subcontractor to the same consequences as those to which the Contractor is bound.

ARTICLE 6

ARBITRATION

6.1 Any controversy or claim between the Contractor and the Subcontractor arising out of or related to this Subcontract, or the breach thereof, shall be settled by arbitration, which shall be conducted in the same manner and under the same procedure as provided in the Prime Contract with respect to claims between the Owner and the Contractor, except that a decision by the Architect shall not be a condition precedent to arbitration. If the Prime Contract does not provide for arbitration or fails to specify the manner and procedure for arbitration, it shall be conducted in accordance with the Construction Industry Arbitration Rules of the American Arbitration Association currently in effect unless the parties mutually agree otherwise.

6.2 Except by written consent of the person or entity sought to be joined, no arbitration arising out of or relating to the Subcontract shall include, by consolidation or joinder or in any other manner, any person or entity not a party to the Agreement under which such arbitration arises, unless it is shown at the time the demand for arbitration is filed that (1) such person or entity is substantially involved in a common question of fact or law, (2) the presence of such person or entity is required if complete relief is to be accorded in the arbitration, (3) the interest or responsibility of such person or entity in the matter is not insubstantial, and (4) such person or entity is not the Architect, the Architect's employee, the Architect's consultant, or an employee or agent of any of them. This agreement to arbitrate and any other written agreement to arbitrate with an additional person or persons referred to herein shall be specifically enforceable under applicable law in any court having jurisdiction thereof.

6.3 The Contractor shall give the Subcontractor prompt written notice of any demand received or made by the Contractor for arbitration if the dispute involves or relates to the Work, materials, equipment, rights or responsibilities of the Subcontractor. The Contractor shall consent to inclusion of the Subcontractor in the arbitration proceeding whether by joinder, consolidation or otherwise, if the Subcontractor requests in writing to be included within ten days after receipt of the Contractor's notice.

6.4 The award rendered by the arbitrator or arbitrators shall be final, and judgment may be entered upon it in accordance with applicable law in any court having jurisdiction thereof.

6.5 This Article 6 shall not be deemed a limitation of rights or remedies which the Subcontractor may have under Federal law, under state mechanics' lien laws, or under applicable labor or material payment bonds unless such rights or remedies are expressly waived by the Subcontractor.

ARTICLE 7

TERMINATION, SUSPENSION OR ASSIGNMENT OF THE SUBCONTRACT

7.1 TERMINATION BY THE SUBCONTRACTOR

7.1.1 The Subcontractor may terminate the Subcontract for the same reasons and under the same circumstances and procedures with respect to the Contractor as the Contractor may terminate with respect to the Owner under the Prime Contract, or for nonpayment of amounts due under this Subcontract for 60 days or longer. In the event of such termination by the Subcontractor for any reason which is not the fault of the Subcontractor, Sub-subcontractors or their agents or employees or other persons performing portions of the Work under contract with the Subcontractor, the Subcontractor shall be entitled to recover from the Contractor payment for Work executed and for proven loss with respect to materials, equipment, tools, and construction equipment and machinery, including reasonable overhead, profit and damages.

7.2 TERMINATION BY THE CONTRACTOR

7.2.1 If the Subcontractor persistently or repeatedly fails or neglects to carry out the Work in accordance with the Subcontract Documents or otherwise to perform in accordance with this Agreement and fails within seven days after receipt of written notice to commence and continue correction of such default or neglect with diligence and promptness, the Contractor may, after seven days following receipt by the Subcontractor of an additional written notice and without prejudice to any other remedy the Contractor may have, terminate the Subcontract and finish the Subcontractor's Work by whatever method the Contractor may deem expedient. If the unpaid balance of the Subcontract Sum exceeds the expense of finishing the Subcontractor's Work, such excess shall be paid to the Subcontractor, but if such expense exceeds such unpaid balance, the Subcontractor shall pay the difference to the Contractor.

7.3 ASSIGNMENT OF THE SUBCONTRACT

7.3.1 In the event of termination of the Prime Contract by the Owner, the Contractor may assign this Subcontract to the Owner, with the Owner's agreement, subject to the provisions of the Prime Contract and to the prior rights of the surety, if any, obligated under bonds relating to the Prime Contract. If the Work of the Prime Contract has been suspended for more than 30 days, the Subcontractor's compensation shall be equitably adjusted.

7.3.2 The Subcontractor shall not assign the Work of this Subcontract without the written consent of the Contractor, nor subcontract the whole of this Subcontract without the written consent of the Contractor, nor further subcontract portions of this Subcontract without written notification to the Contractor when such notification is requested by the Contractor.

ARTICLE 8
THE WORK OF THIS SUBCONTRACT

8.1 The Subcontractor shall execute the following portion of the Work described in the Subcontract Documents, including all labor, materials, equipment, services and other items required to complete such portion of the Work, except to the extent specifically indicated in the Subcontract Documents to be the responsibility of others:
(Insert a precise description of the Work of this Subcontract, referring where appropriate to numbers of Drawings, sections of Specifications and pages of Addenda, Modifications and accepted Alternates.)

ARTICLE 9
DATE OF COMMENCEMENT AND SUBSTANTIAL COMPLETION

9.1 The Subcontractor's date of commencement is the date from which the Contract Time of Paragraph 9.3 is measured; it shall be the date of this Agreement, as first written above, unless a different date is stated below or provision is made for the date to be fixed in a notice to proceed issued by the Contractor.
(Insert the date of commencement, if it differs from the date of this Agreement or, if applicable, state that the date will be fixed in a notice to proceed.)

9.2 Unless the date of commencement is established by a notice to proceed issued by the Contractor, or the Contractor has commenced visible Work at the site under the Prime Contract, the Subcontractor shall notify the Contractor in writing not less than five days before commencing the Subcontractor's Work to permit the timely filing of mortgages, mechanic's liens and other security interests.

9.3 The Work of this Subcontract shall be substantially completed not later than
(Insert the calendar date or number of calendar days after the Subcontractor's date of commencement. Also insert any requirements for earlier Substantial Completion of certain portions of the Subcontractor's Work, if not stated elsewhere in the Subcontract Documents.)

, subject to adjustments of this Subcontract Time as provided in the Subcontract Documents.
(Insert provisions, if any, for liquidated damages relating to failure to complete on time.)

9.4 Time is of the essence of this Subcontract.

9.5 No extension of time will be valid without the Contractor's written consent after claim made by the Subcontractor in accordance with Paragraph 5.2.

ARTICLE 10
SUBCONTRACT SUM

10.1 The Contractor shall pay the Subcontractor in current funds for performance of the Subcontract the Subcontract Sum of Dollars ($), subject to additions and deductions as provided in the Subcontract Documents.

10.2 The Subcontract Sum is based upon the following alternates, if any, which are described in the Subcontract Documents and have been accepted by the Owner and the Contractor:
(Insert the numbers or other identification of accepted alternates.)

10.3 Unit prices, if any, are as follows:

ARTICLE 11
PROGRESS PAYMENTS

11.1 Based upon applications for payment submitted to the Contractor by the Subcontractor, corresponding to Applications for Payment submitted by the Contractor to the Architect, and Certificates for Payment issued by the Architect, the Contractor shall make progress payments on account of the Subcontract Sum to the Subcontractor as provided below and elsewhere in the Subcontract Documents.

11.2 The period covered by each application for payment shall be one calendar month ending on the last day of the month, or as follows:

11.3 Provided an application for payment is received by the Contractor not later than the day of a month, the Contractor shall include the Subcontractor's Work covered by that application in the next Application for Payment which the Contractor is entitled to submit to the Architect. The Contractor shall pay the Subcontractor each progress payment within three working days after the Contractor receives payment from the Owner. If the Architect does not issue a Certificate for Payment or the Contractor does not receive payment for any cause which is not the fault of the Subcontractor, the Contractor shall pay the Subcontractor, on demand, a progress payment computed as provided in Paragraphs 11.7 and 11.8.

11.4 If an application for payment is received by the Contractor after the application date fixed above, the Subcontractor's Work covered by it shall be included by the Contractor in the next Application for Payment submitted to the Architect.

11.5 Each application for payment shall be based upon the most recent schedule of values submitted by the Subcontractor in accordance with the Subcontract Documents. The schedule of values shall allocate the entire Subcontract Sum among the various portions of the Subcontractor's Work and be prepared in such form and supported by such data to substantiate its accuracy as the Contractor may require. This schedule, unless objected to by the Contractor, shall be used as a basis for reviewing the Subcontractor's applications for payment.

11.6 Applications for payment submitted by the Subcontractor shall indicate the percentage of completion of each portion of the Subcontractor's Work as of the end of the period covered by the application for payment.

11.7 Subject to the provisions of the Subcontract Documents, the amount of each progress payment shall be computed as follows:

11.7.1 Take that portion of the Subcontract Sum properly allocable to completed Work as determined by multiplying the percentage completion of each portion of the Subcontractor's Work by the share of the total Subcontract Sum allocated to that portion of the Subcontractor's Work in the schedule of values, less that percentage actually retained, if any, from payments to the Contractor on account of the Work of the Subcontractor. Pending final determination of cost to the Contractor of changes in the Work which have been properly authorized by Construction Change Directive, amounts not in dispute may be included to the same extent provided in the Prime Contract, even though the Subcontract Sum has not yet been adjusted;

11.7.2 Add that portion of the Subcontract Sum properly allocable to materials and equipment delivered and suitably stored at the site by the Subcontractor for subsequent incorporation in the Subcontractor's Work or, if approved in advance by the Owner, suitably stored off the site at a location agreed upon in writing, less the same percentage retainage required by the Prime Contract to be applied to such materials and equipment in the Contractor's Application for Payment;

11.7.3 Subtract the aggregate of previous payments made by the Contractor; and

11.7.4 Subtract amounts, if any, calculated under Subparagraph 11.7.1 or 11.7.2 which are related to Work of the Subcontractor for which the Architect has withheld or nullified, in whole or in part, a Certificate of Payment for a cause which is the fault of the Subcontractor.

11.8 SUBSTANTIAL COMPLETION

11.8.1 When the Subcontractor's Work or a designated portion thereof is substantially complete and in accordance with the requirements of the Prime Contract, the Contractor shall, upon application by the Subcontractor, make prompt application for payment for such Work. Within 30 days following issuance by the Architect of the Certificate for Payment covering such substantially completed Work, the Contractor shall, to the full extent allowed in the Prime Contract, make payment to the Subcontractor, deducting any portion of the funds for the Subcontractor's Work withheld in accordance with the Certificate to cover costs of items to be completed or corrected by the Subcontractor. Such payment to the Subcontractor shall be the entire unpaid balance of the Subcontract Sum if a full release of retainage is allowed under the Prime Contract for the Subcontractor's Work prior to the completion of the entire Project. If the Prime Contract does not allow for a full release of retainage, then such payment shall be an amount which, when added to previous payments to the Subcontractor, will reduce the retainage on the Subcontractor's substantially completed Work to the same percentage of retainage as that on the Contractor's Work covered by the Certificate.

ARTICLE 12
FINAL PAYMENT

12.1 Final payment, constituting the entire unpaid balance of the Subcontract Sum, shall be made by the Contractor to the Subcontractor when the Subcontractor's Work is fully performed in accordance with the requirements of the Contract Documents, the Architect has issued a Certificate for Payment covering the Subcontractor's completed Work and the Contractor has received payment from the Owner. If, for any cause which is not the fault of the Subcontractor, a Certificate for Payment is not issued or the Contractor does not receive timely payment or does not pay the Subcontractor within three working days after receipt of payment from the Owner, final payment to the Subcontractor shall be made upon demand.
(Insert provisions for earlier final payment to the Subcontractor, if applicable.)

12.2 Before issuance of the final payment, the Subcontractor, if required, shall submit evidence satisfactory to the Contractor that all payrolls, bills for materials and equipment, and all known indebtedness connected with the Subcontractor's Work have been satisfied.

ARTICLE 13
INSURANCE AND BONDS

13.1 The Subcontractor shall purchase and maintain insurance of the following types of coverage and limits of liability:

13.2 Coverages, whether written on an occurrence or claims-made basis, shall be maintained without interruption from date of commencement of the Subcontractor's Work until date of final payment and termination of any coverage required to be maintained after final payment.

13.3 Certificates of insurance acceptable to the Contractor shall be filed with the Contractor prior to commencement of the Subcontractor's Work. These certificates and the insurance policies required by this Article 13 shall contain a provision that coverages afforded under the policies will not be cancelled or allowed to expire until at least 30 days' prior written notice has been given to the Contractor. If any of the foregoing insurance coverages are required to remain in force after final payment and are reasonably available, an additional certificate evidencing continuation of such coverage shall be submitted with the final application for payment as required in Article 12. If any information concerning reduction of coverage is not furnished by the insurer, it shall be furnished by the Subcontractor with reasonable promptness according to the Subcontractor's information and belief.

13.4 The Contractor shall furnish to the Subcontractor satisfactory evidence of insurance required of the Contractor under the Prime Contract.

13.5 Waivers of Subrogation. The Contractor and Subcontractor waive all rights against (1) each other and any of their Subcontractors, Sub-subcontractors, agents and employees, each of the other, and (2) the Owner, the Architect, the Architect's consultants, separate contractors, and any of their subcontractors, sub-subcontractors, agents and employees for damages caused by fire or other perils to the extent covered by property insurance provided under the Prime Contract or other property insurance applicable to the Work, except such rights as they may have to proceeds of such insurance held by the Owner as fiduciary. The Subcontractor shall require of the Subcontractor's Sub-subcontractors, agents and employees, by appropriate agreements, written where legally required for validity, similar waivers in favor of other parties enumerated herein. The policies shall provide such waivers of subrogation by endorsement or otherwise. A waiver of subrogation shall be effective as to a person or entity even though that person or entity would otherwise have a duty of indemnification, contractual or otherwise, did not pay the insurance premium directly or indirectly, and whether or not the person or entity had an insurable interest in the property damaged.

13.6 The Contractor shall promptly, upon request of the Subcontractor, furnish a copy or permit a copy to be made of any bond covering payment of obligations arising under the Subcontract.

13.7 Performance Bond and Payment Bond:
(If the Subcontractor is to furnish bonds, insert the specific requirements here.)

ARTICLE 14
TEMPORARY FACILITIES AND WORKING CONDITIONS

14.1 The Contractor shall furnish and make available to the Subcontractor the following temporary facilities, equipment and services; these shall be furnished at no cost to the Subcontractor unless otherwise indicated below:

14.2 Specific working conditions:
(Insert any applicable arrangements concerning working conditions and labor matters for the Project.)

ARTICLE 15
MISCELLANEOUS PROVISIONS

15.1 Where reference is made in this Agreement to a provision of the General Conditions or another Subcontract Document, the reference refers to that provision as amended or supplemented by other provisions of the Subcontract Documents.

15.2 Payments due and unpaid under this Subcontract shall bear interest from the date payment is due at such rate as the parties may agree upon in writing or, in the absence thereof, at the legal rate prevailing from time to time at the place where the Project is located.
(Insert rate of interest agreed upon, if any.)

(Usury laws and requirements under the Federal Truth in Lending Act, similar state and local consumer credit laws and other regulations at the Owner's, Contractor's and Subcontractor's principal places of business, the location of the Project and elsewhere may affect the validity of this provision. Legal advice should be obtained with respect to deletions or modifications, and also regarding requirements such as written disclosures or waivers.)

ARTICLE 16
ENUMERATION OF SUBCONTRACT DOCUMENTS

16.1 The Subcontract Documents, except for Modifications issued after execution of this Agreement, are enumerated as follows:

16.1.1 This executed Standard Form of Agreement Between Contractor and Subcontractor, AIA Document A401, 1987 Edition;

16.1.2 The Prime Contract, consisting of the Agreement between the Owner and Contractor dated as first entered above and the other Contract Documents enumerated in the Owner-Contractor Agreement; Conditions of the Contract (General, Supplementary and other Conditions), Drawings, Specifications, Addenda and other documents enumerated therein:

16.1.3 The following Modifications to the Prime Contract, if any, issued subsequent to the execution of the Owner-Contractor Agreement but prior to the execution of this Agreement:

Modification **Date**

16.1.4 Other Documents, if any, forming part of the Subcontract Documents are as follows:
(List any additional documents which are intended to form part of the Subcontract Documents. Requests for proposal and the Subcontractor's bid or proposal should be listed here only if intended to be part of the Subcontract Documents.)

This Agreement entered into as of the day and year first written above.

CONTRACTOR	SUBCONTRACTOR
(Signature)	*(Signature)*
(Printed name and title)	*(Printed name and title)*

APPENDIX H

AIA DOCUMENT B141, STANDARD FORM OF AGREEMENT BETWEEN OWNER AND ARCHITECT (1987)

THE AMERICAN INSTITUTE OF ARCHITECTS

AIA Document B141

Standard Form of Agreement Between Owner and Architect

1987 EDITION

THIS DOCUMENT HAS IMPORTANT LEGAL CONSEQUENCES; CONSULTATION WITH AN ATTORNEY IS ENCOURAGED WITH RESPECT TO ITS COMPLETION OR MODIFICATION.

AGREEMENT

made as of the day of in the year of
Nineteen Hundred and

BETWEEN the Owner:
(Name and address)

and the Architect:
(Name and address)

For the following Project:
(Include detailed description of Project, location, address and scope.)

The Owner and Architect agree as set forth below.

TERMS AND CONDITIONS OF AGREEMENT BETWEEN OWNER AND ARCHITECT

ARTICLE 1

ARCHITECT'S RESPONSIBILITIES

1.1 ARCHITECT'S SERVICES

1.1.1 The Architect's services consist of those services performed by the Architect, Architect's employees and Architect's consultants as enumerated in Articles 2 and 3 of this Agreement and any other services included in Article 12.

1.1.2 The Architect's services shall be performed as expeditiously as is consistent with professional skill and care and the orderly progress of the Work. Upon request of the Owner, the Architect shall submit for the Owner's approval a schedule for the performance of the Architect's services which may be adjusted as the Project proceeds, and shall include allowances for periods of time required for the Owner's review and for approval of submissions by authorities having jurisdiction over the Project. Time limits established by this schedule approved by the Owner shall not, except for reasonable cause, be exceeded by the Architect or Owner.

1.1.3 The services covered by this Agreement are subject to the time limitations contained in Subparagraph 11.5.1.

ARTICLE 2

SCOPE OF ARCHITECT'S BASIC SERVICES

2.1 DEFINITION

2.1.1 The Architect's Basic Services consist of those described in Paragraphs 2.2 through 2.6 and any other services identified in Article 12 as part of Basic Services, and include normal structural, mechanical and electrical engineering services.

2.2 SCHEMATIC DESIGN PHASE

2.2.1 The Architect shall review the program furnished by the Owner to ascertain the requirements of the Project and shall arrive at a mutual understanding of such requirements with the Owner.

2.2.2 The Architect shall provide a preliminary evaluation of the Owner's program, schedule and construction budget requirements, each in terms of the other, subject to the limitations set forth in Subparagraph 5.2.1.

2.2.3 The Architect shall review with the Owner alternative approaches to design and construction of the Project.

2.2.4 Based on the mutually agreed-upon program, schedule and construction budget requirements, the Architect shall prepare, for approval by the Owner, Schematic Design Documents consisting of drawings and other documents illustrating the scale and relationship of Project components.

2.2.5 The Architect shall submit to the Owner a preliminary estimate of Construction Cost based on current area, volume or other unit costs.

2.3 DESIGN DEVELOPMENT PHASE

2.3.1 Based on the approved Schematic Design Documents and any adjustments authorized by the Owner in the program, schedule or construction budget, the Architect shall prepare, for approval by the Owner, Design Development Documents consisting of drawings and other documents to fix and describe the size and character of the Project as to architectural, structural, mechanical and electrical systems, materials and such other elements as may be appropriate.

2.3.2 The Architect shall advise the Owner of any adjustments to the preliminary estimate of Construction Cost.

2.4 CONSTRUCTION DOCUMENTS PHASE

2.4.1 Based on the approved Design Development Documents and any further adjustments in the scope or quality of the Project or in the construction budget authorized by the Owner, the Architect shall prepare, for approval by the Owner, Construction Documents consisting of Drawings and Specifications setting forth in detail the requirements for the construction of the Project.

2.4.2 The Architect shall assist the Owner in the preparation of the necessary bidding information, bidding forms, the Conditions of the Contract, and the form of Agreement between the Owner and Contractor.

2.4.3 The Architect shall advise the Owner of any adjustments to previous preliminary estimates of Construction Cost indicated by changes in requirements or general market conditions.

2.4.4 The Architect shall assist the Owner in connection with the Owner's responsibility for filing documents required for the approval of governmental authorities having jurisdiction over the Project.

2.5 BIDDING OR NEGOTIATION PHASE

2.5.1 The Architect, following the Owner's approval of the Construction Documents and of the latest preliminary estimate of Construction Cost, shall assist the Owner in obtaining bids or negotiated proposals and assist in awarding and preparing contracts for construction.

2.6 CONSTRUCTION PHASE—ADMINISTRATION OF THE CONSTRUCTION CONTRACT

2.6.1 The Architect's responsibility to provide Basic Services for the Construction Phase under this Agreement commences with the award of the Contract for Construction and terminates at the earlier of the issuance to the Owner of the final Certificate for Payment or 60 days after the date of Substantial Completion of the Work.

2.6.2 The Architect shall provide administration of the Contract for Construction as set forth below and in the edition of AIA Document A201, General Conditions of the Contract for Construction, current as of the date of this Agreement, unless otherwise provided in this Agreement.

2.6.3 Duties, responsibilities and limitations of authority of the Architect shall not be restricted, modified or extended without written agreement of the Owner and Architect with consent of the Contractor, which consent shall not be unreasonably withheld.

2.6.4 The Architect shall be a representative of and shall advise and consult with the Owner (1) during construction until final payment to the Contractor is due, and (2) as an Additional Service at the Owner's direction from time to time during the correction period described in the Contract for Construction. The Architect shall have authority to act on behalf of the Owner only to the extent provided in this Agreement unless otherwise modified by written instrument.

2.6.5 The Architect shall visit the site at intervals appropriate to the stage of construction or as otherwise agreed by the Owner and Architect in writing to become generally familiar with the progress and quality of the Work completed and to determine in general if the Work is being performed in a manner indicating that the Work when completed will be in accordance with the Contract Documents. However, the Architect shall not be required to make exhaustive or continuous on-site inspections to check the quality or quantity of the Work. On the basis of on-site observations as an architect, the Architect shall keep the Owner informed of the progress and quality of the Work, and shall endeavor to guard the Owner against defects and deficiencies in the Work. *(More extensive site representation may be agreed to as an Additional Service, as described in Paragraph 3.2.)*

2.6.6 The Architect shall not have control over or charge of and shall not be responsible for construction means, methods, techniques, sequences or procedures, or for safety precautions and programs in connection with the Work, since these are solely the Contractor's responsibility under the Contract for Construction. The Architect shall not be responsible for the Contractor's schedules or failure to carry out the Work in accordance with the Contract Documents. The Architect shall not have control over or charge of acts or omissions of the Contractor, Subcontractors, or their agents or employees, or of any other persons performing portions of the Work.

2.6.7 The Architect shall at all times have access to the Work wherever it is in preparation or progress.

2.6.8 Except as may otherwise be provided in the Contract Documents or when direct communications have been specially authorized, the Owner and Contractor shall communicate through the Architect. Communications by and with the Architect's consultants shall be through the Architect.

2.6.9 Based on the Architect's observations and evaluations of the Contractor's Applications for Payment, the Architect shall review and certify the amounts due the Contractor.

2.6.10 The Architect's certification for payment shall constitute a representation to the Owner, based on the Architect's observations at the site as provided in Subparagraph 2.6.5 and on the data comprising the Contractor's Application for Payment, that the Work has progressed to the point indicated and that, to the best of the Architect's knowledge, information and belief, quality of the Work is in accordance with the Contract Documents. The foregoing representations are subject to an evaluation of the Work for conformance with the Contract Documents upon Substantial Completion, to results of subsequent tests and inspections, to minor deviations from the Contract Documents correctable prior to completion and to specific qualifications expressed by the Architect. The issuance of a Certificate for Payment shall further constitute a representation that the Contractor is entitled to payment in the amount certified. However, the issuance of a Certificate for Payment shall not be a representation that the Architect has (1) made exhaustive or continuous on-site inspections to check the quality or quantity of the Work, (2) reviewed construction means, methods, techniques, sequences or procedures, (3) reviewed copies of requisitions received from Subcontractors and material suppliers and other data requested by the Owner to substantiate the Contractor's right to payment or (4) ascertained how or for what purpose the Contractor has used money previously paid on account of the Contract Sum.

2.6.11 The Architect shall have authority to reject Work which does not conform to the Contract Documents. Whenever the Architect considers it necessary or advisable for implementation of the intent of the Contract Documents, the Architect will have authority to require additional inspection or testing of the Work in accordance with the provisions of the Contract Documents, whether or not such Work is fabricated, installed or completed. However, neither this authority of the Architect nor a decision made in good faith either to exercise or not to exercise such authority shall give rise to a duty or responsibility of the Architect to the Contractor, Subcontractors, material and equipment suppliers, their agents or employees or other persons performing portions of the Work.

2.6.12 The Architect shall review and approve or take other appropriate action upon Contractor's submittals such as Shop Drawings, Product Data and Samples, but only for the limited purpose of checking for conformance with information given and the design concept expressed in the Contract Documents. The Architect's action shall be taken with such reasonable promptness as to cause no delay in the Work or in the construction of the Owner or of separate contractors, while allowing sufficient time in the Architect's professional judgment to permit adequate review. Review of such submittals is not conducted for the purpose of determining the accuracy and completeness of other details such as dimensions and quantities or for substantiating instructions for installation or performance of equipment or systems designed by the Contractor, all of which remain the responsibility of the Contractor to the extent required by the Contract Documents. The Architect's review shall not constitute approval of safety precautions or, unless otherwise specifically stated by the Architect, of construction means, methods, techniques, sequences or procedures. The Architect's approval of a specific item shall not indicate approval of an assembly of which the item is a component. When professional certification of performance characteristics of materials, systems or equipment is required by the Contract Documents, the Architect shall be entitled to rely upon such certification to establish that the materials, systems or equipment will meet the performance criteria required by the Contract Documents.

2.6.13 The Architect shall prepare Change Orders and Construction Change Directives, with supporting documentation and data if deemed necessary by the Architect as provided in Subparagraphs 3.1.1 and 3.3.3, for the Owner's approval and execution in accordance with the Contract Documents, and may authorize minor changes in the Work not involving an adjustment in the Contract Sum or an extension of the Contract Time which are not inconsistent with the intent of the Contract Documents.

2.6.14 The Architect shall conduct inspections to determine the date or dates of Substantial Completion and the date of final completion, shall receive and forward to the Owner for the Owner's review and records written warranties and related documents required by the Contract Documents and assembled by the Contractor, and shall issue a final Certificate for Payment upon compliance with the requirements of the Contract Documents.

2.6.15 The Architect shall interpret and decide matters concerning performance of the Owner and Contractor under the requirements of the Contract Documents on written request of either the Owner or Contractor. The Architect's response to such requests shall be made with reasonable promptness and within any time limits agreed upon.

2.6.16 Interpretations and decisions of the Architect shall be consistent with the intent of and reasonably inferable from the Contract Documents and shall be in writing or in the form of drawings. When making such interpretations and initial decisions, the Architect shall endeavor to secure faithful performance by both Owner and Contractor, shall not show partiality to either, and shall not be liable for results of interpretations or decisions so rendered in good faith.

2.6.17 The Architect's decisions on matters relating to aesthetic effect shall be final if consistent with the intent expressed in the Contract Documents.

2.6.18 The Architect shall render written decisions within a reasonable time on all claims, disputes or other matters in question between the Owner and Contractor relating to the execution or progress of the Work as provided in the Contract Documents.

2.6.19 The Architect's decisions on claims, disputes or other matters, including those in question between the Owner and Contractor, except for those relating to aesthetic effect as provided in Subparagraph 2.6.17, shall be subject to arbitration as provided in this Agreement and in the Contract Documents.

ARTICLE 3

ADDITIONAL SERVICES

3.1 GENERAL

3.1.1 The services described in this Article 3 are not included in Basic Services unless so identified in Article 2, and they shall be paid for by the Owner as provided in this Agreement, in addition to the compensation for Basic Services. The services described under Paragraphs 3.2 and 3.4 shall only be provided if authorized or confirmed in writing by the Owner. If services described under Contingent Additional Services in Paragraph 3.3 are required due to circumstances beyond the Architect's control, the Architect shall notify the Owner prior to commencing such services. If the Owner deems that such services described under Paragraph 3.3 are not required, the Owner shall give prompt written notice to the Architect. If the Owner indicates in writing that all or part of such Contingent Additional Services are not required, the Architect shall have no obligation to provide those services.

3.2 PROJECT REPRESENTATION BEYOND BASIC SERVICES

3.2.1 If more extensive representation at the site than is described in Subparagraph 2.6.5 is required, the Architect shall provide one or more Project Representatives to assist in carrying out such additional on-site responsibilities.

3.2.2 Project Representatives shall be selected, employed and directed by the Architect, and the Architect shall be compensated therefor as agreed by the Owner and Architect. The duties, responsibilities and limitations of authority of Project Representatives shall be as described in the edition of AIA Document B352 current as of the date of this Agreement, unless otherwise agreed.

3.2.3 Through the observations by such Project Representatives, the Architect shall endeavor to provide further protection for the Owner against defects and deficiencies in the Work, but the furnishing of such project representation shall not modify the rights, responsibilities or obligations of the Architect as described elsewhere in this Agreement.

3.3 CONTINGENT ADDITIONAL SERVICES

3.3.1 Making revisions in Drawings, Specifications or other documents when such revisions are:

.1 inconsistent with approvals or instructions previously given by the Owner, including revisions made necessary by adjustments in the Owner's program or Project budget;

.2 required by the enactment or revision of codes, laws or regulations subsequent to the preparation of such documents; or

.3 due to changes required as a result of the Owner's failure to render decisions in a timely manner.

3.3.2 Providing services required because of significant changes in the Project including, but not limited to, size, quality, complexity, the Owner's schedule, or the method of bidding or negotiating and contracting for construction, except for services required under Subparagraph 5.2.5.

3.3.3 Preparing Drawings, Specifications and other documentation and supporting data, evaluating Contractor's proposals, and providing other services in connection with Change Orders and Construction Change Directives.

3.3.4 Providing services in connection with evaluating substitutions proposed by the Contractor and making subsequent revisions to Drawings, Specifications and other documentation resulting therefrom.

3.3.5 Providing consultation concerning replacement of Work damaged by fire or other cause during construction, and furnishing services required in connection with the replacement of such Work.

3.3.6 Providing services made necessary by the default of the Contractor, by major defects or deficiencies in the Work of the Contractor, or by failure of performance of either the Owner or Contractor under the Contract for Construction.

3.3.7 Providing services in evaluating an extensive number of claims submitted by the Contractor or others in connection with the Work.

3.3.8 Providing services in connection with a public hearing, arbitration proceeding or legal proceeding except where the Architect is party thereto.

3.3.9 Preparing documents for alternate, separate or sequential bids or providing services in connection with bidding, negotiation or construction prior to the completion of the Construction Documents Phase.

3.4 OPTIONAL ADDITIONAL SERVICES

3.4.1 Providing analyses of the Owner's needs and programming the requirements of the Project.

3.4.2 Providing financial feasibility or other special studies.

3.4.3 Providing planning surveys, site evaluations or comparative studies of prospective sites.

3.4.4 Providing special surveys, environmental studies and submissions required for approvals of governmental authorities or others having jurisdiction over the Project.

3.4.5 Providing services relative to future facilities, systems and equipment.

3.4.6 Providing services to investigate existing conditions or facilities or to make measured drawings thereof.

3.4.7 Providing services to verify the accuracy of drawings or other information furnished by the Owner.

3.4.8 Providing coordination of construction performed by separate contractors or by the Owner's own forces and coordination of services required in connection with construction performed and equipment supplied by the Owner.

3.4.9 Providing services in connection with the work of a construction manager or separate consultants retained by the Owner.

3.4.10 Providing detailed estimates of Construction Cost.

3.4.11 Providing detailed quantity surveys or inventories of material, equipment and labor.

3.4.12 Providing analyses of owning and operating costs.

3.4.13 Providing interior design and other similar services required for or in connection with the selection, procurement or installation of furniture, furnishings and related equipment.

3.4.14 Providing services for planning tenant or rental spaces.

3.4.15 Making investigations, inventories of materials or equipment, or valuations and detailed appraisals of existing facilities.

3.4.16 Preparing a set of reproducible record drawings showing significant changes in the Work made during construction based on marked-up prints, drawings and other data furnished by the Contractor to the Architect.

3.4.17 Providing assistance in the utilization of equipment or systems such as testing, adjusting and balancing, preparation of operation and maintenance manuals, training personnel for operation and maintenance, and consultation during operation.

3.4.18 Providing services after issuance to the Owner of the final Certificate for Payment, or in the absence of a final Certificate for Payment, more than 60 days after the date of Substantial Completion of the Work.

3.4.19 Providing services of consultants for other than architectural, structural, mechanical and electrical engineering portions of the Project provided as a part of Basic Services.

3.4.20 Providing any other services not otherwise included in this Agreement or not customarily furnished in accordance with generally accepted architectural practice.

ARTICLE 4

OWNER'S RESPONSIBILITIES

4.1 The Owner shall provide full information regarding requirements for the Project, including a program which shall set forth the Owner's objectives, schedule, constraints and criteria, including space requirements and relationships, flexibility, expandability, special equipment, systems and site requirements.

4.2 The Owner shall establish and update an overall budget for the Project, including the Construction Cost, the Owner's other costs and reasonable contingencies related to all of these costs.

4.3 If requested by the Architect, the Owner shall furnish evidence that financial arrangements have been made to fulfill the Owner's obligations under this Agreement.

4.4 The Owner shall designate a representative authorized to act on the Owner's behalf with respect to the Project. The Owner or such authorized representative shall render decisions in a timely manner pertaining to documents submitted by the Architect in order to avoid unreasonable delay in the orderly and sequential progress of the Architect's services.

4.5 The Owner shall furnish surveys describing physical characteristics, legal limitations and utility locations for the site of the Project, and a written legal description of the site. The surveys and legal information shall include, as applicable, grades and lines of streets, alleys, pavements and adjoining property and structures; adjacent drainage; rights-of-way, restrictions, easements, encroachments, zoning, deed restrictions, boundaries and contours of the site; locations, dimensions and necessary data pertaining to existing buildings, other improvements and trees; and information concerning available utility services and lines, both public and private, above and below grade, including inverts and depths. All the information on the survey shall be referenced to a project benchmark.

4.6 The Owner shall furnish the services of geotechnical engineers when such services are requested by the Architect. Such services may include but are not limited to test borings, test pits, determinations of soil bearing values, percolation tests, evaluations of hazardous materials, ground corrosion and resistivity tests, including necessary operations for anticipating subsoil conditions, with reports and appropriate professional recommendations.

4.6.1 The Owner shall furnish the services of other consultants when such services are reasonably required by the scope of the Project and are requested by the Architect.

4.7 The Owner shall furnish structural, mechanical, chemical, air and water pollution tests, tests for hazardous materials, and other laboratory and environmental tests, inspections and reports required by law or the Contract Documents.

4.8 The Owner shall furnish all legal, accounting and insurance counseling services as may be necessary at any time for the Project, including auditing services the Owner may require to verify the Contractor's Applications for Payment or to ascertain how or for what purposes the Contractor has used the money paid by or on behalf of the Owner.

4.9 The services, information, surveys and reports required by Paragraphs 4.5 through 4.8 shall be furnished at the Owner's expense, and the Architect shall be entitled to rely upon the accuracy and completeness thereof.

4.10 Prompt written notice shall be given by the Owner to the Architect if the Owner becomes aware of any fault or defect in the Project or nonconformance with the Contract Documents.

4.11 The proposed language of certificates or certifications requested of the Architect or Architect's consultants shall be submitted to the Architect for review and approval at least 14 days prior to execution. The Owner shall not request certifications that would require knowledge or services beyond the scope of this Agreement.

ARTICLE 5
CONSTRUCTION COST

5.1 DEFINITION

5.1.1 The Construction Cost shall be the total cost or estimated cost to the Owner of all elements of the Project designed or specified by the Architect.

5.1.2 The Construction Cost shall include the cost at current market rates of labor and materials furnished by the Owner and equipment designed, specified, selected or specially provided for by the Architect, plus a reasonable allowance for the Contractor's overhead and profit. In addition, a reasonable allowance for contingencies shall be included for market conditions at the time of bidding and for changes in the Work during construction.

5.1.3 Construction Cost does not include the compensation of the Architect and Architect's consultants, the costs of the land, rights-of-way, financing or other costs which are the responsibility of the Owner as provided in Article 4.

5.2 RESPONSIBILITY FOR CONSTRUCTION COST

5.2.1 Evaluations of the Owner's Project budget, preliminary estimates of Construction Cost and detailed estimates of Construction Cost, if any, prepared by the Architect, represent the Architect's best judgment as a design professional familiar with the construction industry. It is recognized, however, that neither the Architect nor the Owner has control over the cost of labor, materials or equipment, over the Contractor's methods of determining bid prices, or over competitive bidding, market or negotiating conditions. Accordingly, the Architect cannot and does not warrant or represent that bids or negotiated prices will not vary from the Owner's Project budget or from any estimate of Construction Cost or evaluation prepared or agreed to by the Architect.

5.2.2 No fixed limit of Construction Cost shall be established as a condition of this Agreement by the furnishing, proposal or establishment of a Project budget, unless such fixed limit has been agreed upon in writing and signed by the parties hereto. If such a fixed limit has been established, the Architect shall be permitted to include contingencies for design, bidding and price escalation, to determine what materials, equipment, component systems and types of construction are to be included in the Contract Documents, to make reasonable adjustments in the scope of the Project and to include in the Contract Documents alternate bids to adjust the Construction Cost to the fixed limit. Fixed limits, if any, shall be increased in the amount of an increase in the Contract Sum occurring after execution of the Contract for Construction.

5.2.3 If the Bidding or Negotiation Phase has not commenced within 90 days after the Architect submits the Construction Documents to the Owner, any Project budget or fixed limit of Construction Cost shall be adjusted to reflect changes in the general level of prices in the construction industry between the date of submission of the Construction Documents to the Owner and the date on which proposals are sought.

5.2.4 If a fixed limit of Construction Cost (adjusted as provided in Subparagraph 5.2.3) is exceeded by the lowest bona fide bid or negotiated proposal, the Owner shall:

.1 give written approval of an increase in such fixed limit;

.2 authorize rebidding or renegotiating of the Project within a reasonable time;

.3 if the Project is abandoned, terminate in accordance with Paragraph 8.3; or

.4 cooperate in revising the Project scope and quality as required to reduce the Construction Cost.

5.2.5 If the Owner chooses to proceed under Clause 5.2.4.4, the Architect, without additional charge, shall modify the Contract Documents as necessary to comply with the fixed limit, if established as a condition of this Agreement. The modification of Contract Documents shall be the limit of the Architect's responsibility arising out of the establishment of a fixed limit. The Architect shall be entitled to compensation in accordance with this Agreement for all services performed whether or not the Construction Phase is commenced.

ARTICLE 6
USE OF ARCHITECT'S DRAWINGS, SPECIFICATIONS AND OTHER DOCUMENTS

6.1 The Drawings, Specifications and other documents prepared by the Architect for this Project are instruments of the Architect's service for use solely with respect to this Project and, unless otherwise provided, the Architect shall be deemed the author of these documents and shall retain all common law, statutory and other reserved rights, including the copyright. The Owner shall be permitted to retain copies, including reproducible copies, of the Architect's Drawings, Specifications and other documents for information and reference in connection with the Owner's use and occupancy of the Project. The Architect's Drawings, Specifications or other documents shall not be used by the Owner or others on other projects, for additions to this Project or for completion of this Project by others, unless the Architect is adjudged to be in default under this Agreement, except by agreement in writing and with appropriate compensation to the Architect.

6.2 Submission or distribution of documents to meet official regulatory requirements or for similar purposes in connection with the Project is not to be construed as publication in derogation of the Architect's reserved rights.

ARTICLE 7
ARBITRATION

7.1 Claims, disputes or other matters in question between the parties to this Agreement arising out of or relating to this Agreement or breach thereof shall be subject to and decided by arbitration in accordance with the Construction Industry Arbitration Rules of the American Arbitration Association currently in effect unless the parties mutually agree otherwise.

7.2 Demand for arbitration shall be filed in writing with the other party to this Agreement and with the American Arbitration Association. A demand for arbitration shall be made within a reasonable time after the claim, dispute or other matter in question has arisen. In no event shall the demand for arbitration be made after the date when institution of legal or equitable proceedings based on such claim, dispute or other matter in question would be barred by the applicable statutes of limitations.

7.3 No arbitration arising out of or relating to this Agreement shall include, by consolidation, joinder or in any other manner, an additional person or entity not a party to this Agreement, except by written consent containing a specific reference to this Agreement signed by the Owner, Architect, and any other person or entity sought to be joined. Consent to arbitration involving an additional person or entity shall not constitute consent to arbitration of any claim, dispute or other matter in question not described in the written consent or with a person or entity not named or described therein. The foregoing agreement to arbitrate and other agreements to arbitrate with an additional person or entity duly consented to by the parties to this Agreement shall be specifically enforceable in accordance with applicable law in any court having jurisdiction thereof.

7.4 The award rendered by the arbitrator or arbitrators shall be final, and judgment may be entered upon it in accordance with applicable law in any court having jurisdiction thereof.

ARTICLE 8

TERMINATION, SUSPENSION OR ABANDONMENT

8.1 This Agreement may be terminated by either party upon not less than seven days' written notice should the other party fail substantially to perform in accordance with the terms of this Agreement through no fault of the party initiating the termination.

8.2 If the Project is suspended by the Owner for more than 30 consecutive days, the Architect shall be compensated for services performed prior to notice of such suspension. When the Project is resumed, the Architect's compensation shall be equitably adjusted to provide for expenses incurred in the interruption and resumption of the Architect's services.

8.3 This Agreement may be terminated by the Owner upon not less than seven days' written notice to the Architect in the event that the Project is permanently abandoned. If the Project is abandoned by the Owner for more than 90 consecutive days, the Architect may terminate this Agreement by giving written notice.

8.4 Failure of the Owner to make payments to the Architect in accordance with this Agreement shall be considered substantial nonperformance and cause for termination.

8.5 If the Owner fails to make payment when due the Architect for services and expenses, the Architect may, upon seven days' written notice to the Owner, suspend performance of services under this Agreement. Unless payment in full is received by the Architect within seven days of the date of the notice, the suspension shall take effect without further notice. In the event of a suspension of services, the Architect shall have no liability to the Owner for delay or damage caused the Owner because of such suspension of services.

8.6 In the event of termination not the fault of the Architect, the Architect shall be compensated for services performed prior to termination, together with Reimbursable Expenses then due and all Termination Expenses as defined in Paragraph 8.7.

8.7 Termination Expenses are in addition to compensation for Basic and Additional Services, and include expenses which are directly attributable to termination. Termination Expenses shall be computed as a percentage of the total compensation for Basic Services and Additional Services earned to the time of termination, as follows:

.1 Twenty percent of the total compensation for Basic and Additional Services earned to date if termination occurs before or during the predesign, site analysis, or Schematic Design Phases; or

.2 Ten percent of the total compensation for Basic and Additional Services earned to date if termination occurs during the Design Development Phase; or

.3 Five percent of the total compensation for Basic and Additional Services earned to date if termination occurs during any subsequent phase.

ARTICLE 9

MISCELLANEOUS PROVISIONS

9.1 Unless otherwise provided, this Agreement shall be governed by the law of the principal place of business of the Architect.

9.2 Terms in this Agreement shall have the same meaning as those in AIA Document A201, General Conditions of the Contract for Construction, current as of the date of this Agreement.

9.3 Causes of action between the parties to this Agreement pertaining to acts or failures to act shall be deemed to have accrued and the applicable statutes of limitations shall commence to run not later than either the date of Substantial Completion for acts or failures to act occurring prior to Substantial Completion, or the date of issuance of the final Certificate for Payment for acts or failures to act occurring after Substantial Completion.

9.4 The Owner and Architect waive all rights against each other and against the contractors, consultants, agents and employees of the other for damages, but only to the extent covered by property insurance during construction, except such rights as they may have to the proceeds of such insurance as set forth in the edition of AIA Document A201, General Conditions of the Contract for Construction, current as of the date of this Agreement. The Owner and Architect each shall require similar waivers from their contractors, consultants and agents.

9.5 The Owner and Architect, respectively, bind themselves, their partners, successors, assigns and legal representatives to the other party to this Agreement and to the partners, successors, assigns and legal representatives of such other party with respect to all covenants of this Agreement. Neither Owner nor Architect shall assign this Agreement without the written consent of the other.

9.6 This Agreement represents the entire and integrated agreement between the Owner and Architect and supersedes all prior negotiations, representations or agreements, either written or oral. This Agreement may be amended only by written instrument signed by both Owner and Architect.

9.7 Nothing contained in this Agreement shall create a contractual relationship with or a cause of action in favor of a third party against either the Owner or Architect.

9.8 Unless otherwise provided in this Agreement, the Architect and Architect's consultants shall have no responsibility for the discovery, presence, handling, removal or disposal of or exposure of persons to hazardous materials in any form at the Project site, including but not limited to asbestos, asbestos products, polychlorinated biphenyl (PCB) or other toxic substances.

9.9 The Architect shall have the right to include representations of the design of the Project, including photographs of the exterior and interior, among the Architect's promotional and professional materials. The Architect's materials shall not include the Owner's confidential or proprietary information if the Owner has previously advised the Architect in writing of the specific information considered by the Owner to be confidential or proprietary. The Owner shall provide professional credit for the Architect on the construction sign and in the promotional materials for the Project.

ARTICLE 10
PAYMENTS TO THE ARCHITECT

10.1 DIRECT PERSONNEL EXPENSE

10.1.1 Direct Personnel Expense is defined as the direct salaries of the Architect's personnel engaged on the Project and the portion of the cost of their mandatory and customary contributions and benefits related thereto, such as employment taxes and other statutory employee benefits, insurance, sick leave, holidays, vacations, pensions and similar contributions and benefits.

10.2 REIMBURSABLE EXPENSES

10.2.1 Reimbursable Expenses are in addition to compensation for Basic and Additional Services and include expenses incurred by the Architect and Architect's employees and consultants in the interest of the Project, as identified in the following Clauses.

10.2.1.1 Expense of transportation in connection with the Project; expenses in connection with authorized out-of-town travel; long-distance communications; and fees paid for securing approval of authorities having jurisdiction over the Project.

10.2.1.2 Expense of reproductions, postage and handling of Drawings, Specifications and other documents.

10.2.1.3 If authorized in advance by the Owner, expense of overtime work requiring higher than regular rates.

10.2.1.4 Expense of renderings, models and mock-ups requested by the Owner.

10.2.1.5 Expense of additional insurance coverage or limits, including professional liability insurance, requested by the Owner in excess of that normally carried by the Architect and Architect's consultants.

10.2.1.6 Expense of computer-aided design and drafting equipment time when used in connection with the Project.

10.3 PAYMENTS ON ACCOUNT OF BASIC SERVICES

10.3.1 An initial payment as set forth in Paragraph 11.1 is the minimum payment under this Agreement.

10.3.2 Subsequent payments for Basic Services shall be made monthly and, where applicable, shall be in proportion to services performed within each phase of service, on the basis set forth in Subparagraph 11.2.2.

10.3.3 If and to the extent that the time initially established in Subparagraph 11.5.1 of this Agreement is exceeded or extended through no fault of the Architect, compensation for any services rendered during the additional period of time shall be computed in the manner set forth in Subparagraph 11.3.2.

10.3.4 When compensation is based on a percentage of Construction Cost and any portions of the Project are deleted or otherwise not constructed, compensation for those portions of the Project shall be payable to the extent services are performed on those portions, in accordance with the schedule set forth in Subparagraph 11.2.2, based on (1) the lowest bona fide bid or negotiated proposal, or (2) if no such bid or proposal is received, the most recent preliminary estimate of Construction Cost or detailed estimate of Construction Cost for such portions of the Project.

10.4 PAYMENTS ON ACCOUNT OF ADDITIONAL SERVICES

10.4.1 Payments on account of the Architect's Additional Services and for Reimbursable Expenses shall be made monthly upon presentation of the Architect's statement of services rendered or expenses incurred.

10.5 PAYMENTS WITHHELD

10.5.1 No deductions shall be made from the Architect's compensation on account of penalty, liquidated damages or other sums withheld from payments to contractors, or on account of the cost of changes in the Work other than those for which the Architect has been found to be liable.

10.6 ARCHITECT'S ACCOUNTING RECORDS

10.6.1 Records of Reimbursable Expenses and expenses pertaining to Additional Services and services performed on the basis of a multiple of Direct Personnel Expense shall be available to the Owner or the Owner's authorized representative at mutually convenient times.

ARTICLE 11
BASIS OF COMPENSATION

The Owner shall compensate the Architect as follows:

11.1 AN INITIAL PAYMENT of Dollars ($)
shall be made upon execution of this Agreement and credited to the Owner's account at final payment.

11.2 BASIC COMPENSATION

11.2.1 FOR BASIC SERVICES, as described in Article 2, and any other services included in Article 12 as part of Basic Services, Basic Compensation shall be computed as follows:

(Insert basis of compensation, including stipulated sums, multiples or percentages, and identify phases to which particular methods of compensation apply, if necessary.)

11.2.2 Where compensation is based on a stipulated sum or percentage of Construction Cost, progress payments for Basic Services in each phase shall total the following percentages of the total Basic Compensation payable:

(Insert additional phases as appropriate.)

Schematic Design Phase:	percent (%)
Design Development Phase:	percent (%)
Construction Documents Phase:	percent (%)
Bidding or Negotiation Phase:	percent (%)
Construction Phase:	percent (%)
Total Basic Compensation:	one hundred percent (100%)

11.3 COMPENSATION FOR ADDITIONAL SERVICES

11.3.1 FOR PROJECT REPRESENTATION BEYOND BASIC SERVICES, as described in Paragraph 3.2, compensation shall be computed as follows:

11.3.2 FOR ADDITIONAL SERVICES OF THE ARCHITECT, as described in Articles 3 and 12, other than (1) Additional Project Representation, as described in Paragraph 3.2, and (2) services included in Article 12 as part of Basic Services, but excluding services of consultants, compensation shall be computed as follows:

(Insert basis of compensation, including rates and/or multiples of Direct Personnel Expense for Principals and employees, and identify Principals and classify employees, if required. Identify specific services to which particular methods of compensation apply, if necessary.)

11.3.3 FOR ADDITIONAL SERVICES OF CONSULTANTS, including additional structural, mechanical and electrical engineering services and those provided under Subparagraph 3.4.19 or identified in Article 12 as part of Additional Services, a multiple of ( ) times the amounts billed to the Architect for such services.

(Identify specific types of consultants in Article 12, if required.)

11.4 REIMBURSABLE EXPENSES

11.4.1 FOR REIMBURSABLE EXPENSES, as described in Paragraph 10.2, and any other items included in Article 12 as Reimbursable Expenses, a multiple of ( ) times the expenses incurred by the Architect, the Architect's employees and consultants in the interest of the Project.

11.5 ADDITIONAL PROVISIONS

11.5.1 IF THE BASIC SERVICES covered by this Agreement have not been completed within ( ) months of the date hereof, through no fault of the Architect, extension of the Architect's services beyond that time shall be compensated as provided in Subparagraphs 10.3.3 and 11.3.2.

11.5.2 Payments are due and payable ( ) days from the date of the Architect's invoice. Amounts unpaid ( ) days after the invoice date shall bear interest at the rate entered below, or in the absence thereof at the legal rate prevailing from time to time at the principal place of business of the Architect.

(Insert rate of interest agreed upon.)

(Usury laws and requirements under the Federal Truth in Lending Act, similar state and local consumer credit laws and other regulations at the Owner's and Architect's principal places of business, the location of the Project and elsewhere may affect the validity of this provision. Specific legal advice should be obtained with respect to deletions or modifications, and also regarding requirements such as written disclosures or waivers.)

11.5.3 The rates and multiples set forth for Additional Services shall be annually adjusted in accordance with normal salary review practices of the Architect.

ARTICLE 12
OTHER CONDITIONS OR SERVICES

(Insert descriptions of other services, identify Additional Services included within Basic Compensation and modifications to the payment and compensation terms included in this Agreement.)

This Agreement entered into as of the day and year first written above.

OWNER	ARCHITECT
(Signature)	*(Signature)*
(Printed name and title)	*(Printed name and title)*

APPENDIX I

AMERICAN ARBITRATION ASSOCIATION, CONSTRUCTION INDUSTRY ARBITRATION RULES (AS AMENDED, AND IN EFFECT JANUARY 1, 1991)

American
Arbitration
Association

CONSTRUCTION INDUSTRY ARBITRATION RULES

as amended and in effect
January 1, 1991

National Construction Dispute Resolution Committee

Representatives of the fourteen organizations listed below constitute the National Construction Dispute Resolution Committee (NCDRC). This committee is the sponsor of the arbitration procedure specially designed for the construction industry by the American Arbitration Association (AAA).

AMERICAN CONSULTING ENGINEERS COUNCIL

AMERICAN INSTITUTE OF ARCHITECTS

AMERICAN SOCIETY OF CIVIL ENGINEERS

AMERICAN SOCIETY OF INTERIOR DESIGNERS

AMERICAN SOCIETY OF LANDSCAPE ARCHITECTS

AMERICAN SUBCONTRACTORS ASSOCIATION

ASSOCIATED BUILDERS AND CONTRACTORS, INC.

ASSOCIATED GENERAL CONTRACTORS

ASSOCIATED SPECIALTY CONTRACTORS, INC.

BUSINESS ROUNDTABLE

CONSTRUCTION SPECIFICATIONS INSTITUTE

NATIONAL ASSOCIATION OF HOME BUILDERS

NATIONAL SOCIETY OF PROFESSIONAL ENGINEERS

NATIONAL UTILITY CONTRACTORS ASSOCIATION

Quick Filing Directions

BY DEMAND FOR ARBITRATION:

If the contract has an arbitration clause providing for arbitration under the rules of the American Arbitration Association, the following is necessary to commence arbitration:

1. Five Demand for Arbitration forms are completed and signed. These forms can be obtained from the AAA at no charge. (The use of these forms is not mandatory; a dispute may be filed with the Association by correspondence, providing that all of the necessary information is provided.)

2. Attach a copy of the arbitration clause contained in the contract (or insurance policy) to the demand form, or quote the clause on the demand form in the space provided.

3. The original is sent by the claimant to the respondent. In some states, it is required that the demand be forwarded by Certified Mail-Return Receipt Requested. Check with the regional office where the demand is to be filed if you are uncertain regarding this requirement.

4. Three copies are sent to the regional office of the AAA referred to in the contract or, if no locale is specified, to the nearest AAA regional office. Regardless of the amount in dispute, a case may be initiated with a $300 filing fee.* The remainder of the fee should be paid within sixty days of filing the case or prior to the first hearing, whichever occurs first.

BY SUBMISSION TO ARBITRATION:

If there is an existing dispute with no arbitration clause, or an arbitration clause that does not provide for arbitration under the AAA rules, a Submission to Dispute Resolution is filed.

1. Five copies of the submission form are completed and signed by both parties.

2. A copy of the submission form is retained by each party.

3. Follow step number four above.

* The fee schedule as well as the locations of the AAA regional offices can be found in this booklet.

Introduction

Each year, thousands of transactions take place in the construction industry. Occasionally, disagreements develop over these business transactions. Many of these disputes are resolved by arbitration, the voluntary submission of a dispute to a disinterested person or persons for final and binding determination. Arbitration has proven to be an effective way to resolve these disputes privately, promptly, and economically.

The American Arbitration Association (AAA) is a public-service, not-for-profit organization offering a broad range of dispute resolution services to business executives, attorneys, individuals, trade associations, unions, management, consumers, families, communities, and all levels of government. Services are available through AAA headquarters in New York City and through offices located in major cities throughout the United States. Hearings may be held at locations convenient for the parties and are not limited to cities with AAA offices. In addition, the AAA serves as a center for education and training, issues specialized publications, and conducts research on all forms of out-of-court dispute settlement.

The AAA assesses an administrative fee on the claim asserted by each party. This fee, which is a percentage of the amount claimed, ranges from a minimum of three percent of the smallest claims (up to $25,000) to about one percent of claims of $250,000, to about one tenth of one percent of claims of

$50,000,000 or more. The fee schedule is contained on the last pages of this pamphlet. This fee covers the AAA administrative services; it does not cover arbitrator compensation or expenses, court reporting services, or any post-award charges incurred by the parties in enforcing the award.

The AAA does not act as arbitrator. Its function is to administer arbitrations in accordance with the agreement of the parties and to maintain panels from which arbitrators may be chosen by the parties. Once designated, the arbitrator decides the issues and renders a final and binding award.

The AAA has established and maintains as members of its National Panel of Arbitrators individuals competent to hear and determine disputes administered under the Construction Industry Arbitration Rules. The AAA considers for appointment to the construction industry panel persons recommended by the National Construction Dispute Resolution Committee as qualified to serve by virtue of their experience in the construction field.

When an agreement to arbitrate is included in a construction contract, it may expedite peaceful settlement without the necessity of going to arbitration at all. Thus, the arbitration clause is a form of insurance against loss of good will.

The parties can provide for the arbitration of future disputes by inserting the following clause into their contracts:

Standard Arbitration Clause
Any controversy or claim arising out of or relating to this contract, or the breach thereof, shall be settled by arbitration in accordance with the Construction Industry Arbitration Rules of the American Arbitration Association, and judgment upon the award rendered by the arbitrator(s) may be entered in any court having jurisdiction thereof.

The arbitration of existing disputes may be accomplished by the use of the following:

> We, the undersigned parties, hereby agree to submit to arbitration under the Construction Industry Arbitration Rules of the American Arbitration Association the following controversy: (cite briefly). We further agree that the above controversy be submitted to (one) (three) arbitrator(s) selected from the panels of arbitrators of the American Arbitration Association. We further agree that we will faithfully observe this agreement and the rules, and that we will abide by and perform any award rendered by the arbitrator(s) and that a judgment of the court having jurisdiction may be entered upon the award.

Where the parties are nationals or residents of different countries, the Supplementary Procedures for International Commercial Arbitration shall, by agreement of all parties, apply as well.

Mediation

In appropriate cases, the parties may wish to submit their dispute to mediation. In mediation, the neutral mediator assists the parties in reaching a settlement, but does not have the authority to make a binding decision or award. Mediation is administered by the AAA in accordance with the Construction Industry Mediation Rules, which are available through all of the AAA's regional offices. There is no additional administrative fee where parties to a pending arbitration attempt to mediate their dispute under the AAA's auspices.

The services of the AAA are generally concluded with the transmittal of the award. Although there is voluntary compliance with the majority of awards, judgment on the award can be entered in a court having appropriate jurisdiction if necessary.

Table of Contents

Construction Industry Arbitration Rules

1. Agreement of Parties

The parties shall be deemed to have made these rules a part of their arbitration agreement whenever they have provided for arbitration by the American Arbitration Association (hereinafter AAA) or under its Construction Industry Arbitration Rules. These rules and any amendment of them shall apply in the form obtaining at the time the demand for arbitration or submission agreement is received by the AAA. The parties, by written agreement, may vary the procedures set forth in these rules.

2. Name of Tribunal

Any tribunal constituted by the parties for the settlement of their dispute under these rules shall be called the Construction Industry Arbitration Tribunal.

3. Administrator and Delegation of Duties

When parties agree to arbitrate under these rules, or when they provide for arbitration by the AAA and an arbitration is initiated under these rules, they thereby authorize the AAA to administer the arbitration. The authority and duties of the AAA are prescribed in the agreement of the parties and in these rules, and may be carried out through such of the AAA's representatives as it may direct.

4. National Panel of Arbitrators

In cooperation with the National Construction Dispute Resolution Committee, the AAA shall establish and maintain a National Panel of Construction Industry Arbitrators and shall appoint arbitrators therefrom as hereinafter provided.

5. Regional Offices

The AAA may, in its discretion, assign the administration of an arbitration to any of its regional offices.

6. Initiation under an Arbitration Provision in a Contract

Arbitration under an arbitration provision in a contract shall be initiated in the following manner:

(a) The initiating party (hereinafter claimant) shall, within the time period, if any, specified in the contract(s), give written notice to the other party (hereinafter respondent) of its intention to arbitrate (demand), which notice shall contain a statement setting forth the nature of the dispute, the amount involved, if any, the remedy sought, and the hearing locale requested, and

(b) Shall file at any regional office of the AAA three copies of the notice and three copies of the arbitration provisions of the contract, together with the appropriate administrative fee as provided in the Administrative Fee Schedule.

The AAA shall give notice of the filing to the respondent or respondents. A respondent may file an answering statement in duplicate with the AAA within ten days after notice from the AAA, in which event the respondent shall at the same time send a copy of the answering statement to the claimant. If a counterclaim is asserted, it shall contain a statement setting forth the nature of the counterclaim, the amount involved, if any, and the remedy sought. If a counterclaim is made in the answering statement, the appropriate fee provided in the Administrative Fee Schedule shall be forwarded to the AAA with the answering statement. If no answering statement is filed within the stated time, it will be treated as a denial of the claim. Failure to file an answering statement shall not operate to delay the arbitration.

7. Initiation under a Submission

Parties to any existing dispute may commence an arbitration under these rules by filing at any regional office of the AAA three copies of a written submission to arbitrate under these rules, signed by the parties. It shall contain a statement of the matter in dispute, the amount of money involved, if any, the remedy sought, and the hearing locale requested, together with the appropriate administrative fee as provided in the Administrative Fee Schedule.

8. Changes of Claim

After filing of a claim, if either party desires to make any new or different claim or counterclaim, same shall be made in writing and filed with the AAA, and a copy shall be mailed to the other party, who shall have a period of ten days from the date of such mailing within which to file an answer with the AAA. After the arbitrator is appointed, however, no new or different claim may be submitted except with the arbitrator's consent.

9. Applicable Procedures

Unless the AAA in its discretion determines otherwise, the Expedited Procedures shall be applied in any case where no disclosed claim or counterclaim exceeds $50,000, exclusive of interest and arbitration costs. Parties may also agree to the Expedited Procedures in cases involving claims in excess of $50,000. The Expedited Procedures shall be applied as described in Sections 53 through 57 of these rules, in addition to any other portion of these rules that is not in conflict with the Expedited Procedures.

All other cases shall be administered in accordance with Sections 1 through 52 of these rules.

10. Administrative Conference, Preliminary Hearing, and Mediation Conference

At the request of any party or at the discretion of the AAA, an administrative conference with the AAA and the parties and/or their representatives will be scheduled in appropriate cases to expedite the arbitration proceedings.

In large or complex cases, at the request of any party or at the discretion of the arbitrator or the AAA, a preliminary hearing with the parties and/or their representatives and the arbitrator may be scheduled by the arbitrator to specify the issues to be resolved, to stipulate to uncontested facts, and to consider any other matters that will expedite the arbitration proceedings. Consistent with the expedited nature of arbitration, the arbitrator may, at the preliminary hearing, establish (i) the extent of and schedule for the production of relevant documents and other information, (ii) the identification of any witnesses to be called, and (iii) a schedule for further hearings to resolve the dispute.

With the consent of the parties, the AAA at any stage of the proceeding may arrange a mediation conference under the Construction Industry Mediation Rules, in order to facilitate settlement. The mediator shall not be an arbitrator appointed to the case. Where the parties to a pending arbitration agree to mediate under the AAA's rules, no additional administrative fee is required to initiate the mediation.

11. Fixing of Locale

The parties may mutually agree on the locale where the arbitration is to be held. If any party requests that the hearing be held in a specific locale and the other party files no objection thereto within ten days after notice of the request has been mailed to it by the AAA, the locale shall be the one requested. If a party objects to the locale requested by the other party, the AAA shall have the power to determine the locale and its decision shall be final and binding.

12. Qualifications of an Arbitrator

Any neutral arbitrator appointed pursuant to Section 13, 14, 15, or 54, or selected by mutual choice of the parties or their appointees, shall be subject to disqualification for the reasons specified in Section 19. If the parties specifically so agree in writing, the arbitrator shall not be subject to disqualification for those reasons.

Unless the parties agree otherwise, an arbitrator selected unilaterally by one party is a party-appointed arbitrator and is not subject to disqualification pursuant to Section 19.

The term "arbitrator" in these rules refers to the arbitration panel, whether composed of one or more arbitrators and whether the arbitrators are neutral or party appointed.

13. Appointment from Panel

If the parties have not appointed an arbitrator and have not provided any other method of appointment, the arbitrator shall be appointed in the following manner: immediately after the filing of the demand or submission, the AAA shall submit simultaneously to each party to the dispute an identical list of names of persons chosen from the panel.

Each party to the dispute shall have ten days from the mailing date in which to cross off any names objected to, number the remaining names in order of preference, and return the list to the AAA. If a party does not return the list within the time specified, all persons named therein shall be deemed acceptable. From among the persons who have been approved on both lists, and in accordance with the designated order of mutual preference, the AAA shall invite the acceptance of an arbitrator to serve. If the parties fail to agree on any of the persons named, or if acceptable arbitrators are unable to act, or if for any other reason the appointment cannot be made from the submitted lists, the AAA shall have the power to make the appointment from among other members of the panel without the submission of additional lists.

14. Direct Appointment by a Party

If the agreement of the parties names an arbitrator or specifies a method of appointing an arbitrator, that designation or method shall be followed. The notice of appointment, with the name and address of the arbitrator, shall be filed with the AAA by that party. Upon the request of any appointing party, the AAA shall submit a list of members of the panel from which the party may, if it so desires, make the appointment.

If the agreement specifies a period of time within which an arbitrator shall be appointed and any party fails to make the appointment within that period, the AAA shall make the appointment.

If no period of time is specified in the agreement, the AAA shall notify the party to make the appointment. If within ten days thereafter an arbitrator has not been appointed by a party, the AAA shall make the appointment.

15. Appointment of Neutral Arbitrator by Party-Appointed Arbitrators or Parties

If the parties have selected party-appointed arbitrators, or if such arbitrators have been appointed as provided in Section 14, and the parties have authorized them to appoint a neutral arbitrator within a specified time and no appointment is made within that time or any agreed extension thereof, the AAA may appoint the neutral arbitrator, who shall act as chairperson.

If no period of time is specified for appointment of the neutral arbitrator and the party-appointed arbitrators or the parties do not make the appointment within ten days from the date of the appointment of the last party-appointed arbitrator, the AAA may appoint the neutral arbitrator, who shall act as chairperson.

If the parties have agreed that their party-appointed arbitrators shall appoint the neutral arbitrator from the panel, the AAA shall furnish to the party-appointed arbitrators, in the manner

prescribed in Section 13, a list selected from the panel, and the appointment of the neutral arbitrator shall be made as prescribed in that section.

16. Nationality of Arbitrator in International Arbitration

Where the parties are nationals or residents of different countries, any neutral arbitrator shall, upon the request of either party, be appointed from among the nationals of a country other than that of any of the parties. The request must be made prior to the time set for the appointment of the arbitrator as agreed by the parties or set by these rules.

17. Number of Arbitrators

If the arbitration agreement does not specify the number of arbitrators, the dispute shall be heard and determined by one arbitrator, unless the AAA, in its discretion, directs that a greater number of arbitrators be appointed.

18. Notice to Arbitrator of Appointment

Notice of the appointment of the neutral arbitrator, whether appointed mutually by the parties or by the AAA, shall be mailed to the arbitrator by the AAA, together with a copy of these rules, and the signed acceptance of the arbitrator shall be filed with the AAA prior to the opening of the first hearing.

19. Disclosure and Challenge Procedure

Any person appointed as neutral arbitrator shall disclose to the AAA any circumstance likely to affect impartiality, including any bias or any financial or personal interest in the result of the arbitration or any past or present relationship with the parties or their representatives. Upon receipt of such information from the arbitrator or another source, the AAA shall communicate the information to the parties and, if it deems it appropriate to do so, to the arbitrator and others. Upon objection of a party to the continued service of a neutral arbitrator, the AAA shall determine whether the arbitrator should be disqualified and shall inform the parties of its decision, which shall be conclusive.

20. Vacancies

If for any reason an arbitrator is unable to perform the duties of the office, the AAA may, on proof satisfactory to it, declare the office vacant.

Vacancies shall be filled in accordance with the applicable provisions of these rules.

In the event of a vacancy in a panel of neutral arbitrators after the hearings have commenced, the remaining arbitrator or arbitrators may continue with the hearing and determination of the controversy, unless the parties agree otherwise.

21. Date, Time, and Place of Hearing

The arbitrator shall set the date, time, and place for each hearing. The AAA shall mail to each party notice thereof at least ten days in advance, unless the parties by mutual agreement waive such notice or modify the terms thereof.

22. Representation

Any party may be represented by counsel or other authorized representative. A party intending to be so represented shall notify the other party and the AAA of the name and address of the representative at least three days prior to the date set for the hearing at which that person is first to appear. When such a representative initiates an arbitration or responds for a party, such notice is deemed to have been given.

23. Stenographic Record

Any party desiring a stenographic record shall make arrangements directly with a stenographer and shall notify the other party of these arrangements in advance of the hearing. The requesting party or parties shall pay the cost of the record. If the transcript is agreed by the parties to be, or determined by the arbitrator to be, the official record of the proceeding, it must be made available to the arbitrator and to the other parties for inspection, at a date, time, and place determined by the arbitrator.

24. Interpreters

Any party wishing an interpreter shall make all arrangements directly with the interpreter and shall assume the costs of the service.

25. Attendance at Hearings

The arbitrator shall maintain the privacy of the hearings unless the law provides to the contrary. Any person having a direct interest in the arbitration is entitled to attend hearings. The arbitrator shall otherwise have the power to require the exclusion of any witness, other than a party or other

essential person, during the testimony of any other witness. It shall be discretionary with the arbitrator to determine the propriety of the attendance of any other person.

26. Postponements

The arbitrator for good cause shown may postpone any hearing upon the request of a party or upon the arbitrator's own initiative, and shall also grant such postponement when all of the parties agree thereto.

27. Oaths

Before proceeding with the first hearing, each arbitrator may take an oath of office and, if required by law, shall do so. The arbitrator may require witnesses to testify under oath administered by any duly qualified person and, if it is required by law or requested by any party, shall do so.

28. Majority Decision

All decisions of the arbitrators must be by a majority. The award must also be made by a majority unless the concurrence of all is expressly required by the arbitration agreement or by law.

29. Order of Proceedings and Communication with Arbitrator

A hearing shall be opened by the filing of the oath of the arbitrator, where required; by the recording of the date, time, and place of the hearing, and the presence of the arbitrator, the parties, and their representatives, if any; and by the receipt by the arbitrator of the statement of the claim and the answering statement, if any.

The arbitrator may, at the beginning of the hearing, ask for statements clarifying the issues involved. In some cases, part or all of the above will have been accomplished at the preliminary hearing conducted by the arbitrator pursuant to Section 10.

The complaining party shall then present evidence to support its claim. The defending party shall then present evidence supporting its defense. Witnesses for each party shall submit to questions or other examination. The arbitrator has the discretion to vary this procedure but shall afford a full and equal opportunity to all parties for the presentation of any material and relevant evidence.

Exhibits, when offered by either party, may be received in evidence by the arbitrator.

The names and addresses of all witnesses and a description of the exhibits in the order received shall be made a part of the record.

There shall be no direct communication between the parties and a neutral arbitrator other than at oral hearings, unless the parties and the arbitrator agree otherwise. Any other oral or written communication from the parties to a neutral arbitrator shall be directed to the AAA for transmittal to the arbitrator.

30. Arbitration in the Absence of a Party or Representative

Unless the law provides to the contrary, the arbitration may proceed in the absence of any party or representative who, after due notice, fails to be present or fails to obtain a postponement. An award shall not be made solely on the default of a party. The arbitrator shall require the party who is present to submit such evidence as the arbitrator may require for the making of an award.

31. Evidence

The parties may offer such evidence as is relevant and material to the dispute and shall produce such evidence as the arbitrator may deem necessary to an understanding and determination of the dispute. An arbitrator or other person authorized by law to subpoena witnesses or documents may do so upon the request of any party or independently.

The arbitrator shall be the judge of the relevance and materiality of the evidence offered, and conformity to legal rules of evidence shall not be necessary. All evidence shall be taken in the presence of all of the arbitrators and all of the parties, except where any of the parties is absent in default or has waived the right to be present.

32. Evidence by Affidavits and Post-hearing Filing of Documents or Other Evidence

The arbitrator may receive and consider the evidence of witnesses by affidavit, but shall give it only such weight as the arbitrator deems it entitled to after consideration of any objection made to its admission.

If the parties agree or the arbitrator directs that documents or other evidence be submitted to the arbitrator after the hearing, the documents or other evidence shall be filed with the AAA for transmission to the arbitrator. All parties shall be afforded an opportunity to examine such documents or other evidence.

33. Inspection or Investigation

An arbitrator finding it necessary to make an inspection or investigation in connection with the arbitration shall direct the AAA to so advise the parties. The arbitrator shall set the date and time and the AAA shall notify the parties. Any party who so desires may be present at such an inspection or investigation. In the event that one or all parties are not present at the inspection or investigation, the arbitrator shall make a verbal or written report to the parties and afford them an opportunity to comment.

34. Interim Measures

The arbitrator may issue such orders for interim relief as may be deemed necessary to safeguard the property that is the subject matter of the arbitration without prejudice to the rights of the parties or to the final determination of the dispute.

35. Closing of Hearing

The arbitrator shall specifically inquire of all parties whether they have any further proofs to offer or witnesses to be heard. Upon receiving negative replies or if satisfied that the record is complete, the arbitrator shall declare the hearing closed and a minute thereof shall be recorded. If briefs are to be filed, the hearing shall be declared closed as of the final date set by the arbitrator for the receipt of briefs. If documents are to be filed as provided in Section 32 and the date set for their receipt is later than that set for the receipt of briefs, the later date shall be the date of closing the hearing. The time limit within which the arbitrator is required to make the award shall commence to run, in the absence of other agreements by the parties, upon the closing of the hearing.

36. Reopening of Hearing

The hearing may be reopened on the arbitrator's initiative, or upon application of a party, at any time before the award is made. If reopening the hearing would prevent the making of the award within the specific time agreed on by the parties in the contract(s) out of which the controversy has arisen, the matter may not be reopened unless the parties agree on an extension of time. When no specific date is fixed in the contract, the arbitrator may reopen the hearing and shall have thirty days from the closing of the reopened hearing within which to make an award.

37. Waiver of Oral Hearing

The parties may provide, by written agreement, for the waiver of oral hearings in any case. If the parties are unable to agree as to the procedure, the AAA shall specify a fair and equitable procedure.

38. Waiver of Rules

Any party who proceeds with the arbitration after knowledge that any provision or requirement of these rules has not been complied with and who fails to state an objection thereto in writing shall be deemed to have waived the right to object.

39. Extensions of Time

The parties may modify any period of time by mutual agreement. The AAA or the arbitrator may for good cause extend any period of time established by these rules, except the time for making the award. The AAA shall notify the parties of any extension.

40. Serving of Notice

Each party shall be deemed to have consented that any papers, notices, or process necessary or proper for the initiation or continuation of an arbitration under these rules; for any court action in connection therewith; or for the entry of judgment on any award made under these rules may be served on a party by mail addressed to the party or its representative at the last known address or by personal service, in or outside the state where the arbitration is to be held, provided that reasonable opportunity to be heard with regard thereto has been granted to the party.

The AAA and the parties may also use facsimile transmission, telex, telegram, or other written forms of electronic communication to give the notices required by these rules.

41. Time of Award

The award shall be made promptly by the arbitrator and, unless otherwise agreed by the parties or specified by law, no later than thirty days from the

date of closing the hearing, or, if oral hearings have been waived, from the date of the AAA's transmittal of the final statements and proofs to the arbitrator.

42. Form of Award

The award shall be in writing and shall be signed by a majority of the arbitrators. It shall be executed in the manner required by law.

43. Scope of Award

The arbitrator may grant any remedy or relief that the arbitrator deems just and equitable and within the scope of the agreement of the parties, including, but not limited to, specific performance of a contract. The arbitrator shall, in the award, assess arbitration fees, expenses, and compensation as provided in Sections 48, 49, and 50 in favor of any party and, in the event that any administrative fees or expenses are due the AAA, in favor of the AAA.

44. Award upon Settlement

If the parties settle their dispute during the course of the arbitration, the arbitrator may set forth the terms of the agreed settlement in an award. Such an award is referred to as a consent award.

45. Delivery of Award to Parties

Parties shall accept as legal delivery of the award the placing of the award or a true copy thereof in the mail addressed to a party or its representative at the last known address, personal service of the award, or the filing of the award in any other manner that is permitted by law.

46. Release of Documents for Judicial Proceedings

The AAA shall, upon the written request of a party, furnish to the party, at its expense, certified copies of any papers in the AAA's possession that may be required in judicial proceedings relating to the arbitration.

47. Applications to Court and Exclusion of Liability

(a) No judicial proceeding by a party relating to the subject matter of the arbitration shall be deemed a waiver of the party's right to arbitrate.

(b) Neither the AAA nor any arbitrator in a proceeding under these rules is a necessary party in judicial proceedings relating to the arbitration.

(c) Parties to these rules shall be deemed to have consented that judgment upon the arbitration award may be entered in any federal or state court having jurisdiction thereof.

(d) Neither the AAA nor any arbitrator shall be liable to any party for any act or omission in connection with any arbitration conducted under these rules.

48. Administrative Fee
As a not-for-profit organization, the AAA shall prescribe an Administrative Fee Schedule and a Refund Schedule to compensate it for the cost of providing administrative services. The schedule in effect at the time the demand for arbitration or submission agreement is received shall be applicable.

The administrative fee shall be advanced by the initiating party or parties, subject to final apportionment by the arbitrator in the award.

When a claim or counterclaim is withdrawn or settled, the refund shall be made in accordance with the Refund Schedule.

The AAA may, in the event of extreme hardship on the part of any party, defer or reduce the administrative fee.

49. Expenses
The expenses of witnesses for either side shall be paid by the party producing such witnesses. All other expenses of the arbitration, including required travel and other expenses of the arbitrator, AAA representatives, and any witness and the cost of any proof produced at the direct request of the arbitrator, shall be borne equally by the parties, unless they agree otherwise or unless the arbitrator in the award assesses such expenses or any part thereof against any specified party or parties.

50. Neutral Arbitrator's Fee
Unless the parties agree otherwise, members of the National Panel of Construction Industry Arbitrators appointed as neutrals will serve without compensation for the first day of service.

Thereafter, compensation shall be based on the amount of service involved and the number of hearings. An appropriate daily rate and other

arrangements will be discussed by the administrator with the parties and the arbitrator. If the parties fail to agree to the terms of compensation, an appropriate rate shall be established by the AAA and communicated in writing to the parties.

Any arrangement for the compensation of a neutral arbitrator shall be made through the AAA and not directly between the parties and the arbitrator. The terms of compensation of neutral arbitrators on a panel shall be identical.

51. Deposits
The AAA may require the parties to deposit in advance of any hearings such sums of money as it deems necessary to defray the expense of the arbitration, including the arbitrator's fee, if any, and shall render an accounting to the parties and return any unexpended balance at the conclusion of the case.

52. Interpretation and Application of Rules
The arbitrator shall interpret and apply these rules insofar as they relate to the arbitrator's powers and duties. When there is more than one arbitrator and a difference arises among them concerning the meaning or application of these rules, it shall be decided by a majority vote. If that is unobtainable, either an arbitrator or a party may refer the question to the AAA for final decision. All other rules shall be interpreted and applied by the AAA.

Expedited Procedures

53. Notice by Telephone
The parties shall accept all notices from the AAA by telephone. Such notices by the AAA shall subsequently be confirmed in writing to the parties. Should there be a failure to confirm in writing any notice hereunder, the proceeding shall nonetheless be valid if notice has, in fact, been given by telephone.

54. Appointment and Qualifications of Arbitrator
Where no disclosed claim or counterclaim exceeds $50,000, exclusive of interest and arbitration costs, the AAA shall submit simultaneously to each party an identical list of five proposed arbitrators drawn from the National Panel of Construction Industry Arbitrators, from which one arbitrator shall be appointed.

Each party may strike two names from the list on a peremptory basis. The list is returnable to the AAA within seven days from the date of the AAA's mailing to the parties.

If for any reason the appointment of an arbitrator cannot be made from the list, the AAA may make the appointment from among other members of the panel without the submission of additional lists.

The parties will be given notice by telephone by the AAA of the appointment of the arbitrator, who shall be subject to disqualification for the reasons specified in Section 19. The parties shall notify the AAA, by telephone, within seven days of any objection to the arbitrator appointed. Any objection by a party to the arbitrator shall be confirmed in writing to the AAA with a copy to the other party or parties.

55. Date, Time, and Place of Hearing

The arbitrator shall set the date, time, and place of the hearing. The AAA will notify the parties by telephone, at least seven days in advance of the hearing date. A formal Notice of Hearing will be sent by the AAA to the parties.

56. The Hearing

Generally, the hearing shall be completed within one day, unless the dispute is resolved by submission of documents under Section 37. The arbitrator, for good cause shown, may schedule an additional hearing to be held within seven days.

57. Time of Award

Unless otherwise agreed by the parties, the award shall be rendered not later than fourteen days from the date of the closing of the hearing.

Administrative Fee Schedule

A filing fee of $300 will be paid when a case is filed. The balance of the administrative fee is based on the amount of each claim or counterclaim as disclosed when the claim or counterclaim is filed. This balance is due and payable sixty days after the AAA's commencement of administration, or prior to the date of the first hearing, whichever occurs first. If at any time a claim or counterclaim is settled or withdrawn, the balance of the administrative fee remains due and the Refund Schedule

shall then be applied. When oral hearings are waived under Section 37, the Administrative Fee Schedule shall still apply.

Amount of Claim or Counterclaim	Fee	
Up to $10,000	$300	
$10,000 to $25,000	3%	
$25,000 to $50,000	$750,	plus 2% of excess over $25,000
$50,000 to $100,000	$1,250,	plus 1% of excess over $50,000
$100,000 to $500,000	$1,750,	plus ½% of excess over $100,000
$500,000 to $5,000,000	$3,750,	plus ¼% of excess over $500,000
$5,000,000 to $50,000,000	$15,000,	plus ⅒% of excess over $5,000,000

Where the claim or counterclaim exceeds $50 million, there is no additional administrative fee.

When no amount can be stated at the time of filing, the administrative fee is $1,000, subject to adjustment in accordance with the above schedule as soon as an amount can be disclosed.

An appropriate administrative fee will be determined by the AAA for claims and counterclaims that are not for a monetary amount.

If there are more than two parties represented in the arbitration, an additional 10% of the administrative fee will be due for each additional represented party.

The minimum administrative fee for a case heard by three arbitrators is $1,500, payable by the party requesting same.

Postponement Fees

Sole-Arbitrator Cases

$100 is payable by a party causing its first postponement of any scheduled hearing.

$200 is payable by a party causing its second or subsequent postponement of any scheduled hearing.

Three-Arbitrator Cases

$150 is payable by a party causing its first postponement of any scheduled hearing.

$300 is payable by a party causing its second or subsequent postponement of any scheduled hearing.

Additional Hearing Fee

$75 is payable by each party for each hearing after the first hearing that is clerked by the AAA.

Hearing-Room Rental

Hearing rooms for second and subsequent hearings are available on a rental basis at AAA offices. Check with your local office for specific availability and rates.

Refund Schedule

The Refund Schedule is based on the administrative fee due on a claim or counterclaim asserted by a party.

If the AAA is notified that a claim or counterclaim has been settled or withdrawn before a list of arbitrators has been sent out, all of the fee in excess of $300 will be refunded.

If the AAA is notified that a claim or counterclaim has been settled or withdrawn after a list of arbitrators has been sent out but before the original due date for the return of the first list, two thirds of the fee in excess of $300 will be refunded.

If the AAA is notified that a claim or counterclaim has been settled or withdrawn after the original due date for the return of the first list but at least two business days before the initial date and time set for the first scheduled hearing, one third of the fee in excess of $300 will be refunded.

There will be no refund after any hearing or mediation conference has been held; where a claim or counterclaim was filed as an undetermined/undisclosed claim and remained so at the time of settlement or withdrawal; where a consent award was issued by the arbitrator; or where a determination is made by the arbitrator resulting in the closing of the file.

Index to Section Numbers

AAA Offices*

Atlanta (30309–3214) • India Johnson
1360 Peachtree Street, NE, Suite 270
(404) 872-3022/881-1134 (Fax)

Boston (02110–1703) • Richard M. Reilly
133 Federal Street
(617) 451-6600/451-0763 (Fax)

Charlotte (28226–8297) • Neil Carmichael
7301 Carmel Executive Park, Suite 110
(704) 541-1367/542-7287 (Fax)

Chicago (60606–1212) • David Scott Carfello
205 West Wacker Drive, Suite 1100
(312) 346-2282/346-0135 (Fax)

Cincinnati (45202–2809) • Philip S. Thompson
441 Vine Street, Suite 3308
(513) 241-8434/241-8437 (Fax)

Cleveland (44115–1632) • Audrey Mendenhall
1127 Euclid Avenue, Suite 875
(216) 241-4741/241-8584 (Fax)

Dallas (75240–6620) • Helmut O. Wolff
Two Galleria Tower, Suite 1440
(214) 702-8222/490-9008 (Fax)

Denver (80264–2101) • Mark Appel
1660 Lincoln Street, Suite 2150
(303) 831-0823/832-3626 (Fax)

Garden City, NY (11530–4789) • Mark A. Resnick
585 Stewart Avenue, Suite 302
(516) 222-1660/745-6447 (Fax)

Hartford (06106–1943) • Karen M. Jalkut
Two Hartford Square West
(203) 278-5000/246-8442 (Fax)

Honolulu (96813–4728) • Keith W. Hunter
810 Richards Street, Suite 641
(808) 531-0541/533-2306 (Fax)

Houston (77002–6707) • Therese Tilley
1001 Fannin Street, Suite 1317
(713) 739-1302/739-1702 (Fax)

Kansas City, MO (64106–2110) • Lori A. Madden
1101 Walnut Street, Suite 903
(816) 221-6401/471-5264 (Fax)

Los Angeles (90020–0994) • Jerrold L. Murase
443 Shatto Place
(213) 383-6516/386-2251 (Fax)

Miami (33131–2501) • René Grafals
99 SE Fifth Street, Suite 200
(305) 358-7777/358-4931 (Fax)

Michigan (Southfield 48034–7405) • Mary A. Bedikian
Ten Oak Hollow Street, Suite 170
(313) 352-5500/352-3147 (Fax)

Minneapolis (55402–1092) • James R. Deye
514 Nicollet Mall, Suite 670
(612) 332-6545/342-2334 (Fax)

Nashville (37219–2111) • Tony Dalton
221 Fourth Avenue North
(615) 256-5857/244-8570 (Fax)

New Jersey (Somerset 08873–4120) • Richard Naimark
265 Davidson Avenue, Suite 140
(201) 560-9560/560-8850 (Fax)

New Orleans (70130–6101) • Deann Gladwell
650 Poydras Street, Suite 1535
(504) 522-8781/561-8041 (Fax)

New York (10020–1203) • Carolyn M. Penna
140 West 51st Street
(212) 484-4000/307-4387 (Fax)

Orange County, CA (Irvine 92714–6220) • Lori S. Markowicz
2601 Main Street, Suite 240
(714) 474-5090/474-5087 (Fax)

Orlando (32801–2742) • Mark Sholander
201 East Pine Street, Suite 800
(407) 648-1185/649-8668 (Fax)

Philadelphia (19102–4121) • James L. Marchese
230 South Broad Street
(215) 732-5260/732-5002 (Fax)

Phoenix (85012–2803) • Deborah A. Krell
3033 North Central Avenue, Suite 608
(602) 234-0950/230-2151 (Fax)

Pittsburgh (15222–1207) • John F. Schano
Four Gateway Center, Room 419
(412) 261-3617/261-6055 (Fax)

St. Louis (63101–1614) • Neil Moldenhauer
One Mercantile Center, Suite 2512
(314) 621-7175/621-3730 (Fax)

Salt Lake City (84111–3834) • Kimberly L. Curtis
645 South 200 East, Suite 203
(801) 531-9748

San Diego (92101–5278) • Dennis Sharp
525 C Street, Suite 400
(619) 239-3051/239-3807 (Fax)

San Francisco (94104–1113) • Charles A. Cooper
417 Montgomery Street
(415) 981-3901/781-8426 (Fax)

Seattle (98104–1455) • Neal M. Blacker
811 First Avenue, Suite 200
(206) 622-6435/343-5679 (Fax)

Syracuse (13202–1376) • Deborah A. Brown
205 South Salina Street
(315) 472-5483/472-0966 (Fax)

Washington, DC (20036–4104) • Garylee Cox
1150 Connecticut Avenue, NW, 6th Floor
(202) 296-8510/872-9574 (Fax)

White Plains, NY (10601–4485) • Marion J. Zinman
34 South Broadway
(914) 946-1119/946-2661 (Fax)

*Hearings may be held anywhere in the United States or abroad.

Commonly Used Rules Available from the AAA:

Commercial Arbitration Rules
Commercial Mediation Rules
Supplementary Procedures for International
  Commercial Arbitration
Construction Industry Arbitration Rules
Construction Industry Mediation Rules
Securities Arbitration Rules
Dispute Resolution Procedures
  for Insurance Claims
Expedited Labor Arbitration Rules
Model Employment Arbitration Procedures
Voluntary Labor Arbitration Rules
  (Including Streamlined Labor Arbitration Rules)

For specialized industry rules, contact the AAA regional office nearest you.

Services Available from the AAA:

Arbitration
Mediation
Mini-trial
Factfinding
Election Services
Education
Training
Insurance ADR
ADR for Business
Court-Referred Mediation

American Arbitration Association
140 West 51st Street, New York, NY 10020-1203
Telephone: (212) 484-4000
Fax: (212) 765-4874

See the inside back cover for a list of AAA regional offices.

AAA47-40M-1/91

APPENDIX J

AMERICAN ARBITRATION ASSOCIATION, CONSTRUCTION CONTRACT DISPUTES: HOW THEY MAY BE RESOLVED

American
Arbitration
Association

Construction Contract Disputes:

How They May Be Resolved

Permission to reprint has been given by the AAA.

NATIONAL CONSTRUCTION DISPUTE RESOLUTION COMMITTEE

CHAIRMAN
J. Robert Mann, Jr.
Associated Specialty Contractors, Inc.

CHAIRMAN, *Subcommittee on Rules and Procedure*
Paul Sieben
American Institute of Architects

SECRETARY
Michael F. Hoellering
American Arbitration Association

MEMBERS

American Consulting Engineers Council
Glen H. Abplanalp
Frank Muller

American Institute of Architects
Donald W. Lehman

American Society of Civil Engineers
Robert F. Borg
Harlan Fair

American Society of Interior Designers
Joan C. Gaulden
Robert Arthur King

American Society of Landscape Architects
Walter Diakow
Bradford M. Greene

American Subcontractors Association
Thomas J. Barfield
Eugene E. Grieve

Associated Builders and Contractors, Inc.
Steven M. Coren
Charlie Hawkins

Associated General Contractors
Robert F. Lathlaen
F. Thomas Westcott

Associated Specialty Contractors, Inc.
David Bevirt

Business Roundtable
Richard A. Coughlin

Construction Specifications Institute
Donald D. Meisel
Philip J. Todisco

National Association of Home Builders
Richard Steiner
Leonard Sylk

National Society of Professional Engineers
Robert L. Damrel
Joseph S. Ward

National Utility Contractors Association
Robert Briant
Joseph D. D'Annunzio

STAFF
John H. Wells, Jr.—ABC
J. Laing Bowles—ACEC
Cheryl Terio—AGC
Dale R. Ellickson—AIA
Chris Stinebert—ASA
Daniel G. Walter—ASC
Grace Waldvogel—ASCE
Jean L. Richmond—ASID
David Bohardt—ASLA
Richard F. Kibben—BR
Walter F. Geisinger—CSI
Campbell Reed—NAHB
Arthur E. Schwartz—NSPE
William G. Harley—NUCA

CONSTRUCTION CONTRACT DISPUTES

Many builders, contractors, engineers, and architects know from painful experience that building contracts sometimes erupt into expensive conflicts.

- Specified materials are not always available when needed. Can equivalent material be substituted for the specified product?
- Changes in plans often result in extra charges. What increased charge for the "extras" is payable?
- Contractors are sometimes subject to penalty clauses for delay. Who caused the delay?
- Estimates of construction costs can later prove to have been unrealistic. Added costs are frequently a source of conflict.

Because parties to such disputes want them resolved by persons familiar with their business and because they want a prompt and final decision, they specify arbitration in their contracts. Although the construction industry has used arbitration for many years, a significant improvement in the system has only recently taken place.

PRIOR SYSTEMS OF ARBITRATION

Until 1966, construction-industry arbitration cases under the existing American Institute of Architects Conditions were either arbitrated informally by each party selecting their own "arbitrator" and these two "arbitrators" selecting a neutral arbitrator or they were administered by the AAA under its commercial rules.

The former system resulted in many complaints because it offered neither rules and nor supervision. Moreover, many cases dragged on and created procedural problems.

On the other hand, the AAA procedures, although well tested, were not specifically designed for the construction industry. And the arbitrators on the national panel of the AAA were not always suitable for construction cases.

A joint committee of engineers and architects comprehensively studied the use of arbitration in the industry and concluded that, although it provided a generally effective method for resolving contractual disputes, the procedure could be greatly improved by creating a nationwide uniform system specifically for the construction industry.

In 1965, the joint committee was enlarged to include the following—the American Consulting Engineers Council, then the Consulting Engineers Council; the American Institute of Architects; Associated General Contractors; Associated Specialty Contractors, Inc., then the Council of Mechanical Specialty Contracting Industries; and the National Society of Professional Engineers.

After a year of study, new rules were adopted, known as the Construction Industry Arbitration Rules, to be administered by the American Arbitration Association. Those rules are now recommended for use by all of the organizations in the industry.

Since that time, nine additional national organizations have provided for the use of this arbitration system in their form documents—the American Society of Civil Engineers; the American Society of Interior Designers; the American Society of Landscape Architects; the American Subcontractors Association; Associated Builders and Contractors, Inc.; the Business Roundtable; the Construction Specifications Institute; the National Association of Home Builders; and the National Utility Contractors Association. Thus, fourteen national construction asociations now use the Construction Industry Arbitration Rules.

THE NATIONAL CONSTRUCTION DISPUTE RESOLUTION COMMITTEE

In 1966, with the object of creating the best possible on-going arbitration system, the National Construction Industry Arbitration Committee was established with representatives of the various industry and professional associations. In 1990, the name was changed to the National Construction Dispute Resolution Committee (NCDRC).

Regional advisory committees were also established to work with the American Arbitration Association's regional offices to improve the available construction panel, to serve as a conduit for information, and to advise the AAA on administrative problems.

To implement these new rules and to meet an anticipated increase in cases, the AAA's National Panel of Construction Industry Arbitrators was enlarged through the addition of construction-industry arbitrators nominated by the regional committees.

HOW TO ARBITRATE UNDER THE CONSTRUCTION INDUSTRY ARBITRATION RULES

Under the Construction Industry Arbitration Rules, arbitration can be provided for in the original contract. This provision is expressed in a *future-dispute arbitration clause* of a contract. A clause reading

> "Any controversy or claim arising out of or relating to this contract, or the breach thereof, shall be settled in accordance with the Construction Industry Arbitration Rules of the American Arbitration Association, and judgment upon the award rendered by the arbitrator(s) may be entered in any court having jurisdiction thereof"

can be used to take advantage of the new procedures.

In the absence of such a clause, parties can bring an existing dispute to arbitration by means of a signed statement in which both parties briefly describe the issue between them and agree to arbitrate under the construction rules.

On receiving the Demand for Arbitration or Submission to Dispute Resolution, the AAA sends to each party a copy of a list of proposed arbitrators to resolve the controversy. In a construction dispute, these could include contractors, engineers, architects, other business people familiar with the construction industry or attorneys who customarily represent such clients. In cases involving lesser sums, one arbitrator is generally appointed. But, in large cases, it might be preferable to have three neutral arbitrators.

Parties are allowed seven days to study the list, cross off any names objected to, and number the remaining names in order of their preference. Where parties want more information about a proposed arbitrator, such information is given upon request.

When these lists are returned, the American Arbitration Association compares them and appoints the arbitrator whom the parties have approved. Where parties are unable to find a mutual choice on a list, additional lists can be submitted at the request of both parties.

If parties cannot agree on an arbitrator, the Association will make administrative appointments, but in no case will an arbitrator whose name was crossed out by either party be appointed.

Arbitrators on AAA panels are generally willing to serve without a fee. They volunteer an occasional day as a public service. But, after spending the first day on a case, the arbitrator must be compensated by the parties. The rate of compensation will then be based upon the amount of service involved and the number of hearings. Any arrangement for the compensation of an arbitrator is made through the AAA, not directly by the arbitrator with the parties.

After the arbitrator is appointed, the AAA consults with the parties to determine a mutually convenient time and place for the hearing. Arrangements are made through the Association, rather than directly between the arbitrator and the parties. The reason for this is twofold: it relieves the arbitrator of routine burdens and it eliminates the danger that, in the course of conversations outside the hearing room, one party will offer arguments on the merits of the case that the other has not had an opportunity to rebut.

Pursuant to the AAA's rules, the AAA and the parties may also use facsimile transmission, telexes, telegrams, or other written forms of electronic communication to give the notices required under the rules.

THE ADMINISTRATIVE CONFERENCE

At the request of any party or when the AAA determines that it would be beneficial to do so, an administrative conference will be set up prior to the selection and appointment of an arbitrator. Particularly with complex cases and when it is evident that more than one day of hearing will be necessary, an informal meeting of the parties and an AAA representative is arranged to discuss such procedural concerns as the composition of the panel, scheduling, the specification of uncontested facts, witness lists, document exchange, on-site inspection, stenographic records, interpreters, and other matters of concern to the parties. Taking the time beforehand to finalize these procedural details avoids unnecessary delay and ensures an orderly, expeditious arbitration.

THE HEARING

Arbitration hearings are less formal than court trials. Arbitrators are not required to follow legal rules of evidence. Rather, they are empowered to listen to all evidence that is relevant and material. Arbitrators often accept evidence that might not be permitted in court, but this does not mean that all evidence is believed or given equal weight.

Each party has a right to be represented by counsel and the hearing is conducted in a businesslike manner. It is customary for the complaining party to proceed first, followed by the respondent. This order may be varied, however, when the arbitrator thinks it advisable. Each party must try to convince the arbitrator of the correctness of its position and the hearing is not closed until each has had a full opportunity to present its case.

THE AWARD

The purpose of the award is to dispose of the controversy finally and conclusively. If the parties have not set other limits, the award must be handed down within thirty days after the closing of the hearing (14 days under the expedited procedures). The power of the arbitrator ends with the making of the award; the decision can be changed only if both parties agree to reopen the case, unless the applicable law provides otherwise.

WHAT IT COSTS TO ARBITRATE

As the American Arbitration Association is a not-for-profit organization, its fees for arbitration services have been set to cover only the costs of administration.

When filing a claim under the Construction Industry Arbitration Rules, the filing party must pay an administrative fee of $300. The balance of the fee is based on the amount of each claim and counterclaim as disclosed when the claim or counterclaim is filed. This balance is due and payable sixty days after the AAA's commencement of administration, or prior to the date of the first hearing, whichever occurs first.

MISCELLANEOUS PROVISIONS

In this booklet, we have indicated the answers to some of the most frequently asked questions about how to proceed with arbitration under the Construction Industry Arbitration Rules. But other matters might also be of interest. The rules—a 28-page booklet—answer other procedural questions:

- How is the place of arbitration determined when the parties are unable to agree?
- May a claim be amended after proceedings begin?
- What must an arbitrator do upon discovering that he or she once did business with one of the parties?
- Who may attend hearings?

Those who would like copies of the full text of these rules are invited to write for free copies to the American Arbitration Association at any of its regional offices (their addresses are listed on the back of this brochure) or to any of the organizations composing the National Construction Dispute Resolution Committee.

MEDIATION

There is now available a supplementary method of dispute settlement that affords parties a chance to resolve their dispute in its early stages. Mediation involves the assistance of one or more individuals who work with the parties to settle a controversy by direct negotiation. The mediator participates impartially, serving as an advisor and a consultant.

Mediation is a voluntary procedure. The sole function of a mediator is to guide the parties in reaching their own terms of settlement. It remains within the discretion of the parties to withdraw from the mediation at any time and it is agreed by them at the outset that nothing that

transpires during the procedure will affect the rights of any of the parties or prejudice their positions in a subsequent arbitration or other proceeding, should such become necessary.

The Construction Industry Mediation Rules have received the full endorsement of the NCDRC. To initiate a mediation under them, all that a party need do is write or place a call to any AAA regional office. Panels of trained construction mediators have been established and the AAA's staff support ensures that the process will be handled in an efficient and professional manner.

Pursuant to Section 10 of the Construction Industry Arbitration Rules, if parties to a pending arbitration agree to mediate under the AAA's mediation rules, no additional administrative fee is required to initiate the mediation.

A successful mediation can eliminate the need to prepare for the more formal and adversarial procedures of arbitration or litigation. Even when full accord is not realized, however, a partial settlement can be achieved that will simplify any arbitration that might follow.

EDUCATION FOR CONSTRUCTION ARBITRATION

The participating associations have also enabled the National Committee to carry out educational programs to inform the industry on the use of arbitration and mediation. Speakers and pamphlets on the resolution of disputes may be obtained by writing to the nearest AAA regional office or to AAA headquarters.

The Punch List is published by the AAA for construction professionals interested in alternative dispute resolution. Free subscriptions are available by writing to the AAA's Department of Publications in New York.

The National Construction Dispute Resolution Committee and the AAA would welcome your assistance and support in their efforts to help the design professions and the construction industry to settle disputes more efficiently and speedily. For information about membership, please write to the Chairman, Mr. J. Robert Mann, Jr., at the American Arbitration Association.

REGIONAL CONSTRUCTION-INDUSTRY ADVISORY COMMITTEES

Alabama

Edward S. Allen, Esq.
P. Lauren Barrett, AIA
Marion L. Bradford, AIA
K. W. Michael Chambers, Esq.
Ralph E. Campagno, ASCE
A. H. Gaede, Jr., Esq.
L. Leon Goddard, AGC
Melvin J. Hines, NAHB
Leonard L. Howard, ASCE, NSPE
Daniel T. Hull, Esq.
Rodney A. Max, Esq.
Holt Rast, NAHB, NSPE, NUCA
E. Mabry Rogers, Esq.

Alaska

Leroy J. Barker, Esq.
Stephen DeLisio, Esq.
Conrad G. B. Frank, AGC
R. Everett Harris, Esq.
Robert N. Hesseltine, AIA
John E. McCool, CSI
Frank E. Nyman, ASCE, NSPE
Albert Swalling, AGC
Leo Walsh, AGC
Donna C. Willard, Esq.

Arizona

Stephen John Andros, CSI
Charles O. Biggs, AIA
Richard Caviness, ASID
Harry G. Costello, Esq.
Charles Daggett

ABC—Associated Builders and Contractors, Inc.
ACEC—American Consulting Engineers Council
AGC—Associated General Contractors
AIA—American Institute of Architects
ASA—American Subcontractors Association
ASC—Associated Specialty Contractors, Inc.
ASCE—American Society of Civil Engineers
ASID—American Society of Interior Designers
ASLA—American Society of Landscape Architects
BR—Business Roundtable
CSI—Construction Specifications Institute
NAHB—National Association of Home Builders
NSPE—National Society of Professional Engineers
NUCA—National Utility Contractors Association

Carl LeMar John, AIA
Donald Maxwell, Esq.
James McDonald, AGC
Robert F. Oberstein
William Wissler, Esq.

Charlotte

John W. Boyd, ASC
Curt Bramblett
Dorrell W. Ford, ASA
Stephen P. Gennett, AGC
Dennis James Hall, AIA, CSI
Martin A. Kupris, CSI
Larry J. Ragland, ASLA
J. H. Spencer, NSPE

Chicago

Donald L. Brenner, ASA
Ronald Bronson
Geraldine Soat Brown, Esq.
Allen E. Bulley, Jr., AGC
Henry M. Chiano
Roy Diamond
Elliot Howes Gage, CSI
Carl A. Jansen
Howard P. Kamin, Esq.
Howard Kane, Esq.
Edward Kenn, Esq.
George A. Kennedy, ASCE
Herbert Kreisman, Esq.
Kent Lawrence, Esq.
Paul M. Lurie, Esq.
William J. Lytle, AGC, ASCE
Mary McElroy, AIA
Monroe McNulty
Jerome Meister, Esq.
Richard Pepper, AGC
Angelo J. Polvere, ASCE
Stanton Schuman, Esq.
Steven G. M. Stein, Esq., CSI
William Tomrose

Cincinnati

John A. Bentley, ASLA
Larry Burkart, AIA
Donald G. Engelhart, CSI
Grant Hesser, AGC
Elmer J. Obermeyer, ACEC
Boyd C. Ringo, ASCE
Donald Stevens, AIA
David Warren, ABC

Cleveland

Robert J. Andrews, Esq.
Michael Benza, ACEC, ASCE
Frank Buddie
Donald Daczko, ASA
Stanley G. Gottsegen, Esq.
Samuel E. Kool, NSPE
Lacy Lott, AGC
Barry Joseph Miller, Esq., AIA
Joseph J. Nemeth, ASC
David A. Schaefer, Esq.
Ralph M. Vara

Columbus, Ohio

Ed Edwards, ACEC
Alfred R. Gallucci, NSPE
James H. Holroyd, AIA
Theodore J. Kauer, ASCE
Thomas Morgan, ASC
Ralph E. Setterlin, Sr., AGC
George A. Van Niel, CSI
Richard F. Ziska, ASA

Connecticut

Joseph Anderson, AGC, ASCE, NSPE
Robert Daddona, AGC
Raymond Garcia, Esq.
Roger Galliher, AIA, CSI
Charles Groves
Maurice Hoben
S. Edward Jeter, AIA
C. James Lawler, AIA
James Miller
Louis Pepe, Esq.
John Reid, Esq., ABC, AGC
James N. Royster, Esq.
Austin Wolf, Esq.
John C. Yavis, Esq.

Dallas

C. DeWitt Brown, Jr., ASA
E. Vernon Brown, AGC
James Richard Bullock, AIA
Joe F. Canterbury, Jr., Esq.
David F. Fleming
Terrell R. Harper, AIA
James E. Laughlin, ASCE
Ronald E. Marabito, CSI
Raleigh K. Roussell, AGC
John E. Reuler
Raymond G. Smerge, Esq.
Pat Y. Spillman, AIA, CSI

Jackson D. Wilson II, Esq.
John P. Zimmer, ABC

Denver

Robert S. Appel, Esq.
William L. Bredar, ACEC, ASCE
Fu Hua Chen, ACEC
Jack R. Janney, ACEC, ASCE, NSPE
R. James Noone, CSI
Richard O'Brecht, AGC
W. David Pantle, Esq.
George O. Thomas, ASCE, NSPE

Garden City, NY

Milton Alpern, ASCE, CSI, NSPE
H. G. Baier, ASA, CSI
James Baranello, AGC
Harold A. Brandt, AGC, ASCE
Ronald D. Goodman, AIA
Bert K. Leffert, Esq.
Sayward Mazur, Esq.
Raymond J. Rice, NSPE
Sigmund S. Semon, Esq.

Georgia

John Barmeyer, Esq.
Donald C. Bricker, AIA, CSI
Kelley David Carey, Esq., NSPE
Ray E. Cumrine, AIA
Overton A. Currie, Esq.
James Fincher, ASCE, NSPE
James P. Groton, Esq.
W. Hensell Harris, Jr., Esq., NSPE
William J. Hart, ABC
George Head, Jr., ABC, ASA
John W. Hinchey, Esq.
Paul J. Kennedy, ASCE
John R. Kohnke, AGC
William Laubmann, AIA, ASLA
Seymour Liebmann, NSPE
H. Fielder Martin, Esq.
Clifford Nahser, AIA
William D. Reeves, NUCA
C. Philip Richards, AGC
Leland S. Riggs, ASCE
Bernard B. Rothschild, AIA, CSI
J. Ben Shapiro, Jr., Esq.
Paul Yantis, Jr., AGC,

Hawaii

Gerald S. Clay, Esq.
E. Alan Holl, AIA

H. Murray Hohns, ASCE, NSPE
George Hutton
Michael J. Leinweber, AIA, CSI

Houston

William K. Andrews, Esq., AGC
Richard Avery, Esq.
R. Stanley Bair, AIA, CSI
Jack B. Buckley, CSI, NSPE
John Carson, ABC, AGC
Anthony A. Cucolo, AGC
E. Jerry Dietrich, ACEC
Steven M. Franklin, Esq.
Ben F. Greenwood, AIA, CSI
Larry B. Horrigan, BR
Alvin L. Jensen, AGC
Richard V. Lewis, AGC
Walter D. Murphy, AGC, ASCE, NSPE
Brisco Parker
George F. Pierce, Jr., AIA
Peter Way, ASA, ASC, CSI
Gary Wood, Esq., NAHB

Indianapolis

Max D. Boots, AIA, CSI
Charles G. Fuller, ACEC
Dana L. Hannan, NSPE
Merlin R. Kirchoff, CSI
Richard K. Lennox, AIA
Donald T. MacRae, ASCE
James C. May, AGC

Kansas City, Mo.

Michael T. Callahan, ABC
Frederick P. Chael, AIA, CSI
James E. Conway, Esq.
John Lawrence Daw, AIA
J. Emerson Finney, ASCE, NSPE
Stuart M. Hutchison, AIA
R. C. Oliver, ACEC, NSPE
Robert G. Scott, Esq.
Robin E. Walker, CSI

Los Angeles

James Acret, Esq.
Yoshiro Befu, ASLA
Milton J. Brock, Jr.
Cyril Chern, Esq., AIA
Scott E. Christensen, Esq.
I. V. Cohen, NSPE
LeRoy Crandall, ACEC

Stephen J. Densmore, Esq.
Warren Driver
Randall L. Erickson, Esq.
Garner V. Grayson III, Esq.
R. M. Holmes, NAHB
Ernest E. Johnson, Esq.
Bernard S. Kamine, Esq.
Donald W. Killian, Jr., Esq.
Jerome I. Orland, CSI
Edward M. Pozzo, AGC, ASCE
Edward K. Takahashi, AIA
Raymond Utin, Esq.

Louisiana

Jerome H. Alciatore, AIA, CSI
W. Paul Andersson, Esq.
Stephen G. Benton, AGC
Charles G. Favret, ASA
Ralph L. Kaskell, Jr., Esq.
Morton Katz, Esq.
Roy S. Kelley, Esq., AGC, ASCE
Harvey C. Koch, Esq.
Ed Mathes, AIA
Kathy McGooey, ASID
Lane Meltzer
Fred Myers, ABC
Larry Schmidt
Lloyd N. Shields, Esq.
Knox H. Tumlin, CSI
Fred H. Vanderbrook, ACEC

Maine

W. John Amerling, Jr., Esq.
Robert Armitage, AIA, CSI
Paul Atwood, AIA, ASCE, NAHB
John Ballou, Esq.
Curt Beal, ASCE
Wilbur Bell, AGC, ASA
Stephen Cole, AGC, ASCE, NSPE
John Demer, AGC
George DiMatteo, AGC, CSI
Hon. Sumner Goffin
Jerry Haynes, AGC, ASCE, CSI
Robert Knight, AIA
Edward Millimet, AIA
John Albert Mitchell, Esq.
Jotham D. Pierce, Jr., Esq.

Massachusetts

Anthony Battelle, Esq.
Frank V. Bonzagni, Esq.
Eliot D. Canter, AGC

Philip Cronin, Esq.
Ellen Craig Davey, Esq.
Richard Donovan, ASCE
Thomas I. Elkind, Esq.
John Fieldsteel, Esq.
Susan G. L. Glovsky, Esq.
Emmett F. Glynn, AIA, CSI
Charles P. Heuer, Esq., AIA
Francis W. Madigan, Jr., AGC
Alphonse Mancuso, ASCE
Donald M. Manzelli, AGC, CSI, NSPE
James J. Marden, AGC, NSPE
Roger Marshall, AIA, CSI
Roland W. Martineau, ASCE
Jon C. Mazuy, Esq.
William G. Meserve, Esq.
Robert B. Michelson, ASCE
John B. Miller, Esq., ASCE, NSPE
James J. Myers, Esq.
John C. Neudorfer, ASCE, NSPE
Jerrold A. Olanoff, Esq.
William H. Pineo, ASCE
Charles H. Resnick, Esq.
Carl M. Sapers, Esq.
Sidney Shuman, ASCE, NSPE
John J. Spignesi, Esq.
Sidney Stamell, AGC
David P. Sullivan, ASCE
Dudley A. Weiss, Esq.

Miami

Richard J. Hays, Esq.
Larry R. Leiby, Esq.
Winston Lippert, ASID
David F. Ludovici, ASCE, NSPE
Richard D. Lyle, ASA
Emmanuel N. Nicolaides, ACEC, CSI
David P. Snyder, CSI

Michigan

A. Arnold Agree, AIA
James M. Alexander, Esq.
Jeffrey M. Bauer, Esq.
Victor J. Baum, Esq.
William M. Brenan, CSI
John K. Grylls, Esq., AIA
Joey E. Lebovic, NSPE
George Segal, ASC
Eugene E. Twitchell, AGC

Minneapolis

John C. Anderson, CSI
Archie N. Carter, ASCE

Marvin Fabyanske, Esq.
John Finn, Esq.
Victor C. Gilbertson, AIA
B. C. Hart, Esq.
Thomas G. Hewitt, ASLA
Thomas G. Johnson, ACEC
Bruce Kelley, ACEC
Ben Mayeron, NSPE
James A. Sweitzer, AGC
G. Fritz Weber, Jr.

Mississippi

Phil Abernathy, Esq.
Robert Canizaro, AIA
Lynton B. Cooper, AIA, CSI
A. Dewitt Day, ASLA
Richard M. Dean, AIA
William Grace
Alton H. Harvey
William R. Bob Henry, Jr., AIA, CSI
Charles E. Hughes, Esq.
Robert C. Lunardini, NSPE
James C. Mingee, Esq.
Roy N. Moore
Mark L. Mullin
Samuel J. Nicholas, Jr., Esq.
Jerry Oaks, AIA
Fred H. Pilgrim, Jr.
Charles Porter
William E. Ready, Esq.
David L. Reynolds, Esq.
Lloyd Spivey, ABC
Eugene C. Stone III, Esq.
H. T. Ware, AGC
Charles White, ABC
Robert Wise, Esq.
Terrell E. Wise, AGC

New Hampshire

Richard Baybutt, AGC
John E. Bowker
Wyman P. Boynton, Esq., NSPE
Paul Cameron, AGC
Courtland Cross, AGC
Woodbury Fogg, AGC, NSPE
Delnoce Whitney Goubert, AIA
Jay Leland Hodes, Esq.
Thomas McCartney, AGC
Peter Millham, Esq.
Maurice O'Donnell
R. F. O'Neill, ASCE
Thomas Richards, Esq.

Donald Snyder, Jr., AGC
Robert A. Wells, Esq.

New Jersey

Glen Abplanalp, ACEC, ASCE
John Best, Jr., ASA
Richard Bonamo, Esq.
Robert A. Briant, NUCA
Anthony P. Bruno, AIA, CSI
Robert G. Buie, ASA
Gerald B. Curtis, NSPE
Frank D. Harmuth, ABC
Philip S. Inglis, AGC
Ingvald L. Moe, NSPE
Joseph A. Natoli, ABC
Maynard C. Nugent, ASCE
Robert Peckar, Esq.
Eleanor Petterson, AIA
Fred M. Van Gaasbeek, AIA

New Mexico

Gordon E. Bosl, AIA, CSI
Carl Calvert, Esq., ABC, AGC
Harry Connelly, Jr., Esq.
Arthur Greenfield, Esq., ABC, AGC, CSI
Katherine J. Meadows, National Association of Women in Construction
Dennis W. Roberts, AGC, CSI
Stanley Sager, Esq.
Timothy Sheehan, Esq.
G. W. Stuckman, AGC
Robert Uhl, ACEC, CSI

Oklahoma

Bill Skeith, AGC

Philadelphia

Robert B. Adcock, NSPE
Walter C. Applegate, ASLA
Charles B. Broudy, AIA
Donald D. Degus, CSI
Joseph R. Farrell, Jr., AGC
John E. Healy II, AGC
William J. Knecht, ASC
Roger T. McDonald, CSI
Frank J. McGettigan, AIA
Donald D. Meisel, CSI
John A. Miller, ACEC
William W. Olmsted, AGC
Celestino R. Pennoni, NSPE
F. Spencer Roach, AIA
Michael Stern, ASA

Pittsburgh

A. Ralph Ellis, Esq.
John J. Fuchs, ASA
Samuel J. Green, ACEC
H. Edgar Lore, ASCE
Silvio J. Lorenzi, ASCE, NSPE
Kenneth T. Nagie, CSI
Robert L. O'Rourke, ASA
Dahlen K. Ritchey, AIA
Leonard L. Scheinholtz, Esq.
Lawrence C. Wolfe, AIA

Portland, Oregon

Robert L. Bloodworth, AIA
Edward F. Ceccacci, AGC
William A. Hammond, AGC
Norman L. Lindstedt, Esq.
John T. Merrifield, ASCE
William E. Merritt, CSI
Guy J. Rappleyea, Esq.
Miles Sweeney, Esq.
Gordon Todd, CSI

Rhode Island

Ronald Ash, ASCE
Frank H. Christoph, Jr., AIA
Stephen Fanning, Esq.
George Geisser, ASCE
Thomas Heald, Esq.
Charles Hirsh, Esq.
J. Michael Keating, Esq.
Peter L. Kennedy, Esq.
Thomas Marcaccio, Esq.
Julius Michaelson, Esq.
Victor Minardi, AGC, ASCE
Robert Lee Pare, AIA
William Sheehan, Esq.
Joshua Teverow, Esq.
Gerry Visconti, Esq.
J. Kenneth Wilbur, Jr., AIA
Frank Williams, Esq.
Melvin Zurier, Esq.

Saint Louis

J. Hunt Benoist, AGC
Dennis A. Bolazina, AIA
Melvin E. Becker, AGC
George E. Crow III, CSI
Alfred J. Fleischer, AGC
Allen M. Gallup, Esq., AGC, ASCE, NSPE
Louis Schwartz, American Society of Certified Engineering Technicians

D. Lynn Whitt, Esq.
Michael E. Wilson, Esq.
Kenneth E. Wischmeyer, AIA
Denis Zavadil, ASCE

San Diego

Robert K. Adams, ABC
Thomas G. Atkinson, ACEC, ASCE, CSI
Robert Baumgarten, Esq.
Dean B. Butterfield, ABC
Frank Drachman, Esq.
John Fulbright, AGC
Terry D. Harper, Esq.
Carl F. Ingwalson, Esq.
Peter J. Ippolito, Esq.
Oscar F. Irwin, Esq.
Frank H. Kawasaki, ASLA
Lynn R. McDougal, Esq.
Charles A. McMahon, ASCE, CSI, NSPE
C. J. Paderewski, AIA
Donald J. Reeves, AIA, CSI
Michael J. Roberts, Esq.
James R. Sternberg, Esq.
Jonathan C. Tibbitts, ASCE
Howard G. Weaver, AGC, NAHB
Joyce A. Wharton, Esq., National Association of Women in Construction

San Francisco

Eugene S. Arntz, AGC
William B. Baker, ASCE
Robert L. Balliet, AGC
Robert C. Field, Esq.
Edward T. Gray, CSI
Thomas R. Hogan, Esq.
James Ingwersen, Esq.
Ronald H. Kahn, Esq., AGC, AIA
Aldo Lera, National Electrical Contractors Association
James R. Madison, Esq., ASCE
William H. McInerney, Sr., Esq., AGC
Charles J. Pankow, AGC
James W. Polk, AGC
William B. Reiner, AIA, CSI
Michael B. Shane, Esq.
Stephen G. Walker, Esq.
Gerald G. Weisbach, Esq., AIA

San Juan

Manuel E. Ares, AGC, ASCE, NSPE
José M. Biaggi Junquera, Esq.
David Carrion Fuentes, Esq.

Adriane J. Dudley, Esq.
Armando López, AGC, ACEC
Pedro Juan Ortíz Santiago, ACEC, ASCE, NSPE
Jorge A. Pierluisi Díaz, AGC, ASCE, NAHB
Emiliano H. Ruiz, AGC, ASCE, NSPE
Eric A. Tulla, Esq.
Enrique Umpierre Suárez, Esq., AGC, ASCE, NAHB
Francisco M. Vázquez Santoni
Armando Vivoni, Esq.

Seattle

Don Paul Badgley, Esq.
John D. Blankinship, Esq.
Richard A. Carothers, Sr., ASLA
Michael Lee Cole, ABC, AGC
James P. Donaldson
James A. Frodesen, AGC
Richard D. Harris, Esq.
David D. Hoff, Esq.
Donald L. Logerwell, Esq.
Gilbert H. Mandeville, ASCE
J. Richard Manning, Esq.
Dan Mortimer, AGC
Donald O. Norman
Robert P. Piper, Esq.
Pinckney M. Rohrback, Esq.
Arden C. Steinhart, AIA
James Waldowski, AIA
Ann Whitacre, CSI
Andy Wick, AGC

Syracuse

Wallace P. Beardsley, Jr., AIA, CSI
Lionel A. Biagiotti, ABC, AGC, ASA
Walter G. Breakell, Esq., AGC, NSPE
John C. Bullock, Jr., AGC, NSPE
James D. Curtin, AIA
Raymond A. DiPasquale, AIA, CSI
Alan R. Elia, ABC, AGC
J. William Ernstrom, Esq., ABC, AGC
George N. Freeman
Frank T. Gaglione, Esq., AGC
Gene Irish, AGC
Mark H. Jordan, ACEC, ASCE, CSI, NSPE
James M. Kernan, Esq., AGC
Joseph Lawton, Jr., Esq., ASA
Thomas J. Lynch, Esq.
Douglas Lyon, AIA
Charles M. Manheim, Esq., CSI

Luther C. Nadler, Esq., AGC
James A. Norris, AGC, ASCE
LaVerne F. O'Dell, Esq.
Martha G. Rayle, ASID
Thomas S. Richards, Esq.
Harold Eugene Schumm, ASLA
Loretta Simmons, NUCA
Warren N. Wittek, AIA, CSI
Jeffrey J. Zogg, AGC

Tennessee

Robert J. Ames, CSI
Lee Askew III, AIA
Stephen C. Betts, ASCE
Antonio R. Bologna, AIA
Eugene E. Burr, AIA
William S. Cifers, AGC
E. Gerry Eastman, ABC, AIA, CSI
Jef Feibelman, Esq.
Joseph W. Fortunato, ASCE, NSPE
W. L. Goodson, Sr., American Society of Heating, Refrigeration & Air-Conditioning Engineers
L. Frank Gower, AIA, CSI
Charles D. Hutsell, AGC, CSI
Robert J. Jones, AGC, ASCE, NSPE
James Wiley Johnson, ABC, AGC, CSI
Gerald B. Kirksey, Esq., ASCE, NSPE
Sam L. Lee, AGC
Michael I. Less, Esq.
Lewis M. Steele, AGC

Utah

Robert F. Babcock, Esq.
Peter M. A. Moyes, AIA
Sandra Lee Peterson, ACEC
Richard J. Thorn, AGC

Vermont

Adam Franklin Bortz III, CSI
Jonathan Bump, Esq.
Kimberly Cheney, Esq.
K. William Clauson, Esq.
Hon. Frank Conrad
Jeremy Dworkin, Esq.
David B. Firestone, Esq.
Samuel Eugene Johnson, Esq.
William J. Keogh, AGC
Philip A. Kolvoord, Esq.
Anthony B. Lamb, Esq.
Martin K. Miller, Esq.

Robert M. Paolini, Esq.
Benjamin F. Richards, ASCE

Washington, DC

Harold F. Blasky, Esq.
J. Slater Davidson, Jr., ASCE
H. Peter Guttmann, ACEC, NSPE
Richard G. Howard, CSI
Robert Bleakley James, Jr., Esq.
John W. Johnson, Esq., ASA
Francis D. Lethbridge, AIA
Nash M. Love, NSPE
Richard H. Norair, ASC
Spencer E. Sanders, ASLA
Charles M. Stover, CSI
Raymond Wycoff, ASC

White Plains, NY

John Cricco, ASCE
George Frank, NAHB
Michael Margulies
Sy J. Schulman
Charles Weinberg, ASCE, NSPE
Alfred Wilner, ASCE, NSPE

AAA Offices

Atlanta (30309–3214) • India Johnson
1360 Peachtree Street, NE, Suite 270 • (404) 872-3022/881-1134 (Fax)
Boston (02110–1703) • Richard M. Reilly
133 Federal Street • (617) 451-6600/451-0763 (Fax)
Charlotte (28226–8297) • Neil Carmichael
7301 Carmel Executive Park, Suite 110 • (704) 541-1367/542-7287 (Fax)
Chicago (60606–1212) • David Scott Carfello
205 West Wacker Drive, Suite 1100 • (312) 346-2282/346-0135 (Fax)
Cincinnati (45202–2809) • Philip S. Thompson
441 Vine Street, Suite 3308 • (513) 241-8434/241-8437 (Fax)
Cleveland (44115–1632) • Audrey Mendenhall
1127 Euclid Avenue, Suite 875 • (216) 241-4741/241-8584 (Fax)
Dallas (75240–6620) • Helmut O. Wolff
Two Galleria Tower, Suite 1440 • (214) 702-8222/490-9008 (Fax)
Denver (80264–2101) • Mark Appel
1660 Lincoln Street, Suite 2150 • (303) 831-0823/832-3626 (Fax)
Garden City, NY (11530–4789) • Mark A. Resnick
585 Stewart Avenue, Suite 302 • (516) 222-1660/745-6447 (Fax)
Hartford (06106–1943) • Karen M. Jalkut
Two Hartford Square West • (203) 278-5000/246-8442 (Fax)
Honolulu (96813–4728) • Keith W. Hunter
810 Richards Street, Suite 641 • (808) 531-0541/533-2306 (Fax)
Houston (77002–6707) • Therese Tilley
1001 Fannin Street, Suite 1317 • (713) 739-1302/739-1702 (Fax)
Kansas City, MO (64106–2110) • Lori A. Madden
1101 Walnut Street, Suite 903 • (816) 221-6401/471-5264 (Fax)
Los Angeles (90020–0994) • Jerrold L. Murase
443 Shatto Place • (213) 383-6516/386-2251 (Fax)
Miami (33131–2501) • René Grafals
99 SE Fifth Street, Suite 200 • (305) 358-7777/358-4931 (Fax)
Michigan (Southfield 48034–7405) • Mary A. Bedikian
Ten Oak Hollow Street, Suite 170 • (313) 352-5500/352-3147 (Fax)
Minneapolis (55402–1092) • James R. Deye
514 Nicollet Mall, Suite 670 • (612) 332-6545/342-2334 (Fax)
Nashville (37219–2111) • Tony Dalton
221 Fourth Avenue North • (615) 256-5857/244-8570 (Fax)
New Jersey (Somerset 08873–4120) • Richard Naimark
265 Davidson Avenue, Suite 140 • (201) 560-9560/560-8850 (Fax)
New Orleans (70130–6101) • Deann Gladwell
650 Poydras Street, Suite 1535 • (504) 522-8781/561-8041 (Fax)
New York (10020–1203) • Carolyn M. Penna
140 West 51st Street • (212) 484-4000/307-4387 (Fax)
Orange County, CA (Irvine 92714–6220) • Lori S. Markowicz
2601 Main Street, Suite 240 • (714) 474-5090/474-5087 (Fax)
Orlando (32801–2742) • Mark Sholander
201 East Pine Street, Suite 800 • (407) 648-1185/649-8668 (Fax)
Philadelphia (19102–4121) • James L. Marchese
230 South Broad Street • (215) 732-5260/732-5002 (Fax)
Phoenix (85012–2803) • Deborah A. Krell
3033 North Central Avenue, Suite 608 • (602) 234-0950/230-2151 (Fax)
Pittsburgh (15222–1207) • John F. Schano
Four Gateway Center, Room 419 • (412) 261-3617/261-6055 (Fax)
St. Louis (63101–1614) • Neil Moldenhauer
One Mercantile Center, Suite 2512 • (314) 621-7175/621-3730 (Fax)
Salt Lake City (84111–3834) • Kimberly L. Curtis
645 South 200 East, Suite 203 • (801) 531-9748
San Diego (92101–5278) • Dennis Sharp
525 C Street, Suite 400 • (619) 239-3051/239-3807 (Fax)
San Francisco (94104–1113) • Charles A. Cooper
417 Montgomery Street • (415) 981-3901/781-8426 (Fax)
San Juan (00917–4532) • Lillian A. Reyes
431 Ponce de Leon Avenue, Suite 1501 • (809) 764-8515/764-8848 (Fax)
Seattle (98104–1455) • Neal M. Blacker
811 First Avenue, Suite 200 • (206) 622-6435/343-5679 (Fax)
Syracuse (13202–1376) • Deborah A. Brown
205 South Salina Street • (315) 472-5483/472-0966 (Fax)
Washington, DC (20036–3169) • Garylee Cox
1730 Rhode Island Avenue, NW, Suite 509 • (202) 296-8510/872-9574 (Fax)
White Plains, NY (10601–4485) • Marion J. Zinman
34 South Broadway • (914) 946-1119/946-2661 (Fax)

AAA76-5M-11/90

APPENDIX K

AMERICAN ARBITRATION ASSOCIATION, A GUIDE FOR CONSTRUCTION INDUSTRY ARBITRATORS

A Guide for CONSTRUCTION INDUSTRY ARBITRATORS

American Arbitration Association
140 West 51st Street, New York, NY 10020-1203
Telephone: (212) 484-4000
Fax: (212) 765-4874

National Construction Industry Arbitration Committee Participating Industry Groups

AMERICAN CONSULTING ENGINEERS COUNCIL

AMERICAN INSTITUTE OF ARCHITECTS

AMERICAN SOCIETY OF CIVIL ENGINEERS

AMERICAN SOCIETY OF INTERIOR DESIGNERS

AMERICAN SOCIETY OF LANDSCAPE ARCHITECTS

AMERICAN SUBCONTRACTORS ASSOCIATION

ASSOCIATED BUILDERS AND CONTRACTORS, INC.

ASSOCIATED GENERAL CONTRACTORS

ASSOCIATED SPECIALTY CONTRACTORS, INC.

BUSINESS ROUNDTABLE

CONSTRUCTION SPECIFICATIONS INSTITUTE

NATIONAL ASSOCIATION OF HOME BUILDERS

NATIONAL SOCIETY OF PROFESSIONAL ENGINEERS

NATIONAL UTILITY CONTRACTORS ASSOCIATION

A Guide for Construction Industry Arbitrators

Every year, thousands of busy members of the construction industry put aside their own concerns for a day or two to act as arbitrators of construction controversies. They sit as private judges, selected by the parties to the dispute. The awards they render are binding and enforceable.

Arbitrators come from many occupations, but all have this in common: they are experts in the construction field, they are known for their good judgment, they are respected for their fairness, and they are honored for putting time and talent at the disposal of others without thought of personal gain. The arbitrator's service is appreciated because the prompt and knowledgeable determination of disputes encourages good business relations and relieves the courts.

What It Takes to Be a Good Arbitrator

The arbitrator should be a person of integrity, sound judgment, and specialized knowledge. He or she must be able to decide cases in accordance with the contractual agreement of the parties and the applicable rules of procedure.

The purpose of this guide is to supplement the arbitrator's substantive qualifications with helpful information about arbitration procedure and how to effectively manage and control it.

The Nature of the Arbitrator's Office

The arbitrator is expected to decide the issues presented. Under prevailing arbitration laws, courts will not review awards (arbitrators' decisions) on their merits. This has long been a settled principle in the relationship between arbitration and the law.

Federal and state laws apply to agreements to arbitrate and to arbitrators' awards. The enforceability of an award could depend on how the hearings were conducted. If the arbitrator has acted in accordance with the agreement of the parties and the applicable rules of the American Arbitration Association, the award will meet the standards prescribed by law.

The Importance of Impartiality

An arbitrator should have no interest, financial or otherwise, in the outcome of the case. This is set forth in Section 19 of the Construction Industry Arbitration Rules.

If you discover, upon being asked to serve, some prior or present business connection with one of the parties and the contact is so close as to be disqualifying, you should decline to serve. Not every business relationship casts doubt on an arbitrator's impartiality. Often, it is enough for an arbitrator to disclose the connection before accepting the appointment. Arbitrators are advised, whenever a question as to such potentially disqualifying information arises, *to err in favor of disclosing same to the parties.*

When parties and their witnesses assemble in the hearing room, you might recall for the first time an association with a person involved. Prompt disclosure gives the parties an opportunity to waive their objections. If this happens, you should stop the proceedings and contact the AAA representative. Such a waiver will bar any subsequent objection to the award on grounds of bias.

An arbitrator is on safest ground when communications with the parties are avoided except in the presence of both parties. The arbitrator should avoid luncheon invitations and similar unilateral contacts during the pendency of the case.

Arbitrators must be impartial in both fact and appearance. You should be careful not to comment unfavorably about any party or witness. Feelings about the merits of a case should not be disclosed by an arbitrator. As an arbitrator, you do, however, have the responsibility to determine the relevance and materiality of any evidence or proof offered. You also have the discretion, in the interest of conducting the arbitration proceeding expeditiously, not to receive irrelevant, immaterial, or unnecessarily repetitive proof.

Occasionally, a neutral arbitrator will serve on a case with arbitrators appointed by a party. For example, a panel may be composed of an arbitrator appointed by the claimant, an arbitrator appointed by the respondent, and the neutral arbitrator.

It is recommended that the neutral arbitrator ascertain from the party-appointed arbitrators the nature and extent of any relationship between that arbitrator and the party that appointed the arbitrator, and whether there will be any direct communication between such arbitrators and the party that appointed them.

AAA Staff Will Help

When arbitration is conducted under the Construction Industry Arbitration Rules, the Association assigns a tribunal administrator to help you in your duties as an arbitrator. The administrator will discover the wishes of the parties as to the time and place of hearings and report their preferences to you. In this way, you can then set a convenient date without having to be in personal contact with the parties. The administrator will provide you with a briefing sheet at the outset of the hearing; please read this briefing sheet carefully. A facsimile of it appears on pages 6 and 7. The same procedure applies when a party asks for a postponement, when letters or documents are exchanged, or when you are asked to issue a subpoena. The administrator can answer questions about the rules at any time. Finally, after the hearings are closed, the administrator will encourage you to decide the case and to render an award.

The Arbitrator's Authority

The arbitrator's authority is created by contract, subject to the applicable arbitration law. In effect, the parties breathe life into the arbitrator.

The arbitration clause is the provision of a contract by which parties agree that disputes arising in the future will be submitted to arbitration. A typical arbitration clause might read: "Any controversy or claim arising out of or relating to this contract, or any breach thereof, shall be settled in accordance with the Construction Industry Arbitration Rules of the American Arbitration Association, and judgment upon the award may be entered in any court having jurisdiction thereof."

Most arbitrations are initiated when one party to a contract containing such a clause demands arbitration, asking the arbitrator to order a remedy for the alleged violation. You must examine the contract to see whether the issue is subject to arbitration and whether the relief sought is within the granted authority.

The party against whom a demand for arbitration is served may deny the claim or file a counterclaim.

Disputes between parties not having a prior agreement to arbitrate may be submitted to arbitration. This is done by means of a submission agreement, signed by both parties.

Facsimiles of the demand for arbitration and submission form appear on pages 10 and 11.

Parties may refer to the Construction Industry Arbitration Rules of the American Arbitration Association. The effect of such a reference is to provide a comprehensive, self-executing method for proceeding.

Expedited Procedures

Under the Construction Industry Arbitration Rules, unless the AAA in its discretion determines otherwise, expedited pro-

CHECK LIST FOR AAA COMMERCIAL AND CONSTRUCTION ARBITRATORS

Arbitrators should conduct hearings in the most efficient way possible while providing each side a full and fair opportunity to present its case. The following is a brief outline designed to assist you in this task:

1. You should receive from the AAA a copy of the following prior to the hearing:
   (a) Demand for Arbitration or Submission to Arbitration including contract(s) in dispute
   (b) Answering Statement and/or Counterclaim (if any) with contract(s) in dispute attached

2. Prior to the hearing, you should introduce yourself and request counsel and/or parties and witnesses to do the same. You should announce that you have executed your oath of office and proceed with the swearing of witnesses using the following oath: *"Do you swear (affirm) that the testimony you are about to give in this arbitration is the truth, the whole truth, and nothing but the truth?"*

3. **If at the hearing any arbitrator recognizes in any capacity an attorney, party, and/or witness, or if any party raises an objection to the continued service of any arbitrator, the arbitrator shall declare a recess and make this disclosure known to the tribunal administrator outside of the presence of the parties.** All communication between any party and the arbitrator, except at oral hearings, must be directed though the AAA. Ex-parte contact may be grounds for nullification of the award. If a party contacts you for any reason during the proceedings or after the case has been concluded, please notify the AAA office immediately.

4. Rulings at the hearing:
   (a) No new or different claims may be submitted without the arbitrators' consent. If a change in claim is approved by the arbitrators, the tribunal administrator should be so advised.
   (b) When it is clear that evidence or testimony being introduced is irrelevant or repetitive, the arbitrators have the authority to so rule and direct the party to move forward with its case.
   (c) Arbitrators must rule on objections as they are made.
   (d) On three arbitrator panels, arbitrators should consult each other prior to issuing rulings, sustaining objections, and so on. The chair should announce the panel's rulings.

5. Occasionally, a neutral arbitrator will serve on a case with arbitrators appointed by a party. For example, a panel may be composed of an arbitrator appointed by the claimant, an arbitrator appointed by the respondent, and the neutral arbitrator. It is recommended that the neutral arbitrator ascertain from the party-appointed arbitrators the nature and extent of any relationship between that arbitrator and the party that appointed the arbitrator, and whether there will be any direct communication between such arbitrators and the party that appointed them.

6. The arbitrators must first determine whether or not an absent party has been given proper notice of the arbitration and the hearing. Additionally, arbitrators must require the party that is present to submit evidence supporting its claim. Advise the tribunal administrator that a party failed to appear and that the hearing proceeded in the absence of that party.

7. As the filing party, the claimant usually makes an opening statement first. The arbitrators can vary this procedure. Upon completion of the claimant's opening statement, the respondent is given the same opportunity. The claimant usually presents its case first, through the testimony of witnesses and/or the introduction of exhibits. The respondent may cross-examine the claimant's witnesses. The respondent then presents its case and the claimant is given the same opportunity to cross-examine the respondent's witnesses. The arbitrator has the authority to vary this procedure when appropriate. Encourage parties to disclose the amount of their claims, if they have not already done so, and advise the tribunal administrator of the amount of the claims in dispute. All documents introduced as exhibits must first be presented to the opposing party for examination before being given to the arbitrators.

8. The arbitrators may ask questions to clarify the issue in dispute. They should, however, be careful not to interrupt the presentation of either side's case in asking such questions.

9. If the arbitrators determine that an on-site inspection is necessary, they should notify the tribunal administrator, who will ensure that all parties are duly notified prior to the inspection.

10. The arbitrators should not reveal their opinions regarding the merits of the case and are further cautioned not to suggest specific terms of settlement to the parties. **Any settlement discussions which the parties decide to enter into must be out of the presence of the arbitrators.**

11. When both parties have completed the presentation of their respective cases, the respondent is requested to make a closing statement, if desired. The claimant is then given the opportunity to make a final closing statement.

12. The arbitrator chairing the panel is requested to read the following statement to the parties at the close of hearing:"*The parties are requested to advise us if any further evidence or testimony is to be presented at this time. Both parties have affirmed that the presentation of their case has been completed. We, the arbitrators, declare these hearings closed.*"

13. A Record of Hearing should be completed by the arbitrator and returned to the tribunal administrator after each day of hearing. Please "X" the box marked "Oath Administered"; your signed notice of appointment serves as your oath of office.

14. The award is due within thirty days of the close of the hearing or receipt of the transcript or briefs, if any. If the parties' contract contains some other time limit for rendering the award or **if the case is heard under the AAA Expedited Procedures, which allow fourteen business days to render the award, that time limit is to be observed.** We request that you advise us of your award, either by mail or telephone, within ten days of the hearing, if possible. We must type the award on our award forms, return it to your for your signature, and have it mailed to the parties no later than thirty days from the date the hearing is closed. A late award may be vacated.

15. An opinion is not required. Your award may take the form of a simple, short statement. It should include the apportionment of AAA's administrative fees and of your fees (i.e. -- equally, all by one party, etc.) Do not give a verbal award at the oral hearing, and avoid any commitment as to your decision on points brought forth at the hearing.

16. The AAA is empowered to request deposits from the parties for arbitrator compensation prior to the commencement of the hearings. However, you should review and be aware of the compensation arrangements as they relate to the parties, the AAA, and the arbitrators. **Do not discuss the matter of your compensation directly with the parties**. Your out-of-pocket expenses will be forwarded promptly by our Central Office, while your compensation must be collected by our Central Office from the parties. Therefore, be sure to advise us of all your out-of-pocket expenses incurred prior to the mailing of the final award. We close the case when we transmit the award to the parties and we may not be able to collect any costs you incur after that date or which you notify us of after that date.

17. Pursuant to the Code of Ethics for Arbitrators in Commercial Disputes, you are required to wait a "reasonable" period following the conclusion of the case before entering into any relationship with individuals involved in the arbitration.

The arbitrators may wish to should consult **A Guide for Commercial / Construction Industry Arbitrators**, the **Commercial / Construction Industry Arbitration Rules**, or **Code of Ethics for Arbitrators in Commercial Disputes** for more information on the authority and responsibilities of an arbitrator.

Form 9AAA-10M-8/88

cedures of construction arbitration shall be applied in any case where no disclosed claim or counterclaim exceeds $25,000, exclusive of interest and arbitration costs. These rules were designed to help expedite the settlement of construction arbitration cases of this size.

Parties in larger, more complicated cases may also agree to the expedited procedures, even if claims and counterclaims exceed $25,000, by jointly requesting that their arbitration be processed under the expedited procedures. The expedited procedures, Sections 53 through 57 of the Construction Industry Arbitration Rules, provide for notice by telephone, the appointment of an arbitrator after the submission of one list, and the fixing of the time and place of the arbitration hearing and the time within which the arbitrator has to render the award, usually not later than 14 days from the date of the closing of the hearing.

Administrative Conferences

At the request of any party or at the discretion of the AAA, an administrative conference with the AAA and the parties and/or their representatives will be scheduled in accordance with Section 10 of the rules. The purpose of the administrative conference is to arrange for an exchange of information, the stipulation of any uncontested facts, and whatever other administrative details are necessary to help expedite the arbitration proceedings. Ask the tribunal administrator whether an administrative conference was held and, if so, what the outcome was.

Guidelines for Expediting Larger, Complex Cases

To facilitate the arbitration of larger, complex cases and to best address the unique procedural problems that sometimes accompany them, the arbitrator may refer to the Guidelines for Expediting Larger, Complex Construction Cases. Although portions of these guidelines require the parties' agreement, many of the sections will provide valuable information on how the arbitration can be conducted as efficiently and expeditiously as possible.

Preliminary Hearing

In larger, more complex cases, at the request of any party or at the discretion of the arbitrator of the AAA, a preliminary hearing between the parties and/or their representatives and the arbitrator may be scheduled by the arbitrator. This is also covered in Section 10 of the rules.

You, as arbitrator, first explain the purpose of the preliminary hearing to the parties, including your indication as to what the parties are expected to accomplish at the hearing. This includes establishing the extent of and schedule for the production of relevant documents and other information, the identification of any witnesses to be called, and a schedule for further hearings to resolve the dispute. The preliminary hearing worksheet is reproduced on pages 14 and 15. Most important, you should make a statement about your intention not only to accomplish these preliminary matters, but also to manage the entire arbitration case in a just and efficient manner consistent with the commonly accepted principle that arbitration is an economical and expeditious way to resolve disputes fairly. You should consider the following:

1. A brief written statement, not to exceed five pages, setting forth each party's outline of its claims—you should give the parties the opportunity to briefly describe what the dispute is about, what issues are expected to be resolved, and other brief comments that will educate you about the issues to be decided.

2. Specification of claims and counterclaims—you should encourage the parties to specify the amount involved in claims and counterclaims and advise the AAA of same.

3. Stipulation of uncontested facts—whatever facts the parties can stipulate to should be reduced to writing, and each party should sign and receive a copy before the conclusion of the preliminary hearing. You should also keep a copy of the stipulated agreement and have one placed in the AAA's file.

4. Exchange of information and schedule for the exchange of information (including any reports from experts)—you should inform the parties that they will be required to cooperate in committing to, conducting, and completing the exchange of information concerning the other's documents and witnesses. If the parties do not agree to exchange particular information, you should hear their disagreement and make a ruling on the issue. When the information to be exchanged has been specified, you should establish a schedule for such exchange.

While arbitrators do not have the authority to hold a party in contempt, most parties are reluctant to antagonize

American Arbitration Association

MEDIATION. Please consult the Construction Industry Mediation Rules regarding mediation procedures. If you want the AAA to contact the other party and attempt to arrange a mediation, please check this box. ☐

CONSTRUCTION INDUSTRY ARBITRATION RULES

DEMAND FOR ARBITRATION

DATE: ____________________

TO: (Name) ____________________
(of party upon whom the demand is made)

(Address) ____________________

(City and State) ____________________

(Telephone) ____________________

Named claimant, a party to an arbitration agreement contained in a written contract, dated ____________________, providing for arbitration under the Construction Industry Arbitration Rules, hereby demands arbitration thereunder.

(Attach arbitration clause or quote hereunder.)

NATURE OF DISPUTE:

CLAIM OR RELIEF SOUGHT: (amount, if any)

Please indicate industry category for each party.

Claimant:

☐ Owner ☐ Architect ☐ Engineer ☐ Contractor ☐ Subcontractor, Specify __________,
Other __________

Respondent:

☐ Owner ☐ Architect ☐ Engineer ☐ Contractor ☐ Subcontractor, Specify __________,
Other __________

HEARING LOCALE REQUESTED: ____________________
(City and State)

You are hereby notified that copies of our arbitration agreement and of this demand are being filed with the American Arbitration Association at its ____________________ regional office, with the request that it commence the administration of the arbitration. Under Section 7 of the arbitration rules, you may file an answering statement within seven days after notice from the administrator.

Signed ____________________ Title ____________________
(may be signed by attorney)

Name of Claimant ____________________

Address (to be used in connection with this case) ____________________

City and State ____________________ ZIP Code __________

Telephone ____________________

Name of Attorney ____________________

Address ____________________

City and State ____________________ ZIP Code __________

Telephone ____________________

To institute proceedings, please send three copies of this demand and the arbitration agreement, with the filing fee, as provided in Section 48 of the rules, to the AAA. Send the original demand to the respondent.

Form C2CR-AAA-7/87

American Arbitration Association

SUBMISSION TO ARBITRATION

Date:

The named Parties hereby submit the following dispute to arbitration under the CONSTRUCTION INDUSTRY ARBITRATION RULES of the American Arbitration Association:

Amount of money involved: ______________________

Place of hearing: ______________________

We agree that we will abide by and perform any Award rendered hereunder and that a judgment may be entered upon the Award.

Name of Party ______________________

Telephone ______________________

Address ______________________

Signed by ______________________

Name of Party ______________________

Telephone ______________________

Address ______________________

Signed by ______________________

PLEASE FILE TWO COPIES

Consult counsel about valid execution

FORM AAA-CIC

an arbitrator by refusing to obey his or her directive. Where the arbitrator's request is reasonable, no court will be inclined to disturb it. In some cases, arbitrators have enforced their orders by directing the tribunal administrator to advise a recalcitrant party that failure to obey will result in their claim or counterclaim being stricken. Obviously, this should be a last resort and should first be discussed with the tribunal administrator.

5. Lists of witnesses with outlines of testimony (including biographies of expert witnesses)—you should inform the parties that they will be required to provide each other with a list of witnesses they intend to call, in the order in which they will be called. A summary by name for each witness, along with the subject matter of the anticipated testimony of each, should also be submitted.

6. Advance filing and advance identification of exhibits—you should inform the parties that they will be required to identify and exchange exhibits in advance of the evidentiary hearing.

7. Estimated length of case, schedule of hearings—the parties should attend the preliminary hearing prepared to give a realistic estimate as to the number of hearing days needed to present their respective cases, and should be ready to agree to a schedule for the evidentiary hearings as soon as feasible. Keeping in mind that a prompt award is an important part of arbitration, you should encourage the scheduling of consecutive hearing days, including evenings or weekends if necessary (where permitted by law). Full hearing days should be scheduled with minimal lapses for lunch and breaks. Requests to postpone scheduled hearings will be granted only for just cause.

8. Number of copies of exhibits to be made—it will save time and effort if the parties know how many copies of exhibits and other evidence to provide (enough for each arbitrator and each party). You should raise this item and establish agreement.

9. Briefs—you may discuss whether briefs will be required. When briefs are to be filed, the arbitrator will establish a schedule for their production.

10. Award—it is not required or customary for arbitrators to render written opinions with the awards; however, in some cases, findings of fact, including a breakdown of amounts awarded, may be requested by the parties. The arbitrator should accommodate this need if it exists.

11. Conduct of the preliminary hearing and concluding remarks—at the conclusion of the preliminary hearing, you should reiterate the opening remarks, namely, the goal to move the arbitration to a fair and speedy conclusion. Making the parties aware of their respective responsibilities will set the right tone and facilitate management of a case. This is the time for you to speak about how the evidentiary hearing is to be conducted. This is also the time to announce that documents and/or pictures will be self identifying wherever possible and that repetitious testimony and evidence will be cut off. Whatever other time-saving devices you wish to impose should be indicated at this time.

12. Arbitrators' directive—directions to exchange information can be confirmed in a directive issued by the arbitrators following the preliminary hearing. Any controversies regarding exchanges of information will be resolved by the arbitrators.

"Paper" Preliminary Hearing

If you do not feel that a preliminary hearing is appropriate, but still desire the information requested on the preliminary hearing worksheet, you may direct the parties to complete some or all of the worksheet, and issue directives regarding any item that is necessary. This should be discussed in advance with the tribunal administrator.

Mediation

It is important for arbitrators to remember the availability of mediation. Mediation is a process in which the parties submit their dispute to an impartial third party—the mediator—who attempts to draw them together to produce a settlement. A sample of a Request for Mediation is on page 17. Unlike an arbitrator, a mediator cannot impose a settlement on the parties. Section 10 of the rules provides for a voluntary mediation conference without payment of an additional administrative fee.

The AAA has published Construction Industry Mediation Rules. Where a case bogs down, it might be appropriate for you to suggest mediation. While it is proper for an appointed arbitrator to serve as a mediator, bear in mind that, if the media-

American Arbitration Association
Preliminary Hearing Worksheet

Case # : ______________________________ Date: ________________

Arbitrators : Case Caption (parties):

______________________________ ______________________________

______________________________ ______________________________

______________________________ ______________________________

1. Explanation by Arbitrators of the purpose of the Preliminary Hearing (see instructions attached and note that these worksheet numbers are keyed to instruction paragraph numbers in the attached) .

2. Brief written statement, not to exceed one page, setting forth the outline of the each party's claims:

Claimant:

Respondent:

3. Specification of claims and counterclaims (monetary amounts).

Claims: __________________________

Counterclaims: ______________________

4. Stipulation to uncontested facts.

5. Exchange of information and schedule for exchange of information (including any reports from experts).

Claimant:

Respondent:

6. Lists of witnesses with outlines of testimony (including biographies and reports of expert witnesses).

Claimant:

Respondent:

7. Advance filing and pre-identification of exhibits (schedule for filing of affidavits or for conducting deposition).

Claimant:

Respondent:

8. Estimated length of case.

Claimant:

Respondent:

9. Schedule of hearings.

10. Number of copies of exhibits to be made.

11. Briefs Required? (schedule)

12. Form of Award.

13. Conduct of hearing and concluding remarks (arbitrators' directive, if any).

[phgws-9/87]

tion fails, you more than likely would not be able to continue on as an arbitrator. It is probably wiser to have the Association appoint one of its trained mediators in such instances.

The Arbitrator at the Hearing

If the arbitration involves three neutral arbitrators, it is suggested that the chairperson be the arbitrator with the most experience with construction arbitration and the best knowledge of the subject matter in dispute. If one of the arbitrators is an attorney, that fact alone does not automatically qualify that arbitrator to be the chairperson.

When the parties, their attorneys, and their witnesses convene at the hearing, the arbitrator is in charge.[1] After the parties are introduced and witnesses sworn, you should require that the case move ahead expeditiously. It is important that you exercise firm control.

In some states, it is obligatory for arbitrators to take an oath of office. In others, it is optional with the parties. Many experienced arbitrators prefer to be sworn even when this is not required.[2]

When parties so desire, arbitration awards may be rendered on the basis of documents only, without oral hearing. The material from both parties is sent to the AAA, which then transmits all of the information to the arbitrator for consideration and a determination.

Compensation and Expenses of the Arbitrator

Construction arbitrators generally contribute their services in the smaller cases that take only a day of their time. In larger, more prolonged cases, the parties will agree to payment of a fee as provided for in Section 50 of the rules. The administrator will advise you of the parties' estimates as to the number of hearings expected. These are only estimates. The actual number of hearings required could differ.

You are cautioned not to raise the issue of compensation with the parties. The rules provide that any arrangements for compensation shall be made through the AAA, not directly between the parties and the arbitrator. All negotiations as to fees must be carried out between the AAA and the parties.

Where compensation is to be paid, the Association will secure a compensation agreement from the parties. You will be asked for your written consent. If it is impossible to obtain agreement, the AAA may establish an appropriate rate of com-

American Arbitration Association

CONSTRUCTION INDUSTRY MEDIATION RULES

SUBMISSION TO MEDIATION

Date:____________________

The named Parties hereby submit the following dispute to mediation under the Construction Industry Mediation Rules of the American Arbitration Association:

Amount of Claim in Dispute: ____________________

Place of Mediation: ____________________

Name of Party	Name of Party
Address	Address
Telephone	Telephone
Signed by	Signed by
Industry Category*	Industry Category*
Name of Party	Name of Party
Address	Address
Telephone	Telephone
Signed by	Signed by
Industry Category*	Industry Category*

*Please indicate industry category for each party: Owner, Architect, Engineer, Contractor, Subcontractor, Other (please specify).

PLEASE FILE TWO COPIES

Consult counsel about valid execution

FORM CM1C-AAA-1/86

pensation in accordance with its rules.

Arbitrators are usually invited to serve on cases where the hearing is to be held near their business or residence. In accordance with Section 49 of the rules, out-of-pocket expenses are reimbursed in those cases where an arbitrator is asked to travel to a hearing. Such expenses are generally reimbursed for travel at coach or business rates and for reasonable accommodations for lodging and meals.

Filing New Claims

Sometimes, a party will want to modify a claim in some respect or add a new one. In accordance with Section 8 of the Construction Industry Arbitration Rules, new and different claims may be filed by the parties at any time before the appointment of the arbitrator. After an arbitrator has been appointed, such changes may be made only with the arbitrator's consent.

There are various factors that you should take into account before deciding whether new issues should be consolidated into a single arbitration. Because arbitration is a voluntary process, it is improper for an arbitrator to rule on matters that the parties excluded from the scope of the agreement to arbitrate. When you believe, however, that the new claims relate to the dispute that gave rise to the original demand and that it would be consistent with the intention of the parties as expressed in their arbitration clause, new claims and counterclaims may be considered along with those contained in the original papers.[3] Before coming to a decision on the question of new claims, you should ask both sides to state their positions.

Who Goes First?

Arbitration hearings follow a logical pattern, with opening statements, introduction of the initiating documents, examination of witnesses, presentation of exhibits, and final summations.[4] It is customary for the complaining party to be heard first. Section 29 of the rules gives arbitrators authority over the order of the proceedings; the order may therefore be varied when you think it necessary. Witnesses are usually sworn and are always subject to cross examination.[5]

A party may wish to have a stenographic record made of the proceeding or may require the services of an interpreter. Such arrangements are made directly between that party and the stenographer or interpreter, and the opposing party should be so notified. The requesting party or parties shall assume the costs of such service. At the hearing, a party may realize that they desire the presence of a stenographer, but failed to make arrangements in advance, and request that the hearing be postponed pending the arrival of the stenographer or interpreter. After hearing the parties' positions and considering whether any party will be severely prejudiced by the delay, you must decide whether further testimony will be postponed pending the arrival of the reporter or whether the hearing will proceed in the absence of same. If the parties agree to temporarily suspend the hearing or if you decide to do so, it would be proper for you to inquire whether there are any procedural matters that can be discussed in the interim. Any arrangements arising out of these discussions may then be placed on the record when the case proceeds.

Parties to an arbitration case often have important economic interests at stake; it is proper that they assert their positions energetically. A vigorous tone of argument and a strong objection to evidence hotly resisted by the responding party are common in arbitration.

You should have no difficulty maintaining order. A reminder that certain behavior serves no useful purpose is usually sufficient.

Some General Rules

One of the reasons that parties resort to arbitration is their desire for privacy. You should therefore maintain the privacy of proceedings, unless both parties agree to open the hearings or unless a statute requires otherwise.

Persons directly concerned in an arbitration have the right to attend hearings. You may permit others to attend, with due regard to the right of privacy. You may require the retirement of a witness from the hearing room during the testimony of other witnesses. Refer to Section 25 of the rules regarding attendance at hearings.

Parties are usually represented by attorneys, although that is not required. In either event, the parties "bring the case to the arbitrator," presenting such evidence and arguments as they deem pertinent. You may directly question witnesses, but most experienced arbitrators let the parties develop the facts in their own way and then ask questions or call for the production of additional evidence as required.

At times, parties will become repetitious or wander from

essential matters. You should insist on an expeditious presentation. Asking the disputants to stipulate to agreed-upon facts is often an effective way to save time and clarify the issues.

How to Deal with Objections

After a hearing gets under way and the flow of testimony begins, you might be asked to rule on the admissibility of evidence. This is addressed in Section 31 of the Construction Industry Arbitration Rules. Courtroom rules of evidence do not apply in arbitration, but you might still have to decide whether to permit a witness to continue after a party objects. One guiding principle must be kept in mind: *everything that could further understanding of the case should be heard.*

Because arbitration awards can be subject to attack when arbitrators refuse to hear material testimony, arbitrators may accept doubtful material but give it little weight. You need not accept everything offered. When in doubt, you should ask the parties to present arguments. When one party has stated reasons for its objection and the other has answered, the arbitrator can decide whether the testimony is relevant or material. Even when evidence of doubtful relevance is accepted, the arbitrator can judge how much weight any piece of evidence is worth. Where two or more arbitrators are serving, rulings should be made by a majority of the panel.

Affidavits and Subpoenas

The industry generally prefers arbitration because of its convenience. When witnesses are in a distant city and it would be costly to bring them to the hearing, parties may ask you to accept testimony in the form of affidavits. You may agree if convinced that there is good reason for the request. If there is an objection, you should hear arguments from both sides before making a decision. In evaluating affidavits, you should take into account the fact that witnesses were not subject to cross examination, as they would have been had they appeared in person.

Occasionally, one party will wish to put into the record a document held by the other party. The decision as to whether the document should be produced is made by you. Again, the parties should be asked to comment. If you are convinced that the document is essential, the party should be directed to produce it. This usually results in compliance because few parties in arbitration wish to risk the adverse conclusions that might be drawn from a refusal. When, as rarely happens, there is no compliance, you may either issue a subpoena or draw a negative inference. Pursuant to Section 31 of the rules, the arbitrator or other person authorized by law to subpoena witnesses or documents may do so upon the request of any party or independently.

As the time approaches for you to reach a decision, a legal issue might require further clarification. You should not do independent research, but should ask counsel for each side to explain the law.

Investigations outside the Hearing Room

When the issue turns on an expert's judgment of workmanship or quality of materials, you might want to resolve the matter by personal examination, not just hearing it described. No difficulties are encountered when examination of material can be conducted in the hearing room, with both parties present. But, when it is necessary to visit a building site, for instance, or go to a warehouse to look at the materials, you are not permitted to make an unannounced investigation.

An on-the-scene investigation is an extension of a hearing. An arbitrator should not accept testimony from one side that the other has had no opportunity to hear and comment on. Nor may an arbitrator receive evidence of any kind outside the hearing room without protecting the rights of both parties. An arbitrator who must make an inspection should inform both parties of the time, place, and purpose of that inspection or investigation. The parties are then free to attend. If they choose not to do so, they waive their right to object at a later date. When the parties are not present, you are expected to make a verbal or written report and allow them an opportunity to comment.

Interim Relief

On occasion, a party will request that you grant interim relief to safeguard the subject matter of the arbitration, in accordance with Section 34 of the rules. Interim orders may be used for a wide variety of disputes.

Having considered the parties' respective positions, whether at an oral hearing or in written form, you then make a ruling that may be made a part of the stenographic record, if any, or reduced to writing and transmitted to the parties through the tribunal administrator.

Postponements

Under Section 26 of the rules, you must accept postponements where the parties agree to them. You may postpone any hearing, either on your own initiative or, for good cause shown, at a party's request. In practice, requests for postponement must first be communicated by a party to its adversary. Should the adversary object to the postponement request, it becomes necessary for you to rule on the request through the tribunal administrator. If time permits, the parties will indicate their positions in writing, which will be forwarded to you for review and consideration. Where there is insufficient time for this, the administrator will attempt to arrange a conference call involving counsel for the parties, the arbitrators, and the administrator, who monitors the call. A majority ruling of the arbitrators will prevail. This ruling should be communicated to the administrator, not to the parties directly.

The law requires that arbitrators act reasonably in considering requests for postponements. There are several factors to consider when ruling on postponements. Is this the party's first request? Can the hearings proceed in the absence of an unavailable witness (who can be brought back at a later date) or is that witness vital to the case? Will any party suffer extreme prejudice by the granting or denial of the postponement request?

A possible cure for numerous postponements is the "peremptory hearing," where the arbitrator sets a date that is simply not subject to postponement.

Delegation of Authority Is Not Permitted

Arbitrators have broad powers to determine issues of fact, law, or procedure. This authority must be exercised by you; it may not be delegated to others. Occasionally, a question will arise that requires the services of an outside agency, such as a testing company or a firm of public accountants. Parties may permit you to engage such help. The cost of such an outside service will be subject to allocation in the award unless the parties agree otherwise. This is not a delegation of authority. The ultimate decision is still yours.

The Arbitrator Should Not Participate in Settlement Discussions

During an arbitration, parties sometimes ask for postponement to permit further negotiations. If they resolve the dispute, they may ask you to incorporate the settlement into your award. This is known as a consent award. This is permissible and can be a collateral advantage of arbitration.

You should not participate in settlement discussions. If the parties wish to discuss settlement, you should be excused from the room. If you are a party to settlement discussions or attempt to mediate unsuccessfully, a party might later challenge your impartiality on that basis. In appropriate situations, the AAA may appoint a mediator to conduct mediation under the Construction Industry Mediation Rules.

The Closing of Hearing

After all witnesses have been heard, counsel for each side should be given an opportunity for summation. You should then close the hearing. Unless briefs are to be filed, the time period for delivery of the award begins to run when the oral hearings are closed. The time limits in the contract are binding on the arbitrators. If parties have not set other limits, you have thirty days under AAA rules within which to render the award (14 days under the Expedited Procedures). It is vital that the award be rendered before time runs out. Otherwise, the parties could reject the award or even involve the arbitrators in litigation. Don't procrastinate!

At times, parties will want to provide you with briefs or with additional material not immediately available. You may agree to receive such material at a later date. You do not close the hearing until the date that has been set for such receipt, at which point the time limit starts running.

The procedure for briefs is simple. The parties deliver them to the AAA, which transmits them to you. Parties should be encouraged to exchange all correspondence, including post-hearing briefs, directly. This procedure serves a double function: it makes it unnecessary for you to be in personal contact with the parties, and it ensures that each side receives the other's brief.

On occasion, a party will send an unsolicited response to a post-hearing brief. It is wise practice to advise the parties in advance that this will not be permitted and that the administrator will be instructed to return such a response to its sender.

Reopening the Hearing

Once the hearing is closed, it may not be reopened without the arbitrators' permission. You may reopen a hearing upon a showing of good cause by one of the parties, as long as the award has not yet been rendered. You may not reopen a hearing if it would delay the delivery of the award beyond the time in the contract, unless both parties agree to extend that limit. This rule of procedure, when arbitration is administered by the AAA, tends to ensure that the award will be rendered promptly. Once again, it is vital that the award be rendered on time.

The Award Must Be in Writing

The award must be signed by the arbitrators and, under the laws of some states, be notarized or witnessed. An award form is reproduced on page 25. If two or more arbitrators are involved, the award must have the signatures of a majority. Arbitrators usually meet after the last hearing to agree on the award. They might disagree on some points. Their conversations should be kept confidential. No arbitrator should disclose what any other member of the panel said during these conferences. The majority rules, but any arbitrator who disagrees may note a dissent on the award form or elect not to sign the award.

After deciding the award, you should communicate its substance to the tribunal administrator, who will put it into correct form. This should be done early enough to allow time for typing the award and having it circulated for signature. Some experienced arbitrators are willing to prepare the award on their own. This is acceptable to the AAA, provided that the award contains the necessary information. At a minimum, it must contain

(1) a preamble,
(2) the body of the award (addressing all issues raised, including an allocation of administrative fees and any arbitrator compensation), and
(3) a closing and affirmation.

Your tribunal administrator can provide model language for these basic elements. The closing must state in some form or another that the award is in full settlement of all claims (and counterclaims) submitted in the arbitration. The award must be affirmed by you or notarized, depending on the jurisdiction of the arbitration.

When the award is signed, the AAA will deliver it to the parties.[6]

American Arbitration Association

CONSTRUCTION INDUSTRY ARBITRATION TRIBUNAL

In the Matter of the Arbitration between

CASE NUMBER:

AWARD OF ARBITRATOR(S)

I, (WE), THE UNDERSIGNED ARBITRATOR(S), having been designated in accordance with the Arbitration Agreement entered into by the above-named Parties, and dated
and having been duly sworn and having duly heard the proofs and allegations of the Parties, AWARD as follows:

DATED:

STATE OF
} ss.:
COUNTY OF

On this day of , 19 , before me personally came and appeared
to me known and known to me to be the individual(s) described in and who executed the foregoing instrument and he acknowledged to me that he executed the same.

FORM AAA-C3C

Three Tests of a Good Award

An arbitration award should have three chief characteristics: it should be clear and definite, leaving no doubt as to what the parties must do to comply; it should decide every issue put to the arbitrators; and it should not rule on anything outside the scope of the arbitrators' authority.

When so requested by a party, you may include a breakdown of items awarded. As an example, a general contractor might need an award specified or broken down by items to be able to deal more directly and competently with the subcontractors.[7]

Money Damages and Other Relief

The award must be consistent with the agreement of the parties. For instance, ruling on a claim by a building owner against a contractor who failed to do certain work according to specifications, you might award monetary damages or direct the contractor to do the work over again. You might also dismiss the claim. An award directing the contractor to reimburse the owner by landscaping the property would be improper, however, where landscaping was not contemplated in the contract.

Remedies do not necessarily take a monetary form. For example, you might direct specific performance of the contract. You might also grant injunctive relief, barring a person from continuing a violation of the contract.

The award should also include the arbitrators' final apportionment of administrative fees, expenses, and arbitrator compensation, if any. In some cases, you are authorized by the parties' contract to award attorney fees.

No Written Opinion Is Required

Construction arbitrators are not required to explain the reasons for their decisions. As a general rule, the award consists of a brief direction to the parties on a single sheet of paper. One reason for brevity is that written opinions might open avenues for attack on the award by the losing party.

Courts will not review arbitrators' decisions on the merits of the case, even where the conclusions are different from those that a court might reach. But a carelessly expressed thought in a written opinion could afford an opportunity to delay enforcement of the award. The obligations to the parties are better fulfilled when the award leaves no room for attack. In situations where you feel it necessary to write such an opinion, it should be contained in a separate document.

No Compromises

You should not compromise unless the dispute clearly calls for this result. Parties generally expect a decision on the issues. This does not mean that the award should grant "all or nothing." If it is your judgment that some percentage of a claim is justified, the award may so provide.

For example, an owner of a construction project may ask for a price allowance of an amount of money because of some alleged inferior quality of workmanship. The general contractor may deny that any allowance is warranted. If the arbitrator should find that the workmanship was substandard and that the deficiency justified a reduced price, that would be an appropriate award. But arbitrators are not expected to award half of what a claimant sought in an effort to equalize the burden on the parties. That would be a compromise. What distinguishes an award from a compromise is not its size, but whether it represents your judgment on the contractual rights of the parties.

The Arbitrator's Task Is Completed with the Signing of the Award

When you sign the award and return it to the AAA for delivery to the parties, your task is completed. The arbitrator need have no further concern with the case. You become, in most jurisdictions, "functus officio" upon making an award. Arbitrators should not participate in further litigation in the matter. In fact, you should not discuss the award or respond to a request for clarification unless requested to do so by both parties.[8] In some jurisdictions, arbitrators may modify awards in which there have been technical or clerical errors. Your administrator will bring this to your attention where appropriate.

Your obligation to maintain the confidentiality of business affairs of the parties disclosed in the arbitration continues. It would be a breach of ethics for you not to keep such information confidential.

Post-award Communications from the Parties and/or Their Representatives

On occasion, a party or its representative will call an arbitrator after the award is issued, seeking an explanation of the decision. It is a poor idea to talk at all, Instead, refer the party or attorney to the AAA. One reason why an arbitrator should not answer these questions is that the award could still be subject to enforcement proceedings in court. A party might try to impeach the award by getting the arbitrator to divulge his or her reasoning or thought processes. Another reason is that, in rare instances, all or part of a case is remanded to the arbitrator for further consideration. You could be "tainted" by having had direct ex-parte communications with one party. Pursuant to the *Code of Ethics for Arbitrators in Commercial Disputes*, you are required to wait a "reasonable" period following the conclusion of the case before entering into any relationship with individuals involved in the arbitration.

Conclusion

In this guide, we have called attention to some problems that you might encounter. The problems are usually few. Those that do arise can be readily resolved by you in a businesslike manner.

No attempt is made to impose a rigid formula. The proposed solutions represent methods that many arbitrators have found effective after years of experience.

You can usually rely on the cooperation of the parties. Parties come to the hearing prepared to accept the arbitrators' decisions on procedural and substantive matters, recognizing in advance that some of those decisions might not be to their liking. As an arbitrator, you should be firm, but fair.

The arbitrator who agrees to hear and decide a case finds that the task is mostly an agreeable one. The honor and respect accorded an arbitrator are pleasurable. There is, too, the personal satisfaction of having made a positive contribution to the construction community, avoiding the necessity of going to court.

Volunteer service offers few outlets for the exercise of talent and judgment for the public good to equal those available to the arbitrator of construction disputes.

Footnotes

1. When the parties so desire, arbitration awards may be rendered without oral hearing. The rules then provide for written statements and documents to be transmitted by the parties to the AAA. The material from both parties is sent to the arbitrator by the AAA at the same time.

2. As administered by an AAA tribunal administrator, the oath reads: "Do you solemnly swear (affirm) that you will faithfully and fairly hear and examine the matters in controversy and that you will make a just award to the best of your understanding?"

3. This problem arises when arbitration is initiated by one party in a demand for arbitration, on the basis of a future-dispute arbitration clause of a contract. Where arbitration is initiated by the parties jointly through a submission agreement, the issues before you are those agreed to by both parties in writing.

4. Arbitrators may proceed even in the absence of a party, if proper notice of the hearing was given. This does not mean, of course, that the absent party will lose the case by default. The other party must still produce proof to support the claim and you may not find in favor of the claimant solely by reason of the respondent's failure to appear.

5. As administered by the AAA, the oath reads, "Do you solemnly swear (affirm) that the evidence you are about to give in this arbitration will be the truth, the whole truth, and nothing but the truth?"

6. Delivery consists of placing of the award in the mail by the AAA, addressed to the parties at their last known addresses, personal service, or any other method permitted by law. If the state law requires that a copy of the award be deposited with a court, this will be done by the AAA.

7. Unless the agreement of the parties provides otherwise, an arbitrator rendering a monetary award may require interest to be paid. In that case, the award should instruct the parties clearly on how interest is to be computed. It could be from the time the obligation was incurred to the date of the award, from the date of the award to the date payment is made, or between any other pair of dates that you think equitable. In some states, the rate of interest is specified by law.

8. In some states, the arbitration law provides the parties with an opportunity to resubmit the decision to the arbitrator for modification or correction. In such situations, the parties will correspond with the AAA, which will then forward the matter to the arbitrator for a decision.

Index

The American Arbitration Association

The administrative services described in this booklet, as well as administration of labor-management and international-trade arbitration, represent one aspect of AAA activities. Its other chief functions are education, promotion, and research into the uses of arbitration and other methods for settling all types of dispute.

For these purposes, the AAA was founded in 1926 as a private, not-for-profit organization "to foster the study of arbitration, to perfect its techniques and procedures under arbitration law, and to advance generally the science of arbitration." In line with this purpose, the AAA conducts programs at universities and law schools, bar associations, labor unions, trade associations, and civic organizations of all kinds. Special research projects are also undertaken, often at the request of member organizations, to help adapt arbitration procedures to the needs of particular industries.

Among the AAA's publications are a quarterly magazine, a quarterly letter on legal issues, a quarterly bulletin for members, specialized books, videotapes, and pamphlets on arbitration practice and procedure, and outlines for teaching arbitration and arbitration law courses.

The educational aspects of AAA work are supported by tax-deductible contributions and membership fees. Its membership rolls include companies, labor unions, trade associations, civic groups, foundations, and organizations of all kinds, as well as individuals who believe in arbitration.

There are many advantages in joining the Association. Members of the AAA receive certain regular and special publications, and have access to the AAA's research and educational facilities. For many member organizations, these services are a valuable asset, often making it possible for corporation counsel and other staff members to use their time much more profitably. But, above all, membership makes it possible for the AAA to assist the business community at large with many stand-by services that cannot be paid for on a case-by-case basis. Business people who would like more information about how arbitration serves them, and on how they may participate, are invited to address their inquiries to the AAA's Membership Department.

ATLANTA (30309-3214) • INDIA JOHNSON •
1360 Peachtree Street, NE, Suite 270 • (404) 872-3022/881-1134 (Fax)
BOSTON (02110-2409) • RICHARD M. REILLY •
230 Congress Street • (617) 451-6600/451-0763 (Fax)
CHARLOTTE (28226-8297) • MARK SHOLANDER •
7301 Carmel Executive Park, Suite 110 • (704) 541-1367/542-7287 (Fax)
CHICAGO (60606-1212) • DAVID SCOTT CARFELLO •
205 West Wacker Drive, Suite 1100 • (312) 346-2282/346-0135 (Fax)
CINCINNATI (45202-2809) • PHILIP S. THOMPSON •
441 Vine Street, Suite 3308 • (513) 241-8434/241-8437 (Fax)
CLEVELAND (44115-1632) • EARLE C. BROWN •
1127 Euclid Avenue, Suite 875 • (216) 241-4741/241-8584 (Fax)
DALLAS (75240-6620) • HELMUT O. WOLFF •
Two Galleria Tower, Suite 1440 • (214) 702-8222/490-9008 (Fax)
DENVER (80203-4318) • MARK APPEL •
1775 Sherman Street, Suite 1717 • (303) 831-0823/832-3626 (Fax)
GARDEN CITY, NY (11530-4789) • MARK A. RESNICK •
585 Stewart Avenue, Suite 302 • (516) 222-1660/745-6447 (Fax)
HARTFORD (06106-1943) • KAREN M. JALKUT •
Two Hartford Square West • (203) 278-5000/246-8442 (Fax)
HONOLULU (96813-4728) • KEITH W. HUNTER •
810 Richards Street, Suite 641 • (808) 531-0541/533-2306 (Fax)
HOUSTON (77010-3025) • THERESE TILLEY •
1331 Lamar Street, Suite 1459 • (713) 739-1302
KANSAS CITY, MO (64106-2110) • LORI A. MADDEN •
1101 Walnut Street, Suite 903 • (816) 221-6401/471-5264 (Fax)
LOS ANGELES (90020-0994) • JERROLD L. MURASE •
443 Shatto Place • (213) 383-6516/386-2251 (Fax)
MIAMI (33131-2501) • RENE GRAFALS •
99 SE Fifth Street, Suite 200 • (305) 358-7777/358-4931 (Fax)
MICHIGAN (SOUTHFIELD 48034-7405) • MARY A. BEDIKIAN •
Ten Oak Hollow, Suite 170 • (313) 352-5500/352-3147 (Fax)
MINNEAPOLIS (55402-1092) • JAMES R. DEYE •
514 Nicollet Mall, Suite 670 • (612) 332-6545/342-2334 (Fax)
NASHVILLE (37219-2412) • TONY DALTON •
162 Fourth Avenue North, Suite 106 • (615) 256-5857/256-7819 (Fax)
NEW JERSEY (SOMERSET 08873-4002) • RICHARD NAIMARK •
265 Davidson Avenue, Suite 140 • (201) 560-9560/560-8850 (Fax)
NEW ORLEANS (70130-6101) • DEANN GLADWELL •
650 Poydras Street, Suite 2035 • (504) 522-8781/561-8041 (Fax)
NEW YORK (10020-1203) • CAROLYN M. PENNA •
140 West 51st Street • (212) 484-4000/765-4874 (Fax)
PHILADELPHIA (19102-4121) • JAMES L. MARCHESE •
230 South Broad Street • (215) 732-5260/732-5002 (Fax)
PHOENIX (85012-2803) • DEBORAH A. KRELL •
3033 North Central Avenue, Suite 608 • (602) 234-0950/230-2151 (Fax)
PITTSBURGH (15222-1207) • JOHN F. SCHANO •
Four Gateway Center, Room 221 • (412) 261-3617/261-6055 (Fax)
ST. LOUIS (63101-1614) • NEIL MOLDENHAUER •
One Mercantile Center, Suite 2512 • (314) 621-7175/621-3730 (Fax)
SALT LAKE CITY (84111-3834) • KIMBERLY L. CURTIS •
645 South 200th Street, East, Suite 203 • (801) 531-9748
SAN DIEGO (92101-5278) • DENNIS SHARP •
525 C Street, Suite 400 • (619) 239-3051/239-3807 (Fax)
SAN FRANCISCO (94104-1113) • CHARLES A. COOPER •
417 Montgomery Street • (415) 981-3901/781-8426 (Fax)
SAN JUAN (00918-3628) • LILLIAN A. REYES •
Esquire Building, Suite 800 • (809) 764-8515/764-8848 (Fax)
SEATTLE (98104-1455) • NEAL M. BLACKER •
811 First Avenue, Suite 200 • (206) 622-6435/343-5679 (Fax)
SYRACUSE (13202-1838) • DEBORAH A. BROWN •
State Tower Building, Room 720 • (315) 472-5483/472-0966 (Fax)
WASHINGTON, DC (20036-3169) • GARYLEE COX •
1730 Rhode Island Avenue, NW, Suite 509 • (202) 296-8510/872-9574 (Fax)
WHITE PLAINS, NY (10601-4485) • MARION J. ZINMAN •
34 South Broadway • (914) 946-1119/946-2661 (Fax)

AAA147-5M-11/88

APPENDIX L

AMERICAN ARBITRATION ASSOCIATION, A COMMERCIAL ARBITRATION GUIDE FOR BUSINESS PEOPLE

American
Arbitration
Association

A COMMERCIAL ARBITRATION GUIDE

FOR BUSINESS PEOPLE

Headquarters at
140 West 51st Street, New York, NY 10020-1203

Offices in

Atlanta • Boston • Charlotte • Chicago • Cincinnati • Dallas • Denver • Garden City, NY • Hartford • Honolulu • Houston • Irvine, Calif. • Kansas City, Mo. • Los Angeles • Miami • Middleburg Heights, Ohio • Minneapolis • Nashville • New Orleans • New York • Orlando • Philadelphia • Phoenix • Pittsburgh • Providence • St. Louis • Salt Lake City • San Diego • San Francisco • Seattle • Somerset, NJ • Southfield, Mich. • Syracuse • Washington, DC • White Plains, NY

Tribunals for

Accidents • Alternative Dispute Resolution • Commercial • Computers • Construction • Insurance • International Commerce • Labor • Securities

Introduction

People want to *do* business, not argue about it.

But, in the world of trade and commerce, disputes are inevitable. Parties might disagree as to their individual rights and obligations no matter how carefully a contract is written. This can lead to delayed shipments, complaints about the quality of merchandise, claims of nonperformance of contracts, and similar misunderstandings. And, even with the best of intentions, parties often perform less than they promise.

These controversies seldom involve great legal issues. On the contrary, they concern the same evaluation of facts and interpretation of contract terms that people and their attorneys are accustomed to dealing with every day. Consequently, when differences arise out of day-to-day commercial affairs, parties often prefer to settle them privately and informally, in the kind of businesslike way that encourages continued business relationships. That is what commercial arbitration is for.

Arbitration is the referral of a dispute to one or more impartial persons for final and binding determination. It is private and informal, designed for quick, practical, and economical settlements. Parties can exercise additional control over the arbitration process by adding specific provisions to their contracts' arbitration clause, or, when a dispute arises, by modifying certain of the arbitration rules to suit a particular dispute. Stipulations may be made regarding confidentiality of proprietary information used; evidence, locale, number of arbitrators, and issues subject to arbitration, for example. The parties may also provide for expedited arbitration procedures, including the rendering of the award, if they anticipate the need for hearings to be scheduled on short notice. All such mutual agreements will be binding on the AAA and the arbitrator.

Prior to the initial hearing in a case, the AAA may schedule either an administrative conference with the parties or a preliminary hearing with the arbitrator(s) and the parties to arrange for such matters as the production of relevant documents and the identification of witnesses, and for discussion of and agreement by the parties to any desired rule modifications. AAA administration is guided by the decisions that the parties make about how to handle such sensitive issues as privacy of proceedings, confidentiality, trade secrets, evidence, proprietary information, and injunctive relief.

The rules and procedures for using the Commercial Arbitration Tribunal of the American Arbitration Association are described on the following pages.

Mediation can be used as an alternative to arbitration in appropriate cases. Parties may submit their dispute to an impartial mediator who assists the parties in reaching a settlement but does not have the authority to make a binding decision or award. Submission forms and the Commercial Mediation Rules are available for this purpose from any AAA office. Also, to facilitate an expeditious settlement, the AAA offers parties to a pending arbitration the opportunity to mediate their dispute under its mediation rules with no additional administrative fee.

An AAA Glossary of Dispute Resolution Terms

Some of the commonly used terms follow.

Arbitration submission of a dispute to one or more impartial persons for a final and binding decision

Award An award is the decision of an arbitrator. Awards are made in writing and are enforceable in court under state and federal statutes. Such enforcement actions, when necessary, are brought by the parties to the arbitration.

Case Administrator the AAA staff person assigned to administer cases. This person is responsible for the general management of a particular case, including panel selection, scheduling and exchange of information among the parties, and all of the other administrative details involved in moving cases through the system.

Caucus a meeting in which a mediator talks with the parties individually to discuss the issues

Claimant a filing party, otherwise known as a plaintiff

Counterclaim a counterdemand made by a respondent in his or her favor against a claimant. It is not a mere answer or denial of the claimant's allegation.

Demand for Arbitration the unilateral filing of a claim in arbitration, based on a contractual or statutory right; also, the form used.

Factfinding the process by which parties present the arguments and evidence to a neutral person who then issues a nonbinding report on the findings, usually recommending a basis for settlement

Hearing a proceeding wherein evidence is taken for the purpose of determining the facts of a dispute and reaching a decision based on the evidence.

Mediation a process in which a neutral assists the parties in reaching their own settlement but does not have the authority to make a binding decision

Mediation–Arbitration (Med–Arb) Sometimes, a neutral is selected to serve as both mediator and arbitrator in a dispute. Med–arb combines the voluntary techniques of persuasion and discussion, as in mediation, with an arbitrator's authority to issue a final and binding decision, when necessary.

Minitrial A minitrial is a confidential, nonbinding exchange of information, intended to facilitate settlement. The goal of the minitrial is to encourage prompt, cost-effective resolution of complex litigation. The minitrial seeks to narrow the areas of controversy, dispose of collateral issues, and encourage a fair and equitable settlement.

Negotiation In negotiation, disputants communicate their differences to one another and with this knowledge try to resolve them.

Parties the disputants.

Respondent a responding party, also known as a defendant.

Submission the filing of a dispute to a dispute resolution process after the controversy arises.

How to Initiate Arbitration

The First Step—the Agreement to Arbitrate

The most important step in initiating arbitration is the agreement to arbitrate. This agreement can be of one of two kinds: it could take the form of a *future-dispute arbitration clause* in a contract or, where the parties did not provide in advance for arbitration, it might take the form of a *submission* of an existing dispute to arbitration. The AAA will, without charge, attempt to get all parties to agree to arbitration of such a dispute.

The parties can provide for the arbitration of future disputes by inserting the following clause into their contracts:

Standard Arbitration Clause

Any controversy or claim arising out of or relating to this contract, or the breach thereof, shall be settled by arbitration in accordance with the Commercial Arbitration Rules of the American Arbitration Association, and judgment upon the award rendered by the arbitrator(s) may be entered in any court having jurisdiction thereof.

Arbitration of existing disputes may be accomplished by the use of the following:

> We, the undersigned parties, hereby agree to submit to arbitration under the Commercial Arbitration Rules of the American Arbitration Association the following controversy: (cite briefly). We further agree that the above controversy be submitted to (one)(three) arbitrator(s) selected from the panels of arbitrators of the American Arbitration Association. We further agree that we will faithfully observe this agreement and the rules, and that we will abide by and perform any award rendered by the arbitrator(s) and that a judgment of the court having jurisdiction may be entered upon the award.

Regardless of how the agreement to arbitrate has been reached, notification of the claim to the AAA and the defending party is all that is required to set machinery for arbitration in motion. On receiving the initiating papers together with the appropriate administrative fee, the Association assigns to the case one of its staff members, whose official title is case administrator and who, from that point on, is at the disposal of the parties, expediting administration and assisting both sides in all procedural matters until the award is rendered. Pursuant to the rules, the parties and the AAA may use facsimile transmission, telexes, telegrams, or other written forms of electronic communication to give the notices required by the rules.

> See page 11 for a facsimile of a Demand for Arbitration (to be signed by the demanding party). A Submission to Dispute Resolution (to be signed by *both* parties) is shown on page 9. The American Arbitration Association will supply these forms free of charge on request but arbitration may also be initiated through ordinary correspondence, provided all of the essential information is included.

Special attention is sometimes required to determine in which state and city hearings are to take place. If the place of arbitration has not been designated in the contract or the Submission to Dispute Resolution, or if the parties have not otherwise notified the Association of their agreement on locale, the AAA will designate the city in accordance with its rules. Among the factors considered are

locations of the parties,
locations of witnesses and documents,
locations of sites or places of materials,
relative costs to the parties,
the place of performance of the contract,

A Checklist for Initiating Arbitration

	By Demand for Arbitration	By Submission to Arbitration
Disposition of the Original	Mailed to the Respondent	Filed with the AAA in Duplicate
Copies Needed by the AAA	Three	Two
Copies Retained by the Parties	The demanding party retains one.	Each party retains one.
Signatures Required	An authorized person for the demanding party signs and lists his or her title.	Authorized persons for both parties sign, listing their titles.
Identification of Parties	The responding party should be clearly identified by official name and address.	Official names and addresses of both parties should appear, with signatures and titles.
Contract Clauses	Arbitration clauses should be quoted in full (may be attached separately if more convenient). Include date of the document.	Not Applicable
The Filing Fee	$300 must be advanced by the demanding party. The balance is due in sixty days. The arbitrator later apportions fees. See the schedule on page 17.	The fee may be shared equally. The arbitrator later apportions fees. See the schedule on page 17.
Statement of Dispute	It should be brief but clear and include the amount claimed, if any, and the relief sought.	Claims and answers should be brief but clear and include the amount claimed, if any, and the relief sought.
Answering Statement	The respondent may mail the answering statement to the claimant and file two copies with the AAA.	See above.
Composition of the Arbitration Panel	The AAA will determine the number of arbitrators unless composition is stated in the arbitration clause.	The number of arbitrators desired may be stated. If not stated, the AAA will determine the composition of the panel.
Locale of Arbitration	If not provided for in arbitration clause, the demanding party should indicate its preference.	Locale should be indicated, if possible.

laws applicable to the contract,
places of previous court actions, if any, and
the location of the most appropriate panel of arbitrators.

Hearings may be held in any geographical area, not just where the AAA maintains regional offices.

Expedited procedures, outlined in Sections 53–57 of the rules, are applied in any case where no disclosed claim or counterclaim exceeds $50,000, exclusive of interest and arbitration costs. In sum, these procedures provide for a list of five proposed arbitrators from which each party shall have the right to strike two names on a peremptory basis, notice of arbitrator appointment and hearing by telephone, and the award of the arbitrator to be rendered no later than fourteen days from the date of closing of the hearing.

The Second Step—Selection of the Arbitrator

To serve the business community with arbitrators representing all fields of specialization, the American Arbitration Association now maintains a National Panel of Arbitrators numbering 50,000 individuals throughout the United States and the rest of the world. Usually nominated by leading figures in their industries, trades, or professions, arbitrators are added to the panel after careful checking of qualifications and reputations.

Commercial arbitrators generally contribute their services in the smaller cases that take only a day of their time. In larger, more prolonged cases, the parties will agree to payment of a fee as provided in Section 50 of the rules. When called on by the Association, they serve under its Commercial Arbitration Rules and their conduct is guided by the Code of Ethics for Arbitrators in Commercial Disputes, a copy of which is sent to them upon their appointment to a case. Arbitrators deserve the same respect and courtesy given to all who dedicate themselves to the public good. Parties can show their appreciation to the arbitrators, and at the same time serve their own best interests, by presenting their cases in an expeditious and orderly way, thereby facilitating the task of the arbitrator.

Unless the parties have indicated another method, the AAA follows this simple and effective system for selecting the arbitrator:

1. Upon receiving the a Demand for Arbitration or a Submission to Dispute Resolution, the case administrator sends *each* party a copy of the same specially prepared list of proposed arbitrators to resolve the controversy. A sample list appears on page 13. In drawing up the list, the case administrator is guided by the nature of the dispute. Biographical data on each arbitrator accompany the list.

2. Parties are allowed ten days to study the list, strike all names objected to, and number the remaining names in the order of preference. **Additional information about the proposed arbitrators is available through the administrator. It is up to each party to research the persons suggested.** If the case is being administered under the expedited provisions of the rules, parties are allowed seven days to study the list of five proposed arbitrators, strike two names on a peremptory basis, and number the remaining names in order of preference.
3. When these lists are returned to the Association, the case administrator compares indicated preferences and makes note of the mutual choices. Where parties are unable to find a

American Arbitration Association

SUBMISSION TO DISPUTE RESOLUTION

Date: ____________

The named parties hereby submit the following dispute for resolution under the ____________ ____________ Rules* of the American Arbitration Association:

Procedure Selected: ☐ Binding arbitration ☐ Mediation settlement
☐ Other ____________ (Describe)

FOR INSURANCE CASES ONLY:

____________ Policy Number ____________ Effective Dates to ____________ ____________ Applicable Policy Limits

____________ Date of Incident ____________ Location

Insured: ____________ Claim Number: ____________

Name(s) of Claimant(s)	Check if a Minor	Amount Claimed
____________	☐	____________
____________	☐	____________

Nature of Dispute and/or Injuries Alleged (attach additional sheets if necessary):

Place of Hearing: ____________

We agree that, if binding arbitration is selected, we will abide by and perform any award rendered hereunder and that a judgment may be entered on the award.

To Be Completed by the Parties

____
Name of Party
____
Address
____
City, State, and ZIP Code
( ) ____________
Telephone Fax
____
Signature†

____
Name of Party's Attorney or Representative
____
Address
____
City, State, and ZIP Code
( ) ____________
Telephone Fax
____
Signature†

____
Name of Party
____
Address
____
City, State, and ZIP Code
( ) ____________
Telephone Fax
____
Signature†

____
Name of Party's Attorney or Representative
____
Address
____
City, State, and ZIP Code
( ) ____________
Telephone Fax
____
Signature†

Please file three copies with the AAA.

* *If you have a question as to which rules apply, please contact the AAA.*
† Signatures of all parties are required for arbitration.

Form G1-6/91

mutual choice on a list, additional lists may be submitted at the request of both parties.

4. If the parties cannot agree on an arbitrator, the AAA will make an administrative appointment, but in no case will an arbitrator whose name was crossed out by either party be appointed.

Panels with Party-Appointed Arbitrators

Under some arbitration clauses in use, each party to a dispute appoints one arbitrator (who might or might not be a member of the AAA's National Panel of Arbitrators) and the two select a third arbitrator from the Association's panels in accordance with procedures just described in steps 2–4. To avoid the danger that a compromise award might have to be rendered for the sake of a majority, the parties sometimes provide, and the Association recommends, that the third arbitrator be permitted to render the award alone when a unanimous award is not possible. This may be done by the parties in their agreement to arbitrate or in a later stipulation.

It is recommended that the neutral arbitrator ascertain from the party-appointed arbitrators the nature and extent of any relationship between the arbitrators and the parties that appointed the arbitrators and whether there will be any direct communication between such arbitrators and the parties that appointed them.

The Third Step—Preparation for the Hearing

The case administrator consults all parties and arbitrators to determine a mutually convenient day and time for the hearing. If the parties cannot agree, the arbitrator is empowered to set dates.

> ***Note:* In this, as in all other administrative matters, the case administrator manages details and arrangements. This has a twofold advantage: it relieves the arbitrator of the burden and eliminates the necessity of direct communication between the parties and the arbitrator except at the hearing. By specifically forbidding communication with the arbitrator, except in the presence of both parties, AAA rules avoid the danger that one side will offer arguments or evidence that the other has no opportunity to rebut.**

At the request of any party or at the discretion of the AAA, an administrative conference with the AAA and the parties and/or their representatives will be scheduled in appropriate cases to expedite the arbitration proceedings.

In large or complex cases, at the request of any party or at the discretion of the arbitrator or the AAA, a preliminary hearing with the parties and/or their representatives and the arbitrator may be scheduled by the arbitrator to specify the issues to be resolved, to stipulate to uncontested facts, and to consider any other matter that will expedite the arbitration proceedings. Consistent with the expedited nature of arbitration, the arbitrator may, at the preliminary hearing, establish (i) the extent of and a schedule for the production of relevant documents and other information, (ii) the identification of all witnesses to be called, and (iii) a schedule for further hearings to resolve the dispute.

American Arbitration Association

MEDIATION Please consult the Commercial Mediation Rules regarding mediation procedures. If you want the AAA to contact the other party and attempt to arrange a mediation, please check this box. ☐

COMMERCIAL ARBITRATION RULES
DEMAND FOR ARBITRATION

DATE: ____________

TO: Name ____________
(of the party upon whom the demand is made)

Address ____________

City and State ____________ ZIP Code ____________

Telephone ( ) ____________ Fax ____________

Name of Representative ____________
(if known)

Representative's Address ____________

City and State ____________ ZIP Code ____________

Telephone ( ) ____________ Fax ____________

The named claimant, a party to an arbitration agreement contained in a written contract, dated ________ ____________, providing for arbitration under the Commercial Arbitration Rules, hereby demands arbitration thereunder.

(Attach the arbitration clause or quote it hereunder.)

NATURE OF DISPUTE:

CLAIM OR RELIEF SOUGHT (amount, if any):

TYPE OF BUSINESS: Claimant ____________ Respondent ____________

HEARING LOCALE REQUESTED: ____________
(City and State)

You are hereby notified that copies of our arbitration agreement and of this demand are being filed with the American Arbitration Association at its ____________ office, with the request that it commence the administration of the arbitration. Under the rules, you may file an answering statement within ten days after notice from the administrator.

Signed ____________ Title ____________
(may be signed by a representative)

Name of Claimant ____________

Address (to be used in connection with this case) ____________

City and State ____________ ZIP Code ____________

Telephone ( ) ____________ Fax ____________

Name of Representative ____________

Representative's Address ____________

City and State ____________ ZIP Code ____________

Telephone ( ) ____________ Fax ____________

To institute proceedings, please send three copies of this demand with the administrative fee, as provided for in the rules, to the AAA. Send the original demand to the respondent.

Form C2-11/90

11

Since the arbitrator will make the award on the basis of the facts and exhibits presented at the hearing, it is essential that the parties or their representatives prepare for arbitration carefully.

1. Assemble all documents and papers that you will need at the hearing. Always make photocopies for the arbitrator and the other party. If documents that are needed are in the possession of the other party, ask that they be brought to the arbitration. Under some state arbitration laws, the arbitrator or another person has authority to subpoena documents and witnesses. A checklist of documents and exhibits will be helpful toward your orderly presentation.
2. If it will be necessary for the arbitrator to visit a building site or warehouse for on-the-spot investigation, make plans in advance. The arbitrator will have to be accompanied by representatives of both parties, unless they specifically authorize that the investigation be conducted without their presence or unless one party fails to attend after being notified.
3. Interview all of your witnesses. Make certain that each one understands the whole case and particularly the importance of his or her own testimony within it.
4. If there is a possibility that others, not on your regular list of witnesses, might have to appear, alert them to be available on call without delay.
5. Make a written summary of what each witness will prove. This will be useful as a checklist at the hearing and will help make sure that nothing is overlooked.
6. Study the case from the other side's point of view. Be prepared to answer the opposition's evidence.
7. If a transcript of the hearing is needed, the parties are responsible for making the arrangements and notifying the other parties of such arrangements in advance of the hearing.

The right to representation in arbitration by counsel or another authorized person is guaranteed by the rules of the American Arbitration Association. A party who desires to be represented should notify the other side and file a copy of the notice with the case administrator at least three days before the hearing. When arbitration is initiated by a representative or when the respondent replies through a representative, however, such notice is deemed to have been given.

The Fourth Step—Presentation of the Case

Arbitration hearings are conducted somewhat like court trials, except that arbitrations are less formal. Arbitrators are not required to follow strict rules of evidence. They must hear all of the evidence material to an issue but they may determine for themselves what is relevant. Arbitrators are therefore inclined to accept evidence that might not be allowed by judges. This does not mean, however, that all evidence will be considered of equal weight.

> *Direct testimony of witnesses is usually more persuasive than hearsay evidence, and facts will be better established by documents and exhibits than by argument only.*

It is customary for the claimant to proceed first with its case, followed by the respondent. This order may be varied, however, when the arbitrator thinks it necessary. In any event, the "burden

American Arbitration Association

Do not send this form to the other party.

CASE NUMBER: ____________ DATE LIST SUBMITTED: ____________

PARTIES: ____________ AND ____________

LIST FOR SELECTION OF ARBITRATOR(S)

After striking the name of any unacceptable arbitrator(s), please indicate your order of preference by number. We will try to appoint arbitrator(s) mutually acceptable who can hear your case promptly. Leave as many names as possible.

NOTE: Biographical information is attached. Unless your response is received by the Association by ____________, all names submitted may be deemed acceptable.

REQUEST FOR DATES

To enable the arbitrator(s) to avoid fixing an inconvenient hearing date, please cross off the dates that are not acceptable for a hearing, but leave as many days as possible so that the first mutually agreeable date for hearing may be set. If a mutually agreeable date is not available for these two months, the arbitrator(s) is/are empowered under the rules to fix the time and place for each hearing. If this form is not returned by ____________, it will be assumed that any open date is satisfactory to you. The hearing will then be scheduled for a date preferred by the other party.

NOTE: Saturdays, Sundays, and other unavailable days have been marked off.

Month of ____________

1	2	3	4	5	6	7
8	9	10	11	12	13	14
15	16	17	18	19	20	21
22	23	24	25	26	27	28
29	30	31				

Month of ____________

1	2	3	4	5	6	7
8	9	10	11	12	13	14
15	16	17	18	19	20	21
22	23	24	25	26	27	28
29	30	31				

Please note that hearings generally commence at ____________ a.m./p.m.

I anticipate that my case will require ____________ hours/days of hearing.

PARTY: ____________

BY: ____________ TITLE: ____________

Your telephone or fax response to this inquiry will be appreciated and will expedite administration. Please refer to the telephone and fax numbers on the enclosed letterhead.

Form G4-4/89

of proof'' is not on one side more than the other; each party must try to convince the arbitrator of the correctness of its position and no hearing is closed until both have had a full opportunity to do so.

That is why it is equally the responsibility of the claimant and the respondent to present their cases to the arbitrator in an orderly and logical manner. This includes:

1. An opening statement that clearly but briefly describes the controversy and indicates what is to be proved. Such a statement iays the groundwork and helps the arbitrator understand the relevance of testimony to be presented.
2. A discussion of the remedy sought. This is important because the arbitrator's power is conferred by the agreement of the parties. Each party should try to show that the relief that it requests is within the arbitrator's authority to grant.
3. Introduction of witnesses in a systematic order to clarify the nature of the controversy and to identify documents and exhibits. Cross examination of witnesses is important, but each party should plan to establish its case by its *own* witnesses.
4. A closing statement that should include a summary of the evidence and arguments and a refutation of points made by the opposition.

Above all, a cooperative attitude is essential for effective arbitration. Overemphasis or exaggeration, concealing of facts, introduction of legal technicalities with the objective of delaying proceedings, or, in general, disregard of ordinary rules of courtesy and decorum can have an adverse effect on arbitrators.

After both sides have had an equal opportunity to present all of their evidence, the arbitrator declares the hearing closed. Under AAA rules, the arbitrator has thirty days from that time within which to render an award, unless the agreement provides otherwise. If the case was administered under the expedited provisions in the rules, the arbitrator has fourteen days within which to render an award.

The Award

The award is the decision of the arbitrator on the matters submitted to him or her under the arbitration agreement. If the arbitration panel consists of more than one arbitrator, the majority decision, under AAA rules, is binding. The purpose of the award is to dispose of the controversy finally and conclusively. It is made within the limits of the arbitration agreement and it rules on each claim

Procedure for Oral Hearings

	Who Decides	Who Makes Arrangements	Notice
Time	The arbitrator, at the convenience of the parties	The case administrator, who consults the parties and the arbitrator	At least ten days, given by the case administrator unless parties agree otherwise
Representation by Counsel	The individual party	The individual party	Three days' notice to other party unless arbitration was initiated by counsel, in which case notice is deemed to have been given
Stenographic Records and Interpreters	The requesting party	The requesting party	The requesting party notifies the other party in advance of the hearing and may inquire of the other side as to whether it would like to share the cost and get a copy of the record.
Attendance at Hearing	Parties attend and bring witnesses. Arbitrators decide which other interested persons may attend and may require withdrawal of witnesses during the testimony of others.	Parties arrange for attendance of witnesses.	Parties notify their own interested persons.
Affidavits and Documents	The arbitrator decides whether to receive such evidence when it is presented.	Each party arranges to submit its own documents. If they are in the possession of the other party, documents may be requested directly.	None is required.
Subpoenas of Witnesses and Documents	The arbitrator issues subpoenas on showing of need by a party. In New York State, attorneys of record may also issue subpoenas.	The case administrator obtains signature of arbitrator for subpoena supplied by party and returns subpoena to party for service.	Subpoenas are served by parties directly on the witness or the custodian of documents.
Inspection or Investigation	The arbitrator may decide on his or her own initiative or at the request of a party, if the arbitrator deems it necessary.	The case administrator.	Parties are notified of time and place of inspection so that they can be present.
Closing of Oral Hearings	The arbitrator closes hearing after both sides complete proofs and witnesses. If briefs, investigations, or more data are required, the hearings are kept open.	The case administrator arranges for receipt of posthearing matters and makes a record of the closing of hearings on instructions from the arbitrator.	The case administrator notifies parties of all official closing dates.

submitted. Arbitrators are not required to write opinions explaining the reasons for their decisions. As a general rule, AAA commercial awards consist of a brief direction to the parties on a single sheet of paper. Written opinions can generate attacks on the award because they identify targets for the losing party. In some cases, both parties will request an opinion or the arbitration agreement provides for one. The AAA then has no objection. Usually, however, the parties look to the arbitrator for a decision, not an explanation.

> The power of the arbitrator ends with the making of the award. An award may not be changed by the arbitrator, once it is made, *unless the parties agree to restore the power of the arbitrator or unless the law provides otherwise.*

When the parties agree to request a clarification or interpretation of a disputed ruling, the agreement must be in writing. Such an agreement is filed with the Association, which then proceeds to make the necessary arrangements with the arbitrator. In some jurisdictions, the law permits arbitrators to clarify or modify the award upon the request of a party. The administrator will provide copies of the state arbitration law upon request.

Large, Complex Cases

Recognizing that large, complex commercial arbitration cases often present unique procedural problems, the AAA, working with attorneys, arbitrators, and industry advisory groups, developed guidelines for expediting such cases. These guidelines are applied when the disclosed claim of any party exceeds $250,000, unless there is an objection. Parties may agree to use the guidelines in smaller cases as well. The guidelines provide for an early administrative conference and a preliminary hearing before the arbitrators. Arbitrators generally serve on a consecutive-day basis. Under the guidelines, the AAA will make the necessary arrangements to have party-appointed arbitrators serve as neutrals, thereby avoiding problems that have arisen in cases using partisan arbitrators. The guidelines contemplate that adjournments will be granted only for good cause. If desired by the parties, the AAA will make arrangements for the arbitrators to write an opinion.

International Cases

In order to serve the parties in international arbitrations best, the AAA devised the Supplementary Procedures for International

Commercial Arbitration, which may be used in conjunction with various sets of arbitration rules. These procedures do not supersede any provision in the applicable rules but merely codify various procedures that are used in international arbitrations. Among the more interesting features are provisions governing consecutive hearing days, language of the hearings, and opinions. The thrust of the procedures is to expedite international commercial arbitrations and to keep them as economical as possible. In a case involving a panel of U.S. nonnationals, for instance, the AAA attempts to appoint resident foreign nationals in order to minimize travel expenses. Pursuant to the Securities Arbitration Rules, a request for a foreign-national arbitrator must be made by the time set for the appointment of the arbitrator as agreed by the parties or set by the rules. In March, 1991, the AAA also promulgated International Arbitration Rules.

Costs

Administrative Fees

A filing fee of $300 will be paid when a case is filed. The balance of the administrative fee is based on the amount of each claim or counterclaim as disclosed when the claim or counterclaim is filed. This balance is due and payable sixty days after the AAA's commencement of administration, or prior to the date of the first hearing, whichever occurs first. If at any time a claim or counterclaim is settled or withdrawn, the balance of the administrative fee remains due and the Refund Schedule shall then be applied. When oral hearings are waived under Section 37, the Administrative Fee Schedule shall still apply.

Amount of Claim or Counterclaim	Fee	
Up to $10,000	$300	
$10,000 to $25,000	3%	
$25,000 to $50,000	$750,	plus 2% of excess over $25,000
$50,000 to $100,000	$1,250,	plus 1% of excess over $50,000
$100,000 to $500,000	$1,750,	plus ½% of excess over $100,000
$500,000 to $5,000,000	$3,750,	plus ¼% of excess over $500,000
$5,000,000 to $50,000,000	$15,000,	plus 1/10% of excess over $5,000,000

Where the claim or counterclaim exceeds $50 million, there is no additional administrative fee. When no amount can be stated at the time of filing, the administrative fee is $1,000, subject to adjustment in accordance with the above schedule as soon as an amount can be disclosed. An appropriate administrative fee will be determined by the AAA for claims and counterclaims that are not for a monetary amount.

If there are more than two parties represented in the arbitration, an additional 10% of the administrative fee will be due for each additional represented party. The minimum administrative fee for a case heard by three arbitrators is $1,500, payable by the party requesting same.

Postponement Fees

Sole-Arbitrator Cases

$100 is payable by a party causing its first postponement of any scheduled hearing. $200 is payable by a party causing its second or subsequent postponement of any scheduled hearing.

Three-Arbitrator Cases

$150 is payable by a party causing its first postponement of any scheduled hearing. $300 is payable by a party causing its second or subsequent postponement of any scheduled hearing.

Additional Hearing Fee

$75 is payable by each party for each hearing after the first hearing that is clerked by the AAA.

Hearing-Room Rental

Hearing rooms for second and subsequent hearings are available on a rental basis at AAA offices. Check with your local office for specific availability and rates.

Refund Schedule

The Refund Schedule is based on the administrative fee due on a claim or counterclaim asserted by a party. If the AAA is notified that a claim or counterclaim has been settled or withdrawn before a list of arbitrators has been sent out, all of the fee in excess of $300 will be refunded. If the AAA is notified that a claim or counterclaim has been settled or withdrawn after a list of arbitrators has been sent out but before the original due date for the return of the first list, two thirds of the fee in excess of $300 will be refunded. If the AAA is notified that a claim or counterclaim has been settled or withdrawn after the original due date for the return of the first list but at least two business days before the initial date and time set for the first scheduled hearing, one third of the fee in excess of $300 will be refunded.

There will be no refund after any hearing or mediation conference has been held; where a claim or counterclaim was filed as an undetermined/undisclosed claim and remained so at the time of settlement or withdrawal; where a consent award was issued by the arbitrator; or where a determination is made by the arbitrator resulting in closing of the file.

The American Arbitration Association

The AAA provides services in administration of arbitration, educational programs, and publications as well as research into the uses of arbitration for settling all types of dispute.

The educational aspects of Association work are supported by tax-deductible contributions and membership fees. Its membership rolls include companies, labor unions, trade associations, civic groups, foundations, and organizations of all kinds, as well as individuals who believe in arbitration. It is inherent in the impartial nature of the Association that in the conduct of arbitration members and nonmembers are treated equally; no advantage accrues to any party from membership in the AAA, *insofar as case administration is concerned.*

Members of the AAA receive publications in their area of practice and have access to the Association's research and educational facilities. Business people who would like more information about how they may participate are invited to address their inquiries to the AAA's Membership Department.

AAA Offices

Atlanta (30309–3214) • India Johnson
1360 Peachtree Street, NE, Suite 270 • (404) 872-3022/881-1134 (Fax)
Boston (02110–1703) • Richard M. Reilly
133 Federal Street • (617) 451-6600/451-0763 (Fax)
Charlotte (28202–2431) • Neil Carmichael
428 East Fourth Street, Suite 300 • (704) 347-0200/347-2804 (Fax)
Chicago (60606–1212) • David Scott Carfello
205 West Wacker Drive, Suite 1100 • (312) 346-2282/346-0135 (Fax)
Cincinnati (45202–2809) • Philip S. Thompson
441 Vine Street, Suite 3308 • (513) 241-8434/241-8437 (Fax)
Cleveland (Middleburg Heights 44130–3490) • Audrey Mendenhall
17900 Jefferson Road, Suite 101 • (216) 891-4741/891-4740 (Fax)
Dallas (75240–6620) • Helmut O. Wolff
Two Galleria Tower, Suite 1440 • (214) 702-8222/490-9008 (Fax)
Denver (80264–2101) • Mark Appel
1660 Lincoln Street, Suite 2150 • (303) 831-0823/832-3626 (Fax)
Garden City, NY (11530–4789) • Mark A. Resnick
585 Stewart Avenue, Suite 302 • (516) 222-1660/745-6447 (Fax)
Hartford (06106–1943) • Karen M. Jalkut
Two Hartford Square West • (203) 278-5000/246-8442 (Fax)
Honolulu (96813–4714) • Keith W. Hunter
810 Richards Street, Suite 641 • (808) 531-0541/533-2306 (Fax)
In Guam: (671) 477-1845/477-3178 (Fax)
Houston (77002–6707) • Therese Tilley
1001 Fannin Street, Suite 1317 • (713) 739-1302/739-1702 (Fax)
Kansas City, MO (64106–2110) • Lori A. Madden
1101 Walnut Street, Suite 903 • (816) 221-6401/471-5264 (Fax)
Los Angeles (90020–0994) • Jerrold L. Murase
443 Shatto Place • (213) 383-6516/386-2251 (Fax)
Miami (33131–2501) • René Grafals
99 SE Fifth Street, Suite 200 • (305) 358-7777/358-4931 (Fax)
Michigan (Southfield 48034–7405) • Mary A. Bedikian
Ten Oak Hollow Street, Suite 170 • (313) 352-5500/352-3147 (Fax)
Minneapolis (55402–1092) • James R. Deye
514 Nicollet Mall, Suite 670 • (612) 332-6545/342-2334 (Fax)
Nashville (37219–2111) • Tony Dalton
221 Fourth Avenue North • (615) 256-5857/244-8570 (Fax)
New Jersey (Somerset 08873–4120) • Richard Naimark
265 Davidson Avenue, Suite 140 • (908) 560-9560/560-8850 (Fax)
New Orleans (70130–6101) • Deann Gladwell
650 Poydras Street, Suite 1535 • (504) 522-8781/561-8041 (Fax)
New York (10020–1203) • Carolyn M. Penna
140 West 51st Street • (212) 484-4000/307-4387 (Fax)
Orange County, CA (Irvine 92714–6220) • Lori S. Markowicz
2601 Main Street, Suite 240 • (714) 474-5090/474-5087 (Fax)
Orlando (32801–2742) • Mark Sholander
201 East Pine Street, Suite 800 • (407) 648-1185/649-8668 (Fax)
Philadelphia (19102–4106) • Kenneth Egger
230 South Broad Street, Floor 6 • (215) 732-5260/732-5002 (Fax)
Phoenix (85012–2365) • Deborah A. Krell
333 E. Osborn Road, Suite 310 • (602) 234-0950/230-2151 (Fax)
Pittsburgh (15222–1207) • John F. Schano
Four Gateway Center, Room 419 • (412) 261-3617/261-6055 (Fax)
Providence (02903) • Mark Bayliss
115 Cedar Street • (401) 453-3250/453-6194 (Fax)
St. Louis (63101–1614) • Neil Moldenhauer
One Mercantile Center, Suite 2512 • (314) 621-7175/621-3730 (Fax)
Salt Lake City (84111–3834) • Kimberly L. Curtis
645 South 200 East, Suite 203 • (801) 531-9748/531-0660 (Fax)
San Diego (92101–5278) • Dennis Sharp
525 C Street, Suite 400 • (619) 239-3051/239-3807 (Fax)
San Francisco (94104–1113) • Charles A. Cooper
417 Montgomery Street • (415) 981-3901/781-8426 (Fax)
Seattle (98101–2511) • Neal M. Blacker
1325 Fourth Avenue, Suite 1414 • (206) 622-6435/343-5679 (Fax)
Syracuse (13202–1376) • Deborah A. Brown
205 South Salina Street • (315) 472-5483/472-0966 (Fax)
Washington, DC (20036–4104) • Garylee Cox
1150 Connecticut Avenue, NW, 6th Floor • (202) 296-8510/872-9574 (Fax)
White Plains, NY (10601–4485) • Marion J. Zinman
34 South Broadway • (914) 946-1119/946-2661 (Fax)

AAA35-20M-8/91

APPENDIX M

DISPUTE RESOLUTION PROCEDURES IN THE NORTHERN DISTRICT OF CALIFORNIA

DISPUTE RESOLUTION PROCEDURES IN THE NORTHERN DISTRICT OF CALIFORNIA

UNITED STATES DISTRICT COURT

Table of Contents

Dispute Resolution Procedures Available in the Northern District

A Introduction

The United States District Court for the Northern District of California provides in this pamphlet an overview of court-sponsored processes which may result in faster and less costly civil dispute resolution than full-scale, formal litigation. The court urges counsel to share this material with their clients so that together they can explore the full range of dispute resolution and case management methods available.

Traditional litigation may impose substantial costs long before the trial commences. Further, the current congestion of trial calendars in federal courts, caused in part by the substantial criminal docket, contributes significantly to that congestion. In this environment, alternative processes for dispute resolution offer many advantages, including the following:

- Reduced time to disposition;
- Streamlined and less costly discovery;
- More effective case management (for pre- and post-trial matters);
- Increased confidentiality;
- Facilitation of early, direct communication and understanding among the parties of the essential issues on each side of the dispute;
- Preservation of ongoing party relations;
- Savings in trial expenses;
- Providing qualified, neutral experts to hear complex matters (with the parties having a role in selecting them under some processes).

While this pamphlet describes only the programs that the court sponsors, there are other dispute resolution procedures that have been developed in the private sector and that have proven effective in a wide range of cases. A list of some of the private organizations and individuals who provide alternative dispute resolution services is available from the clerk's office in San Francisco or San Jose.

Even though two of the court-sponsored options (court-annexed arbitration and early neutral evaluation) are mandatory for cases which fit the requisite criteria, the parties to *any* type of case may *voluntarily* refer their matter to any of the court programs. Many of these programs have been designed by members of the private bar. They have been endorsed as cost-effective and fair by the vast majority of lawyers and clients whose cases have been submitted to them. All the court-sponsored procedures except consent trials before magistrates are *not binding*, meaning that if they fail to resolve the case, the parties retain their full rights to a trial, with no penalties. Of course, if the parties agree, all these procedures can be binding and appeals can be waived.

Parties and counsel who are considering voluntary participation in alternative dispute resoluton should weigh the comparative benefits of each dispute resolution process before deciding which is most appropriate for their particular case. For example, court-annexed arbitration can be effective for many contract and tort cases, especially when outcome is likely to turn on credibility of witnesses. In contrast, the court appointment of a special master may be warranted for technical or complicated matters involving multiple parties and substantially greater sums. With the consent of the parties, a United States Magistrate may preside over any civil trial, jury or non-jury, in the same manner as an Article III judge, and can usually hear the case much sooner as a result of a lighter trial calendar. Some factors parties should consider in deciding which option to pursue include:

- The cost of the process relative to the amount of the dispute involved;
- The importance of prompt resolution;
- Whether an informal or formal proceeding would be beneficial;
- Whether privacy interests need special protection;
- What role the parties want the third-party neutral to play;
- Whether binding or non-binding options should be pursued;
- The number of issues and parties involved in the dispute;
- Whether a transcript of the proceeding should be made; and
- The desirability of establishing precedent.

B Court-Sponsored Dispute Resolution Processes

The Northern District of California sponsors six special procedures to facilitate resolution of disputes:

I. Early Neutral Evaluation;
II. Court-Annexed Arbitration;
III. Consensual Jury or Court Trial before a United States Magistrate;
IV. Settlement Conferences;
V. Non-binding Summary Jury or Bench Trials; and
VI. Special Masters.

These programs are briefly described below. The referenced authorities and contacts, at the end of each subsection, may be consulted for additional information.

I

Early Neutral Evaluation

The Early Neutral Evaluation (ENE) program offers a *confidential, non-binding* conference where the parties (face-to-face) and their counsel present the factual and legal bases of their case to one another and to an experienced and impartial attorney with expertise in the subject matter of the case. In a two-hour informal session, held within 150 days after the complaint is filed, the neutral evaluator hears both sides. The evaluator then identifies the primary issues in dispute, as well as areas of agreement, explores the possibility of settlement (if the parties desire), helps the parties devise a discovery or motion plan, articulates an assessment of the relative strengths and weaknesses of the parties' positions and the value of the case, and discusses whether a follow-up session would be fruitful.

ENE has been endorsed by many lawyers for a number of reasons. First, the program provides *early* case evaluation by a neutral lawyer with expertise in the relevant subject matter. This early expert assessment may result in significant cost savings and may lead to settlement since the parties are compelled to develop an early understanding of the case and of the other side's position. Second, ENE may be a cost-effective substitute for some formal discovery and pretrial motions. It enables parties to communicate and learn more directly and productively about their case than they would in formal litigation. Third, confidentiality is maintained. The judge to whom the action is assigned does not learn any information communicated in the course of the ENE session by any of the participants (including the evaluator). Additionally, the evaluation session is informal; the Federal Rules of Evidence do not apply and there is no direct or cross-examination of witnesses. Thus an effective ENE session usually results in clarification of issues and the development of a case management plan. For these reasons, past ENE participants have highly praised the program.

While certain categories of cases are compelled to participate in ENE and court-annexed arbitration, litigants in other categories of cases may stipulate to participate in either program. Parties trying to decide between programs should consider several factors. For complex cases, or for matters that are not based on straightforward contract or tort theories, ENE offers the advantage of assuring that the neutral advisor is an expert in the relevant subject matter. Compared to arbitration under Local Rule 500, ENE also offers greater assurance of confidentiality, operates on a faster track, and may be less expensive. ENE might be appropriate even in cases where the principal relief sought is equitable if there is a reasonable chance that, with the aid of a neutral expert, the parties might be able to agree on the terms of an injunction or consent decree. Unlike court-annexed arbitration, however, transcripts may not be made of ENE proceedings and there is less opportunity to assess the relative credibility of key witnesses than in arbitration.

Subject Matter Scope: After two years experience, the court has concluded that the following cases are likely to benefit most from ENE:

> Contract (including business contracts, insurance coverage, Miller Act, negotiable instrument, stockholders suits, and contract product liability); Torts (including motor vehicle, motor vehicle product liability, personal injury, personal injury — product liability, and fraud); Civil

Rights (employment); Intellectual Property; Antitrust; Racketeer Influenced and Corrupt Organizations; and Securities/Commodities Exchange.

It is clear, however, that other cases not falling within these subject matters may also profit from ENE.

ENE is not compelled (but may be available upon consent) for *in propria persona* cases, actions where the primary relief is equitable, and matters submitted to court-annexed arbitration under Local Rule 500.

Invocation: Presently, every even-numbered case which falls within the subject matter classifications set forth above (and in which the principal relief sought is not equitable) is automatically referred to ENE by the clerk of the court.

In addition, civil actions may be assigned to ENE on motion by a party and approval by the court or *sua sponte* by the judge to whom the action is assigned.

Right of Appeal: ENE is non-binding and confidential; therefore, there is no occasion for an appeal from the results of an ENE conference.

Requests for relief from the requirements of ENE must be made in writing and must be presented in the first instance to the ENE Magistrate. Appeals from his rulings must be filed with the assigned judge within ten calendar days.

Authority: Northern District of California, General Order No. 26.

Contact: For further information, call the clerk's office: (415) 556-5742.

II

Court-Annexed Arbitration

The Northern District of California was one of the first federal courts in the nation to experiment with court-annexed arbitration, beginning in 1978. Based on the success of that pilot program and recent congressional authorization, today Local Rule 500 establishes a *compulsory, non-binding* arbitration program primarily for contract or tort cases involving not more than $150,000 (exclusive of punitive or exemplary damages, interest and costs). Local Rule 500-8 also permits *voluntary*, non-binding arbitration for cases involving any amount in controversy or subject matter by stipulation of the parties and permission of the judge assigned the action. Court-annexed arbitration is distinct from private sector arbitration, which is usually voluntary and binding. The parties may stipulate, however, that the award in court-annexed arbitration shall be final and binding.

More than 80% of the lawyers who have had cases in the court's arbitration track endorse this program. Ninety-three percent (93%) agree that the arbitration procedures are fair, and when the variables of cost, time, and fairness are considered, arbitration is preferred by more lawyers than either a court trial or a jury trial. *See* B. Meierhoefer & C. Seron, *Court-Annexed Arbitration in the Northern District of California* 21-40 (1988) (Federal Judicial Center publication).

The cases referred to court-annexed arbitration are heard by qualified individual arbitrators or three-member panels, usually within six months of the filing of the answer. Attendance at the arbitration hearing and the production of documents may be compelled by subpoena. Testimony is given under oath and witnesses may be cross-examined. The Federal Rules of Evidence serve as guidelines but are not rigidly enforced. A party may arrange to have the proceedings reported and transcribed at his or her own expense.

To obtain a trial *de novo*, a party must file a written demand within thirty days of entry of judgment on the arbitration award. If no such demand is filed, the award becomes the final judgment of the court (and is not subject to appellate review). Absent a stipulation, the amount of any arbitration award would not be admissible at a subsequent trial *de novo.*

Numerous benefits can be obtained from court-annexed arbitration. Because arbitration is held relatively early in the case, parties can save time and money. The process is fair, as a neutral assessment of the case is made by a qualified, impartial attorney or by a panel of three attorneys. Parties help select the arbitrators from the list of ten names provided by the clerk's office by striking some of the nominees and listing those remaining in the parties' order of preference. An absolute right to a full trial is preserved. There is no penalty for demanding trial *de novo* or for failing to obtain a judgment at trial that is more favorable than the arbitration award. The arbitration award also may provide a useful starting point for settlement discussions. As important, the imminence of the arbitration hearing often encourages counsel to evaluate their case carefully and to enter serious settlement negotiations before the arbitration is held. In fact, most cases that are assigned to the arbitration track settle before the hearing is held.

For parties whose cases have not been ordered into ENE or arbitration, and who are trying to decide whether to participate voluntarily in one of the court's programs, court-annexed arbitration may have an advantage over ENE or a settlement conference when the case turns on credibility of witnesses. At an arbitration hearing, the witnesses testify under oath and are subject to cross-examination. However, arbitration also has disadvantages in comparison to ENE or an early settlement conference: the arbitration hearing may occur later in the pretrial stage, it may be more formal (and therefore potentially more expensive), and provide less protection of confidentiality. Further, arbitration may lead to early disclosure of trial strategies. Court-annexed arbitration is also generally inappropriate when the principal relief sought is equitable.

Subject Matter Scope: Compelled in cases (1) that involve an amount in controversy not exceeding $150,000 (excluding punitive or exemplary damages, interest and costs) *and,* (2) that are based on personal injury, property damage, or contract. Consult Local Rule 500-2 for the particulars.

Court-annexed arbitration is *not* compelled in actions where the principal relief sought is equitable (except that declaratory relief actions in insurance coverage cases are sent to arbitration where the amount in controversy does not exceed $150,000).

Invocation: Automatic referral by the clerk of court where the subject matter criteria (noted above) are satisfied, or by stipulation of the parties and approval of the court.

Right of Appeal: Right to *de novo* trial if demanded within thirty days of entry of judgment on arbitration award. If timely demand for trial *de novo* is not made, the arbitrator's award becomes the final judgement and is not reviewable by appeal.

Authority: Local Rule 500 and the Judicial Improvements and Access to Justice Act of 1988, Title IX, Pub. L. No. 100-702, §§ 901-907 (codified at 28 U.S.C. §§ 651-658).

Contact: Courtroom deputy clerk for the judge to whom the action is assigned.

III

Consent to Jury or Court Trial Before a Magistrate

By written stipulation, the parties to any civil action may elect to have a magistrate of their choice (instead of the assigned Article III judge) conduct all proceedings in any civil case, including presiding over a jury or a non-jury trial. A trial before a magistrate is governed by exactly the same procedural and evidentiary rules as trial before a district judge, and a right of appeal is automatically preserved directly to the United States Court of Appeals under the same standards which govern appeals from district court judgments.

Parties often consent to resolution of their civil disputes by magistrate *bench* or *jury* trial because: (1) magistrates have less crowded calendars and do not set multiple cases for trial on the same date (as many judges do), thus usually permitting the parties to secure an *earlier* and *firm* trial date; (2) the parties may select any available magistrate within the Northern District of California; (3) jurisdiction extends to any civil matter within the jurisdiction of the federal courts; (4) a formal trial is conducted under the Federal Rules of Evidence and the Federal Rules of Civil Procedure; (5) a full right of appeal is retained; and (6) a transcript of the proceedings may be taken by court reporter or, if the parties prefer, by electronic sound recording.

Subject Matter Scope: Any civil proceeding.

Invocation: Written consent of the parties indicated on a form available from the clerk's office. 28 U.S.C. § 636(c)(2).

The parties may also discuss reference to a magistrate at pretrial conference. *See* Fed. R. Civ. P. 16(c)(6).

Right of Appeal: There is a direct appeal to the U.S. Court of Appeals. *See* 28 U.S.C. § 636(c)(3); Fed. R. Civ. P. 73(c). As an alternative, the parties may choose to preserve a right to appeal to the U.S. District Court. *See* 28 U.S.C. § 636(c)(4); Fed. R. Civ. P. 73(d) & 74-76. *See also* Local Rule 410-2(c).

Authority: 28 U.S.C. § 636(c); Fed. R. Civ. P. 16(c)(6) & 73-76; Local Rules 405(k) & 410-2(c).

Contact: Courtroom deputy clerk for the judge to whom the action is assigned or any magistrate.

IV

Settlement Conferences Conducted by a Judge or Magistrate

A judicially conducted settlement conference may be held at any time during the pendency of a civil case. Most settlement conferences are conducted by magistrates, but district judges also help with settlement negotiations as their calendars permit. Under Local Rule 240-1, the judge or magistrate who would preside at trial does not conduct the settlement conference unless the parties request in writing that he or she do so. Normally, the settlement process is initiated upon the request of a party or on the motion of the judge to whom the action is assigned. However, the parties should be prepared to discuss the possibility of settlement at any pretrial conference, including the initial status conference. A court-appointed special master (discussed in the next subsection) may also be used for settlement purposes.

Counsel who attend the settlement conference are required to be thoroughly familiar with the case and have authority to negotiate a settlement. Many judges have standing orders that require the attendance of the parties unless they reside more than fifty miles from the settlement conference site, in which case they are required to be available by telephone for consultation with the judge or magistrate.

The settlement judge or magistrate acts as a mediator or facilitator at the settlement conference, promoting communication among the parties, holding one-on-one sessions with each side, offering an objective assessment of the case, and suggesting settlement options. A magistrate or judge has no power to impose settlement and does not attempt to coerce a party to accept any proposed terms. If settlement is reached, the parties will sign an agreement, thereby avoiding the cost of trial or other litigation. If no settlement is reached, the case proceeds to trial before the judge to whom the action is assigned.

Confidentiality is maintained at the conference, which fosters frank, open discussions. Written settlement conference statements are submitted directly to the judge or magistrate and are not made part of the case file. The judge who would preside at trial is not told what positions the parties took during the settlement conference (unless by agreement he or she is also the settlement judge). Federal Rules of Evidence 408 and 403 may bar use at trial of communications made during the settlement negotiations.

A settlement conference is suitable for any kind of civil case, from the most straightforward to the most complex. The process is informal. Parties can communicate *ex parte* with the person conducting the conference, who can offer the parties an informed yet neutral view of the case. Often the parties help select the judge or magistrate who will conduct the settlement conference. A special master may be selected when the available judicial officers do not have the time for the negotiations or when specialized subject-matter expertise is required.

The settlement conference is similar to ENE in that it may aid the identification and narrowing of issues in dispute. Creative lawyers also can use settlement conferences to establish streamlined discovery plans. Unlike an ENE session, which usually occurs shortly after the case is filed, a settlement conference may be conducted at any time.

Subject Matter Scope: Any civil action.

Invocation: On motion of the judge to whom the action is assigned or on motion of a party with approval by the court.

The parties may also discuss "the possibility of settlement" at the pretrial conference. Fed. R. Civ. P. 16(c)(7).

Right of Appeal: Not applicable; the case proceeds to trial if settlement is not reached.

Authority: Fed. R. Civ. P. 16(a)(5) & (c)(7); Local Rule 240-1. *See also* Fed. R. Evid. 408 (concerning inadmissibility for certain purposes of statements made in settlement negotiations).

Contact: Courtroom deputy clerk for the judge to whom the action is assigned or the judge or magistrate conducting the settlement conference.

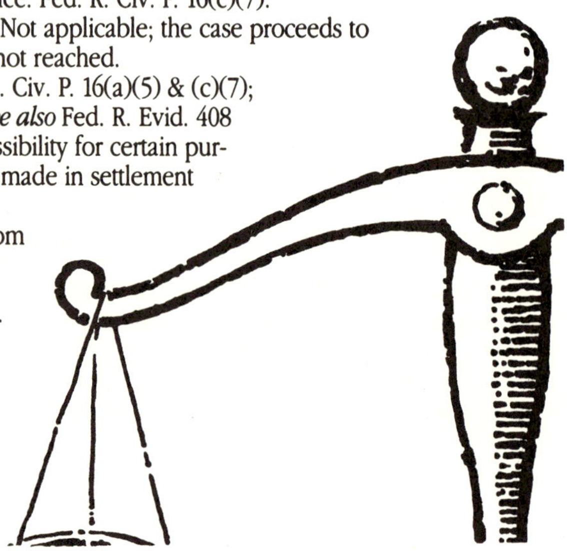

V

Non-Binding Summary Jury or Bench Trials

In cases that would take many court days to try, counsel might consider more elaborate settlement vehicles known as non-binding summary jury or bench trials. These procedures are designed to enable parties and their lawyers to see how, in abbreviated forms, the presentation of their case compares to the presentation of their opponent's case, and to gain insights into how a jury or judge is likely to react to the basic components of the parties' respective positions. While these procedures can be tailored to the specific needs of individual cases or situations, they generally include telescoped presentations of each side's case either to a judge or magistrate (not the judge to whom the case is assigned for trial if no settlement is reached) or to a jury selected from the same venire as the court's other juries. If a jury is used, it is given brief instructions by the court in the relevant legal principles, then each side takes anywhere from an hour to a day to present a well-focused version of its case, using narratives by counsel, documents, deposition transcripts, video-tapes, or live testimony. Rules of evidence are not applied, and cross-examination either is not permitted or is quite limited. After closing arguments, the jury or judge retires to deliberate, then returns to articulate an advisory judgment.

After the judgment is announced, counsel are permitted to ask questions of the judge or jury to explore the reasoning that supported the judgment, to assess how the fact-finder reacted to particular arguments or evidence and whether it failed to understand any important parts of the presentations. Neither party is bound by the advisory judgment and both are free to proceed to full trial without penalty (unless, by stipulation, the parties agree in advance that they should build some disincentive to go to trial into the rules of their special procedure). The parties themselves are required to attend these proceedings and to watch both how the respective cases are presented and how the jury or judge responds to them.

After the questioning of the fact-finder the parties may elect to discuss settlement, with or without the assistance of the judge or magistrate who presided over the summary proceedings. If they fail to settle, the matter proceeds to trial on a normal schedule.

This more elaborate vehicle for exploring settlement possibilities is appropriate only in cases that would require relatively lengthy trials. Such non-binding summary proceedings can be especially helpful to parties who disagree fundamentally about how neutral fact-finders are likely to respond to the basic outlines of each side's story, or how a judge or jury is likely to apply an elusive legal standard to a given set of facts.

Counsel who would like to set up a non-binding jury or bench trial should raise the matter with the assigned judge, asking him or her for permission to contact the magistrate or judge whom the parties would like to preside at the non-binding proceedings. No special fee is charged for this service. There are organizations in the private sector which, for a fee, will help parties design and conduct similar kinds of proceedings outside the court system. One version of such proceedings is sometimes called a "mini-trial". A list of some of the private organizations that can help parties set up these kinds of procedures is available from the clerk's office.

Subject Matter Scope: Any civil suit, but best suited to cases that would require many court days to try formally.

Invocation: By stipulation of the parties and with the permission of the assigned judge, as well as the judge or magistrate who would preside over the non-binding proceedings.

Right of Appeal: Since the "judgment" of the jury or judge in the summary proceeding is non-binding and only advisory, there is no occasion for an appeal.

Authority: Written stipulation by all parties; Fed. R. Civ. P. 16 (by implication); and the inherent powers of trial courts.

Contact: The assigned judge and the magistrate or judge whom the parties would like to preside at the non-binding summary proceeding.

VI

Special Masters

A special master is a private lawyer, retired judge, law professor, or other person who may be appointed to perform any of a wide range of tasks, including case management, discovery resolution, fact-finding, and settlement. A master may be appointed in response to a motion or on the court's own initiative. Special masters have contributed significantly to dispute resolution in several roles, including:

- Managing the case development process;
- Resolving discovery disputes;
- Monitoring discovery (e.g., depositions and document productions);
- Reviewing documents to determine whether they are protected by privilege or fall within the terms of a protective order;
- Performing technical or specialized functions (such as auditing);
- Fact-finding;
- Assisting in settlement negotiations, (e.g., in complex cases and class actions); and
- Monitoring the implementation of equitable decrees after trial.

Compensation for masters is determined by the judge. The court can appoint a master selected by either the court itself or by the parties, and may appoint a magistrate. The parties may be ordered to share equally the master's hourly fee. When a magistrate is designated to serve as a master, no fee is charged.

The authority of the master is fixed by the court's order of reference. Depending on the terms of the appointment, a master may compel the production of evidence, rule on the admissibility of evidence, and examine witnesses or parties under oath. Witnesses may be subpoenaed by the parties to appear before the master. Masters' determinations are subject to district court review. In non-jury trials, a court accepts the findings of fact of the master unless clearly erroneous. Within ten days of service, a party may file an objection to such findings. In jury trials, a master's findings are admissible as evidence. The parties may also stipulate that a master's findings be deemed final.

There can be several advantages to using a special master. The order of reference to a master can be carefully tailored to fit a particular case need. Although the direct cost of a special master can be substantial (depending on the terms of compensation), in appropriate cases the appointment of a special master has resulted in substantial savings to the litigants. Masters can reduce dramatically the time to resolve pretrial disputes. The master's immediate availability and thorough familiarity with the details of the case can discourage litigants from taking disruptive and costly positions. The master can bring expertise in a specialized field to the case that a generalist judge may not have. In addition, parties retain the right to have the court review decisions made by the master. The use of a master is generally appropriate in matters where the amount in controversy is substantial and technical issues are involved. Masters may be used to conduct lengthy settlement negotiations for which a judge or magistrate may not have time. In class actions, masters have been used successfully not only to administer distributions of funds, but also to facilitate negotiations about the substantive terms of settlement, about attorneys' fees, and about setting up claims procedures.

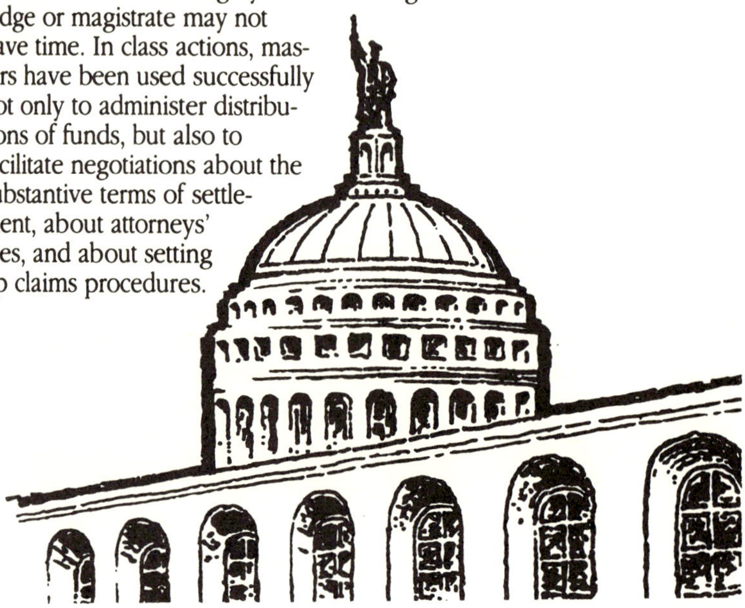

Subject Matter Scope: Any civil matter referred by the district court and subject to the specifications and limitations of the appointment order.

Invocation: *Sua sponte* court appointment or at the request of the parties. Fed. R. Civ. P. 53.

The parties may discuss reference to a master at any pretrial conference. Fed. R. Civ. P. 16(c)(6).

Right of Appeal: Fixed by order of reference or Fed. R. Civ. P. 53. Rulings and findings generally are reversible only if clearly erroneous.

Authority: Fed. R. Civ. P. 16(c)(6) & 53, as well as the Court's inherent authority. Concerning the possibility of a magistrate serving as a special master, see 28 U.S.C. § 636(b)(2); Fed. R. Civ. P. 53(a), (b) & (f).

Contact: Courtroom deputy clerk for the judge to whom the action is assigned.

C Private Options for Alternative Dispute Resolution

Federal Rule of Civil Procedure 16(c)(7) expressly provides that after a civil action has been filed in federal court, the parties may consider "the use of extrajudicial procedures to resolve the[ir] dispute." There are numerous dispute resolution providers in the private sector offering a variety of services, including arbitration, fact-finding, conciliation, mediation, negotiation, and private trials. The role of the "neutral" can be played by retired judges, law professors, former government officials and experienced attorneys with specialized expertise in dispute resolution techniques. Virtually all private sector providers charge fees for their services.

Interested lawyers or parties may obtain from the clerk's office in San Francisco or San Jose a list of private organizations, firms, and individuals who have informed the court that they provide dispute resolution services.

This list is not necessarily complete. Any provider omitted may contact the clerk's office to request inclusion in subsequent editions. By providing this list, the court does not endorse any particular organization and does not imply any opinion about the quality of the services that may be offered.

Special Acknowledgement

The Judges of this Court wish to express their sincere gratitude to Mark L. Krotoski, former law clerk to Chief Judge William A. Ingram, for the many hours of splendid work he committed to the development of this brochure. Working under considerable time pressure, Mr. Krotoski marshalled a great deal of information, coordinated inputs from many sources, prepared several working drafts of this document, and helped publicize the Court's work in this area. His work, and the spirit in which he performed it, were marvelous. We are in his debt.

For more information contact:

Arbitration/ENE Unit
(415) 556-5742

TABLE A: CASES

Case	*Book §*
A.A. Equipment, Inc. v. Farmoil, Inc., 31 Conn. Supp. 322, 330 A.2d 99 (1974)	§§ 5.22, 7.19
Abrams, *In re,* 127 Bankr. 239 (Bankr. 9th Cir. 1991)	§ 2.28
Adobe Building Center, Inc. v. Reynolds, 403 So. 2d 1033 (Fla. Dist. Ct. App. 1981)	§ 7.16
Aerojet General v. Superior Court, 211 Cal. App. 3d 216, 257 Cal. Rptr. 621 (1989), *reh'g denied* 8/10/89, at _____ Cal. App. 3d _____, 258 Cal. Rptr. 684, (1989)	§ 8.17
Aerojet General Corp. v. Transport Indemnity Insurance Co., No. 262425, Calif. Daily, Jan. 14, 1992	§ 8.17
AIU Insurance v. Superior Court, _____ Cal. App. 3d _____, 262 Cal. Rptr. 182 (1989), *rev'd,* 51 Cal. 3d 807, 274 Cal. Rptr. 820 (1990)	§ 8.17
All Bay Mill & Lumber Co. v. Surety Co., 208 Cal. App. 3d 11, 255 Cal. Rptr. 790 (1989)	§ 9.5
Allied Properties v, John A. Blume & Associates, Engineers, 25 Cal. App. 3d 848, 102 Cal. Rptr. 259 (1972)	§ 7.16
Alperin v. National Home Life Assurance Co., 32 Ill. App. 3d 261, 336 N.E.2d 365 (1965)	§ 8.2
American & Foreign Ins. Co. v. Allied Plumbing and Heating Co., 36 Mich. App. 561, 194 N.W.2d 158 (1971)	§ 8.5
Armour & Co. v. Scott, 360 F. Supp. 319 (W.D. Pa. 1972), *aff'd,* 480 F.2d 611 (3d Cir. 1973)	§ 4.19
Baker v. Keller Construction Co., 219 So. 2d 569 (La. Ct. App. 1969)	§ 3.7
Balboa Insurance Co. v. United States, 775 F.2d 1158 (Fed. Cir. 1985)	§ 9.8
Ballinger v. Howell Manufacturing. Co., 407 Pa. 319, 180 A.2d 555 (1962)	§ 4.7
Becker v. Black & Veatch Consulting Engineers, 509 F.2d 42 (8th Cir. 1974)	§ 5.22
Bender-Miller C. V. Thomwood Farms, Inc., 211 Va. 585, 179 S.E. 2d 636 (1971)	§ 5.19
Bergier v. Internal Revenue Service, 83 Bank. Rptr. 324, 110 S. Ct. 2258 (1990)	§ 11.4
Bethlehem Fabricators, Inc. v. British Overseas Airways Corp., 434 F.2d 840 (2d Cir. 1970)	§ 5.10
Bexiga v. Havir Mfg. Corp., 60 N.J. 402, 290 A.2d 281 (1972)	§§ 4.4, 7.16

Case	*Book §*
B-K of Kansas, Inc., *In re,* 69 Bankr. 812 (Bankr. D. Kan. 1987)	§ 11.11
Bos Material Handling, Inc. v. Crown Controls Corp., 137 Cal. App. 3d 99, 186 Cal. Rptr. 740 (1982)	§ 3.4
Boyle v. United States Technologies, Corp., 487 U.S. 500, 108 S. Ct. 2510, 101 L. Ed. 2d 442 (1988), U.S. *reh'g denied,* 109 S. Ct. 1182 (1989)	§ 7.22
Bridgeport Restoration v. Petrucci Construction Co., 211 Conn. 230, 557 A.2d 1263 (1989)	§ 10.10
Brinker v. Superior Court (Page Realty Corp., _____ Cal. App. 3d _____, 1 Cal. Rptr. 2d 358, (1991)	§ 3.5
Broadway Maintenance Corp. v. Rutgers, 90 N.J. 253, 447 A.2d 906 (1982)	§§ 4.8, 5.19
Bulley v. Andrews, Inc. v. Symons Corp., 25 Ill. App. 3d 696, 323 N.E.2d 806 (1975)	§ 7.16
Burton v. Sosinsky, 203 Cal. App. 3d 562, 250 Cal. Rptr. 33 (1988)	§ 6.11
Burton-Dixie Corp. v. Timothy McCarthy Construction Co., 436 F.2d 405 (5th Cir. 1971)	§ 3.4
C.A. Davis, Inc. v. City of Miami, 400 So. 2d 536 (Fla. App. 1981)	§§ 3.7, 7.14, 11.3
C. Norman Petersen Co. Container Corp. of Am., 172 Cal. App. 3d 628, 218 Cal. Rptr. 592 (1985)	§§ 3.14, 4.19
C.H. Leavell & Co. v. Glantz Contracting Corp., 322 F. Supp. 779 (E.D. La. 1971)	§ 6.17
Cannavino & Shea, Inc. v. Water Works Supply Corp., 361 Mass. 363, 280 N.E.2d 147 (1972)	§ 6.16
Capolino Sons, Inc. v. Electronic Missile Facilities, Inc., 364 F. 2d 705 (2d Cir.), *cert. dismissed,* 385 U.S. 924 (1966)	§ 3.6
Cardenas v. Miami-Dade Yellow Cab Co., 538 So. 2d 491 (Fla. App. 1989)	§ 8.14
Carl Bolander & Sons Co. v. City of Minneapolis, 438 N.W.2d 735 (Minn. App. 1989)	§ 10.10
Carrabine Construction Co. v. Chrysler Realty Corp., 25 Ohio St. 222, 495 N.E.2d 952 (1986)	§ 5.19
Christiansen v. Dewor Developments, 33 Cal. 3d 778, 661 P. 2d 1088, 191 Cal. Rptr. 8 (1983)	§ 3.4
Cincinnati Gas & Electric Co. v. Westinghouse Electric, 465 F.2d 1064 (6th Cir. 1972)	§ 5.23
Citadel Properties, Inc., *In re,* 86 Bankr. 275 (Bankr. M.D. Fla. 1988	§ 11.4
City of Los Angeles v. Superior Court (Hutcherson), 176 Cal. App. 3d 856, 222 Cal. Rptr. 562 (1986)	§ 7.11
Coatesville Contractors & Engineers, Inc. v. Borough of Ridley Park, Pa., 509 Pa. 553, 506 A.2d 862 (1986)	§ 5.19
Colard v. American Family Mutual Ins. Co., 709 P.2d 11 (Colo. App. 1985)	§ 8.7

Case	*Book §*
Commonwealth Coatings Corp. v. Continental Casualty Co., 393 U.S. 145, 89 S. Ct. 337 (1968), *petition for reh'g denied,* 21 L. Ed. 2d 301 (1969)	§ 3.10
Comstock Insurance Co. v. Thomas A. Hanson & Associates, Inc., 77 Md. App. 431, 550 A.2d 731 (1988)	§ 8.9
Consulting Engineers v. Dept. of Transportation, 547 N.Y.S. 2d 205 (Sup. 1989)	§ 5.39
Cooper v. Goodwin, 155 D. Ct. App. 449, 478 F.2d 653 (D.C. Cir. 1973)	§ 7.9
Corrinno Civetta Construction v. City of New York, 67 N.Y.2d 297, 493 N.E.2d 905, 502 N.Y.S.2d 681 (1986)	§ 5.10
Costanza Construction Corp. v. Rochester, 147 A.D.2d 929, 537 N.Y.S.2d 394 (1989)	§ 5.26
Crown Construction Co. v. Opelika Mfg. Corp., 343 F. Supp. 1266 (N.D. Ga. 1972), [*modified on appeal on another point,* 480 F.2d 149 (1973)]	§ 6.4
Culligan Corp. v. Transamerica Ins. Co., 580 F.2d 251 (7th Cir. 1978)	§ 5.34
D&L Building Inc. v. Maltby Tank & Barge, 747 P.2d 517 (Wyo. 1987)	§ 9.1
Dalehite v. United States, 346 U.S. 15, 73 S. Ct. 956, 97 L. Ed. 1427 (1953)	§ 7.21
Damora v. Stresscon International, Inc., 324 So. 2d 80 (Fla. 1975)	§ 3.5
Del Mar Beach Club Owners Association v. Imperial Contracting Co., 123 Cal. App. 3d 898, 176 Cal. Rptr. 886 (1981)	§ 7.16
DiSalvatore v. United States, 456 F. Supp. 1079 (E.D. Pa. 1978), 472 F. Supp. 816 (E.D. Pa. 1979), *appeal dismissed,* 622 F.2d 578 (1979)	§ 4.9
Dole v. Dow Chemical Co., 30 N.Y. 2d 143, 282 N.E.2d 288 (1972)	§§ 5.22, 7.19
Domingue v. H&S Construction Co., 546 So. 2d 913 (La. Ct. App. 1989)	§ 8.17
Drennan v. Star Paving, 51 Cal. 2d 409, 333 P. 2d 757 (1958)	§ 4.13
E.C. Ernst, Inc. v. City of Tallahasee, 527 F. Supp. 1141 (N.D. Fla. 1981)	§ 3.4
E.C. Ernst, Inc. v. General Motors Corp., 482 F.2d 1047 (5th Cir. 1973)	§ 6.1
Emasco Ins. Co. v. L&M Development, Inc., 372 N.W.2d 908 (N.D. 1985)	§ 8.7
Empire City Subway Co. v. Greater New York Mutual Insurance Co., 35 N.Y.2d 8, 315 N.E.2d 755, 358 N.Y.S.2d 691 (1974)	§ 8.12

Case	*Book §*
Employer's Insurance v. Construction Management Engineers, 297 S.C. 354, 377 S.E.2d 119 (Ct. App. 1989)	§ 9.6
Episcopal Housing Corp. v. Federal Ins. Co., 273 S.C. 181, 255 S.E.2d 451 (1979)	§ 3.9
Ernest W. Hahn, Inc. v. Armco Steel Co, 601 P.2d 152 (Utah 1979)	§ 7.16
First National Bank of Akron v. Cann, 503 F. Supp. 419 (1980), *aff'd,* 669 F.2d 415 (6th Cir. 1982)	§ 4.9
Foster Wheeler Corp. v. United States, 206 Ct. Cl. 533, 513 F.2d 588 (1975)	§ 3.14
Framingham Trust Co. v. Gould-National Batteries, 307 F. Supp. 1008, 427 F.2d 856 (1st Cir. 1970)	§ 9.8
Gagne v. Bertran, 43 Cal. 2d 481, 275 P.2d 15 (1954)	§ 7.16
Galimi v. Jetco, Inc., 514 F.2d 949 (2d Cir. 1975)	§§ 5.22, 7.19
Galt v. Libbey-Owens-Ford Glass Co., 397 F. 2d 439 (7th Cir.), *cert. denied,* 393 U.S. 925, 895 S. Ct. 258 (1968)	§ 3.7
Garden Grove Community Church v. Pittsburgh-Des Moines Steel Co., 140 Cal. App. 3d 251, 191 Cal. Rptr. 15 (1983)	§ 3.9
Giem v. Searles, 470 S.W.2d 327 (Ky. 1971)	§ 7.16
Glover Construction Co., *In re,* 30 Bank. 873 (W.D. Ky. 1983)	§ 9.8
Goldfarb v. Virginia State Bar, 421 U.S. 773, 95 S. Ct. 2004, 44 L. Ed. 2d 572 (1975)	§ 4.6
Gon v. First State Insurance Co., 871 F.2d 863 (9th Cir. 1989)	§ 8.15
Gray v. Zurich Ins., 65 Cal.2d 263, 419 P.2d 168, 54 Cal. Rptr. 104 (1966)	§ 8.15
Great Lakes Gas Transmission Co. v. Grayco Constructors, Inc., 506 F.2d 498 (6th Cir. 1974), *cert. denied* 2/24/75, 95 S. Ct. 133	§§ 3.7, 7.14, 11.3
Greenman v. Yuba Power Products, Inc. 59 Cal. 2d 57, 377 P.2d 897, 27 Cal. Rptr. 697 (1963)	§ 7.14
Grinnell Fire Protection Systems Co. v. W.C. Ealy & Associates, Inc., 552 S.W.2d 747 (Tenn. App. 1977), *cert. denied,* by Supreme Court, June 6, 1977	§ 4.19
Grover-Diamond Associates v. American Arbitration Association, 297 Minn. 324, 211 N.W. 2d 787 (1973)	§ 3.9
Gruenberg v. Aetna Insurance Co., 9 Cal. 3d 566, 510 P.2d 1032, 108 Cal. Rptr. 480 (1973)	§ 8.14
Gurney Industries, Inc. v. St. Paul Fire and Marine Insurance Co., 467 F.2d 588 (4th Cir. 1972)	§§ 3.7, 4.4, 8.7
Hart v. Aetna Cas. & Surety Co., 756 S.W.2d 27 (Tex. App. 1988)	§ 8.14
Hess v. Zoological Society of Buffalo, 134 A.D.2d 824, 521 N.Y.S.2d 903 (1987)	§ 7.16

Case	*Book §*
Higley South, Inc. v. Park Shore Development Co., 494 So. 2d 227 (Fla. App. 1986)	§ 3.9
Holman v. State, 53 Cal. App. 3d 317, 124 Cal. Rptr. 773 (1975)	§§ 3.14, 7.22
J.J. Brown Co. v. J.L. Simmons Co., 2 Ill. App. 2d 132, 118 N.E.2d 781 (1954)	§ 5.33
James Baird Co. v. Gimbel Brothers, 64 F.2d 344 (2d Cir. 1933)	§ 4.13
Jarchow v. Transamerica Title Ins. Co., 48 Cal. App. 3d 917, 122 Cal. Rptr. 470 (1975)	§ 8.15
Jiminez v. Western Pacific Construction Co., 185 Cal. App. 3d 102, 229 Cal. Rptr. 575 (1986), *reh'g denied,* 11/25/86	§§ 5.22, 7.19
John E. Green Plumbing & Heating Co., Inc. v. Turner Construction Co., 742 F.2d 965 (6th Cir. 1984)	§ 5.19
Jones v. Continental Casualty Co., 123 N.J. Super. 353, 303 A.2d 91 (1973)	§ 8.2
Kalisch-Jarcho, Inc. v. City of New York, 58 N.Y. 2d 377, 448 N.E.2d 413, 461 N.Y.S.2d 746 (1983)	§ 5.19
Kalman Floor Co. v. Joseph L. Muscarelle, Inc., 196 N.J. Super. 16, 481 A.2d 553 (N.J. Sup. A.D. 1984), *aff'd,* 98 N.J. 266, 486 A.2d 334 (1985)	§ 3.4
Knutson Construction Co. v. St. Paul Fire and Marine Ins. Co., 396 N.W.2d 229 (Minn. 1986)	§ 8.7
Kriegler v. Eichler Homes, Inc, 269 Cal. App. 2d 224, 74 Cal. Rptr. 749 (1969)	§§ 4.4, 7.16
Leo F. Piazza Paving Co. v. Bebek & Brkich, 141 Cal. App. 2d 226, 296 P.2d 368 (1956)	§ 6.16
Lozier Brothers, Inc. v. Fremont County School District, 747 P.2d 515 (Wyo. 1987)	§ 9.1
Lough v. Ins. Co. of North America, 789 P.2d 576 (Mont. 1990)	§ 8.14
Loughman Cabinet Co. v. C. Iber & Sons, Inc., 46 Ill. App. 3d 873, 361 N.E.2d 379 (1977)	§ 5.33
Manchester Township Board of Educ. v. Carney, 199 N.J. Super. 266, 489 A.2d 682 (1985)	§ 3.9
Manors of Inverrary XII Condominium Association v. Atreco-Florida, 438 So. 2d 490 (Fla. App. 1983)	§ 7.22
Marioenzi v. Joseph Di Ponte, Inc., 333 A.2d 127 (R.I. 1975)	§ 7.9
Martin Bloom Associates v. Manzie, 389 F. Supp. 848 (D. Nev. 1975)	§ 4.10
Maryland v. Armco, Inc., 822 F.2d 1348 (4th Cir. 1987), *cert. denied,* 484 U.S. 1008, 108 S. Ct. 703, 102 L. Ed. 2d 43 (1988)	§ 8.17
Mascioni v. I.B. Miller, Inc., 261 N.Y. 1, 184 N.E. 473 (1933)	§ 5.34

Case	*Book §*
Maxum Foundations, Inc. v. Salus Corp., 817 F.2d 1086 (4th Cir. 1987)	§ 3.9
McCarter v. State Farm Mutual Ins. Co., 130 Ill. App. 3d 97, 85 Ill. Dec. 416, 473 N.E.2d 1015 (1985)	§ 8.14
McGhee v. Aramco, 871 F. 2d 1412 (9th Cir. 1989)	§ 5.31
Meva Corp. v. U.S., 511 F.2d 548 (Ct. Cl. 1975)	§ 6.19
Midland Engineering Co. v. John A. Hall Construction Co., 398 F. Supp. 981 N.D. Ind. 1975)	§ 5.34
Mile High Fence Co. v. Radovich, 175 Colo. 537, 489 P.2d 308 (1971)	§ 7.9
Mills v. MMM Carpets, Inc., ______ Cal. App. 3d ______, 1 Cal. Rptr. 2d 813 (1991)	§ 7.11
Milwaukee County Pavers Association v. Fiedler, 707 F. Supp. 1016 (W.D. Wis. 1989)	§ 10.10
Mississippi Coast Coliseum Commission v. Stuart Construction Co., 417 So. 2d 541 (Miss. 1982)	§ 3.9
Moore v. Denune & Pipic, Inc., 26 Ohio St.2d 125, 269 N.E.2d 599 (1971)	§ 7.9
Moradi-Shalal v. Fireman's Fund Insurance Cos., 46 Cal. 3d 287, 758 P.2d 58, 250 Cal. Rptr. 116 (1988)	§ 8.14
Morris v. County of Marin, 18 Cal. 3d 901, 559 P.2d 606, 136 Cal. Rptr. 251 (1977)	§§ 3.14, 7.22
Moses H. Cone Memorial Hospital v. Mercury Construction Corp., 460 U.S. 1, 103 S. Ct. 927, 74 L. Ed. 2d 765 (1983)	§ 3.5
Mraz v. Canadian Universal Insurance Co., Ltd., 804 F.2d 1325 (4th Cir. 1986)	§ 8.17
Muskopf v. Corning Hosp. District, 55 Cal. 2d 211, 359 P.2d 457, 11 Cal. Rptr. 89 (1961)	§ 7.22
National Union Indemnity Co. v. G.E. Bass & Co., 369 F.2d 75 (5th Cir. 1966)	§ 9.6
Nelson v. Commonwealth, 235 Va. 228, 368 S.E.2d 239 (1988)	§ 7.16
Nicholas Acoustics & Specialty Co. v. H&M Construction Co., 695 F.2d 839 (5th Cir. 1983)	§ 5.34
O'Neil v. Picillo, 682 F. Supp. 706 (D.R.I. 1988), *aff'd,* 883 F.2d 176 (1st Cir. 1989), *cert. denied,* 110 S. Ct. 1115 (1990)	§ 7.11
Oxford Development Corp. v. Rausauer Builders, Inc., 158 Ind. App. 622, 304 N.E.2d 211 (1973)	§ 6.17
Paul Allison Inc. v. Miniken Storage of Omaha, Inc., 486 F. Supp. 1 (D.C. Neb. 1980)	§ 3.6
Pearlman v. Reliance Insurance Co., 371 U.S. 132, 83 S. Ct. 232 (1962)	§ 9.8
Pettinaro Construction Co. v. Harry C. Partridge et al., 408 A.2d 957 (Del. 1980)	§ 3.4
Player v. Geo. M. Brewster & Son, Inc., 18 Cal. App. 3d 526, 96 Cal. Rptr. 149 (1971)	§ 3.4

Case	*Book §*
Post Tensioned Engineering Corp. v. Fairways Plaza Associates, 412 So. 2d 871 (Fla. Dist. Ct. App. 1982)	§ 3.4
Protective Life Ins. Corp. v. Lincoln Nat'l Life Insurance Corp., 873 F.2d 281 (11th Cir. 1989)	§ 3.9
R. Baker v. Motel 6, Inc., 180 Cal. App. 3d 928, 225 Cal. Rptr. 849 (1986)	§ 3.3
Reliance Ins. Co. v. Mogavero, 640 F. Supp. 84 (D.Md. 1986)	§ 8.7
Ribeira & Lourenco Concrete Construction Co. v. Jackson Health Care Associates, 231 N.J. Super. 16, 554 A.2d 1350, *cert. granted,* 563 A.2d 826 (1989), *aff'd,* 571 A.2d 1311 (1990)	§ 9.4
Richmond v. J.A. Croson, 480 U.S. 469, 109 S. Ct. 706, 102 L. Ed. 2d 854 (1989)	§ 10.10
Riverside Steel Const. Co. V. William H. Simpson Const. Co., 227 Cal. Rptr. 424, *review granted and ordered citable,* 709 P.2d 1309, *cause transferred* 748 P.2d 307, *transferred* to 199 Cal. App. 3d 1051, 245 Cal. Rptr. 477 (1986)	§ 7.11
Rosswog, *In re,* 48 Bankr. 689 (Bankr. M.D. Pa. 1985)	§ 11.11
Rowland v. Christian, 69 Cal. 2d 108, 443 P.2d 561, 70 Cal. Rptr. 97 (1968)	§ 7.9
Royal Globe Insurance Co. v. Superior Court, 23 Cal. 3d 880, 592 P.2d 329, 153 Cal. Rptr. 842 (1979)	§ 8.14
Rudd v. Anderson, 153 Ind. App. 11, 285 N.E.2d 836 (1972)	§§ 3.14, 4.4
Rutter v. Arlington Park Jockey Club, 510 F.2d 1065 (7th Cir. 1975)	§ 5.23
S.J. Groves & Sons Co. v. Warner Co., 576 F.2d 524 (3d Cir. 1978)	§§ 3.9, 7.14, 11.3
S.M. Wilson & Co. v. Prepakt Concrete Co., 23 Ill. App. 3d 137, 318 N.E.2d 722 (1974)	§ 6.16
Saddler v. United States, 152 Ct. Cl. 557, 287 F.2d 411 (1961)	§ 5.33
Schipper v. Levitt & Sons, Inc., 44 N.J. 70, 207 A.2d 314 (1965)	§§ 4.4, 7.16
Schuler-Haas Electric Co. v. Aetna Casualty and Surety Co., 40 N.Y.2d 883, 357 N.E.2d 1003, 389 N.Y.S.2d 348 (1976)	§ 5.34
Security Ins. Co. v. Deshotels, 458 So. 2d 186 (La. App. 1984)	§ 8.14
Service Engineering Co. v. Southwest Marine, Inc., 719 F. Supp. 1500 (1989)	§ 10.10
Severson, Werson, Berke & Melchior v. Bolinger, ____ Cal. App. 3d ____, 1 Cal. Rptr. 2d 531 (1991)	§ 1.17
Sharp Brothers Contracting Co. v. Commercial Restoration, Inc., 334 S.W.2d 248 (Mo. App. 1960)	§ 6.16
Shearer v. Reed, 428 A.2d 635 (Pa. Super. 1981)	§ 8.14

Case	*Book §*
Silver Cross Hospital v. S.N. Nielsen Co., 8 Ill. App. 3d 1000, 291 N.E. 2d 247 (1972)	§ 3.4
Singer Co. v. Superior Court (Cool-Temp Air Conditioning Co.), 179 Cal. App. 3d 875, 225 Cal. Rptr. 159 (1986)	§ 7.11
Sky Group International, Inc., *In re,* 108 Bankr. 86 (Bankr. W.D. Pa. 1989)	§ 11.4
Spence v. Omnibus Industries, 44 Cal. App. 3d 970, 119 Cal. Rptr. 171 (1975)	§ 3.4
Stop and Shop Cos. Inc. v. Gilbane Building Co., 364 Mass. 325, 304 N.E. 2d 429 (1973)	§ 3.9
Sunlake Apartment Residents v. Tonti Development Corp. 522 So. 2d 1298 (La. App. 1988)	§§ 3.14, 7.22
Tech-Bilt v. Woodward-Clyde & Associates, 38 Cal. 3d 488, 698 P.2d 159, 213 Cal. Rptr. 256 (1985)	§ 7.11
Terra Group Inc. v. Sandefur Management, Inc., 527 So. 2d 849 (Fla. Dist. Ct. App. 1988)	§ 9.3
Thomas J. Dyer Co. v. Bishop International Engineering Co., 303 F. 2d 655 (6th Cir. 1962)	§§ 5.34, 6.6
Tolboe Construction Co. v. Staker Paving & Construction Co., 682 P.2d 843 (Utah 1984)	§ 6.15
Travelers Indemnity Co. v. Ewing, Cole, Erdman and Eubank, 711 F. 2d 14 (3d Cir. 1983), *cert. denied,* 464 U.S. 1041, 104 S.Ct. 707 (1983)	§ 4.3
Tunkl v. Regents of the University of California, 60 Cal. 2d 92, 383 P.2d 441, 32 Cal. Rptr. 33 (1963)	§ 5.23
U.S.F. & G. v. Rob Homes, Inc., 323 So. 2d 105 (Miss. 1975)	§ 8.5
United States v. Spearin, 248 U.S. 132, 39 S. Ct. 59, 63 L. Ed. 166 (1918)	§ 5.26
United States v. Young Lumber Co., 376 F. Supp. 1290 (D.S.C. 1974)	§ 5.4
United States Financial v. Sullivan, 37 Cal. App. 3d 5, 112 Cal. Rptr. 18 (1974)	§ 7.16
Vandermark v. Ford Motor Co., 61 Cal. 2d 256, 391 P.2d 168, 37 Cal. Rptr. 896 (1964)	§§ 4.4, 7.16
Varco-Pruden, Inc. v. Hampshire Construction Co., 50 Cal. App. 3d 654, 123 Cal. Rptr. 606 (1975)	§ 6.17
Vernon Williams & Son Construction Inc. v. Continental Ins. Co., 591 S.W.2d 760 (Tenn. 1979)	§ 8.7
Village of Cairo v. Bodine Contracting Co., 685 S.W. 2d 253 (Mo. Ct. App. 1985)	§ 3.7
Village of Euclid v. Ambler Realty Co., 272 U.S. 365, 47 S. Ct. 114, 71 L. Ed. 303 (1926)	§ 10.3
Volt Information Services, Inc. v. Board of Trustees of Leland Stanford Junior University, 489 U.S. 468, 109 S. Ct. 1248, 103 L. Ed. 2d 488 (1989)	§ 3.5

Case	*Book §*
Votaw Precision Tool Co. v. Air Canada, 60 Cal. App. 3d 52, 131 Cal. Rptr. 335 (1976)	§§ 3.7, 8.7
Waldinger Corp. v. CRS Group Engineers, Inc., 775 F.2d 781 (7th Cir. 1985)	§ 4.4
Warburton v. Phoenix Steel Corp., 321 A.2d 345 (Del. Super. 1974)	§ 5.22
Wechsler v. Beall & Assoc. Roofing Corp., 442 So. 2d 1104 (Fla. App. 1987)	§ 8.7
Williams Engineering Inc. v. Goodyear, 496 So. 2d 1012 (La. 1986)	§ 2.6
Willman v. Children's Hospital of Pittsburgh, 74 Pa. Comm. 303, 459 A.2d 855 (1983)	§ 4.9
Windmill Farms, Inc., *In re,* 841 F.2d 1467 (9th Cir. 1988)	§ 11.11
Wright v. Creative Corp., 30 Colo. App. 575, 498 P.2d 1179 (1972)	§§ 4.4, 7.16
Wunderlich Contracting Co. v. United States, 240 F.2d 201 (10th Cir. 1957)	§ 5.33
Yamanishi v. Bleily & Collishaw, Inc., 29 Cal. App. 3d 457, 105 Cal. Rptr. 580 (1972)	§ 5.34

TABLE B: AIA DOCUMENT REFERENCES

Document References	*Book §*
AIA Doc. A101	§§ 3.2, 6.4
AIA Doc. A201	§§ 3.2, 6.4
AIA Doc. A201, Para. 1.1.2	§ 6.12
AIA Doc. A201, Para. 4.3.6	§ 5.26
AIA Doc. A201, Para. 4.4.1	§ 3.2
AIA Doc. A201, Para. 4.4.2	§ 3.2
AIA Doc. A201, Para. 4.4.3	§ 3.2
AIA Doc. A201, Para. 4.4.4	§ 3.2
AIA Doc. A201, Para. 4.5.1	§ 3.7
AIA Doc. A201, Para. 4.5.2	§ 3.7
AIA Doc. A201, Para. 4.5.3	§ 3.7
AIA Doc. A201, Para. 4.5.4	§ 3.8
AIA Doc. A201, Para. 4.5.4.1	§ 3.8
AIA Doc. A201, Para. 4.5.4.2	§ 3.8
AIA Doc. A201, Para. 4.5.5	§ 3.9
AIA Doc. A201, Para. 4.5.6	§ 3.9
AIA Doc. A201, Para. 4.5.7	§ 3.10
AIA Doc. A201, Para. 5.2.1	§ 6.15
AIA Doc. A201, Para. 5.3.1	§ 6.17
AIA Doc. A201, Para. 5.4.1	§ 6.4
AIA Doc. A201, Para. 8.2.1	§ 5.17
AIA Doc. A201, Para. 9.6.2	§ 6.6
AIA Doc. A201, Para. 11.3.1	§ 8.8
AIA Doc. A201, Para. 11.3.10	§ 8.8
AIA Doc. A201, Para. 13.1.1	§ 3.5
AIA, Doc. A312, Performance Bond	§ 9.3
AIA Doc. A312, Payment Bond	§ 9.4
AIA Doc. A401	§ 3.3
AIA Doc. A401	§ 3.7
AIA Doc. A401 Para. 6.2	§ 3.9
AIA Doc. A401, Para. 7.3.1	§ 6.4
AIA Doc. A401, Para. 11.3	§ 6.6
AIA Doc. A401, Para 11.11.1 (1978)	§§ 5.22, 7.14
AIA Doc. A401, Para. 12.1	§ 6.6
AIA Doc. B141	§ 3.7
AIA Doc. B141	§ 4.7
AIA Doc. B141, Para. 9.1	§ 3.5
AIA Doc. B141, Para. 10.1.1	§ 4.10
AIA Doc. B161	§ 4.7
AIA Doc. B511, Para. 1.2	§ 4.7
AIA Doc. B511, Para. 2.6	§ 4.7
AIA Doc. B511, Para. 2.7	§ 4.7
AIA Doc. B801	§ 4.7

TABLE C: NSPE DOCUMENT REFERENCES

Document References	*Book §*
NSPE 1910-8, Para. 6.11	§ 6.9
NSPE 1910-8, Para. 12.3	§ 5.17
NSPE 1910-8, Para. 6.9	§ 6.7

TABLE D: EJCDC DOCUMENT REFERENCES

Document References	*Book §*
EJCDC No. 1910-1, Para. 7.6.1	§ 3.7
EJCDC No. 1910-1, Para. 7.6.2	§ 3.8
EJCDC No. 1910-1, Para. 7.6.3	§ 3.7
EJCDC No. 1910-1, Para. 7.6.4	§ 3.9
EJCDC No. 1910-1, Para. 7.6.4	§ 3.7
EJCDC No. 1910-1, Para. 7.6.6	§ 3.10
EJCDC No. 1910-8 (1990)	§ 3.10
EJCDC No. 1910-8 Exh. GC-A, Para. 16.2	§ 3.8
EJCDC No. 1910-8, Para. 6.8.2	§ 6.15
EJCDC Exhibit OG-B	§ 3.7

INDEX